Springer Optimization and Its Applications

VOLUME 104

Aims and Scope
Optimization has been expanding in all directions at an astonishing rate during the last few decades. New algorithmic and theoretical techniques have been developed, the diffusion into other disciplines has proceeded at a rapid pace, and our knowledge of all aspects of the field has grown even more profound. At the same time, one of the most striking trends in optimization is the constantly increasing emphasis on the interdisciplinary nature of the field. Optimization has been a basic tool in all areas of applied mathematics, engineering, medicine, economics, and other sciences.

The series *Springer Optimization and Its Applications* publishes undergraduate and graduate textbooks, monographs and state-of-the-art expository work that focus on algorithms for solving optimization problems and also study applications involving such problems. Some of the topics covered include nonlinear optimization (convex and nonconvex), network flow problems, stochastic optimization, optimal control, discrete optimization, multi-objective programming, description of software packages, approximation techniques and heuristic approaches.

More information about this series at http://www.springer.com/series/7393

Alexander J. Zaslavski

Turnpike Theory of Continuous-Time Linear Optimal Control Problems

 Springer

Alexander J. Zaslavski
Department of Mathematics
The Technion - Israel Institute of Technology
Haifa, Israel

ISSN 1931-6828   ISSN 1931-6836   (electronic)
Springer Optimization and Its Applications
ISBN 978-3-319-36663-0   ISBN 978-3-319-19141-6   (eBook)
DOI 10.1007/978-3-319-19141-6

Mathematics Subject Classification (2010): 49J15, 49K15, 49K40, 49N05, 49N70, 91A05, 91A23, 93C15

Springer Cham Heidelberg New York Dordrecht London
© Springer International Publishing Switzerland 2015
Softcover reprint of the hardcover 1st edition 2015

Printed on acid-free paper

Springer International Publishing AG Switzerland is part of Springer Science+Business Media (www.springer.com)

Preface

In this monograph, we study the structure of approximate solutions of linear optimal control problems with nonsmooth integrands. These problems are governed by differential equations whose right-hand side is linear with respect to a state and control variables. We establish a number of new results on properties of approximate solutions which are independent of the length of the interval, for all large intervals. As a matter of fact, these results provide a full description of the structure of approximate solutions of linear optimal control problems.

In our book, we study the turnpike phenomenon arising in the optimal control theory. The term was first coined by P. Samuelson in 1948 when he showed that an efficient expanding economy would spend most of the time in the vicinity of a balanced equilibrium path (also called a von Neumann path). To have the turnpike property means, roughly speaking, that the approximate solutions of the problems are determined mainly by the objective function and are essentially independent of the choice of interval and endpoint conditions, except in regions close to the endpoints. The turnpike property discovered by P. Samuelson is well known in the economic literature, where it was studied for various models of economic growth. Usually for these models a turnpike is a singleton.

Now it is well known that the turnpike property is a general phenomenon, which holds for large classes of variational and optimal control problems. In our research, using the Baire category (generic) approach, it was shown that the turnpike property holds for a generic (typical) variational problem [44] and for a generic optimal control problem [52]. According to the generic approach, we say that a property holds for a generic (typical) element of a complete metric space (or the property holds generically) if the set of all elements of the metric space possessing this property contains a G_δ everywhere dense subset of the metric space which is a countable intersection of open everywhere dense sets. This means that the property holds for most elements of the metric space.

Individual (non-generic) turnpike results and sufficient and necessary conditions for the turnpike phenomenon are of great interest because of their numerous applications in engineering and the economic theory. In particular, we are interested in the cases when a turnpike has a simple structure (a singleton or a periodic trajectory).

In this case, it is possible to find the turnpike (or at least its approximations) numerically. In our research which was summarized in [51], we obtained a number of individual (non-generic) turnpike results for variational problems. In our more recent book [53], we studied the turnpike phenomenon for discrete-time optimal control problems, which describe a general model of economic dynamics and for autonomous variational problems with extended-valued integrands. For these problems, the turnpike property was established with the turnpike being a singleton. In [53], for problems which satisfy concavity (convexity) assumption common in the literature, we also studied the structure of approximate solutions in the regions containing end points and obtained a full description of the structure of approximate solutions.

In this monograph, we are also interested in individual turnpike results but for linear optimal control problems which have important applications in engineering. We study two large classes of problems. The first class studied in Chap. 2 consists of linear control problems with periodic nonsmooth convex integrands. We show that for these problems the turnpike property holds and the turnpike is a periodic trajectory-control pair. The second class studied in Chaps. 3–5 consists of linear control problems with autonomous nonconvex nonsmooth integrands. It is shown that for this class of problems the turnpike phenomenon takes place with the turnpike being a singleton. For these two classes of problems, we study the structure of approximate solutions in the regions containing end points and obtain a full description of the structure of approximate solutions. We show that the structure of approximate solutions is stable under small perturbations of integrands. This stability is an important property from the view of practice if we are interested to find a turnpike or its approximations numerically. In the other chapters of the book, we establish a turnpike property for dynamic zero-sum games with linear constraints (see Chap. 6), obtain the description of the structure of variational problems with extended-valued integrands (see Chap. 8), and study the turnpike phenomenon for dynamic games with extended-valued integrands (see Chap. 9).

Haifa, Israel   Alexander J. Zaslavski
February 28, 2015

Contents

Chapter 1
Introduction

The study of the existence and the structure of solutions of optimal control problems defined on infinite intervals and on sufficiently large intervals has recently been a rapidly growing area of research. See, for example, [2–4, 6–11, 13, 14, 16, 19, 20, 22, 25, 27, 32–34, 36, 37, 46–49, 51, 52] and the references mentioned therein. These problems arise in engineering [1, 23, 44, 56, 57], in models of economic growth [12, 13, 17, 22, 26, 31, 35, 39–41, 44, 45, 53, 55], in the game theory [18, 21, 43, 44, 50, 53, 54], in infinite discrete models of solid-state physics related to dislocations in one-dimensional crystals [5, 42], and in the theory of thermodynamical equilibrium for materials [15, 24, 28–30]. In this chapter we explain the turnpike phenomenon for a simple class of variational problems, discuss certain turnpike results obtained in our previous research, and describe the structure of the book.

1.1 The Turnpike Phenomenon

Denote by $|\cdot|$ the Euclidean norm in the n-dimensional Euclidean space R^n which is induced by the inner product $\langle \cdot, \cdot \rangle$ in R^n.

Assume that a function $f : R^n \times R^n \to R^1$ is strictly convex and differentiable and satisfies

$$f(y, z)/(|y| + |z|) \to \infty \text{ as } |y| + |z| \to \infty.$$

In this section we study a family of variational problems

$$\int_0^T f(v(t), v'(t))dt \to \min, \tag{P_0}$$

© Springer International Publishing Switzerland 2015
A.J. Zaslavski, *Turnpike Theory of Continuous-Time Linear Optimal Control Problems*, Springer Optimization and Its Applications 104,
DOI 10.1007/978-3-319-19141-6_1

$v : [0, T] \to R^n$ is an absolutely continuous function

such that $v(0) = y, \; v(T) = z,$

where T is a positive number and $y, z \in R^n$. More precisely, we analyze the structure of minimizers of the problem (P_0) when the values at the endpoints y, z and T vary and T is sufficiently large.

Consider the following optimization problem

$$f(y, 0) \to \min, \; y \in R^n. \tag{P_1}$$

The strict convexity of f and the growth condition imply that the problem (P_1) possesses a unique minimizer $\bar{y}$ which satisfies

$$\partial f / \partial y(\bar{y}, 0) = 0.$$

Define a function $L : R^n \times R^n \to R^1$ by

$$
\begin{aligned}
L(y, z) &= f(y, z) - f(\bar{y}, 0) \\
&\quad - \langle \nabla f(\bar{y}, 0), (y, z) - (\bar{y}, 0) \rangle \\
&= f(y, z) - f(\bar{y}, 0) - \langle (\partial f / \partial z)(\bar{y}, 0), z \rangle
\end{aligned}
$$

for all $y, z \in R^n$. It is not difficult to see that the function $L : R^n \times R^n \to R^1$ is differentiable and strictly convex and satisfies

$$L(y, z) / (|y| + |z|) \to \infty \text{ as } |y| + |z| \to \infty.$$

In view of the strict convexity of the functions f and L we have [38]

$$L(y, z) \geq 0 \text{ for all } (y, z) \in R^n \times R^n$$

and

$$L(y, z) = 0 \text{ if and only if } y = \bar{y}, \; z = 0.$$

Consider an auxiliary variational problem

$$\int_0^T L(v(t), v'(t)) dt \to \min, \tag{P_2}$$

$v : [0, T] \to R^n$ is an absolutely continuous function

such that $v(0) = y, \; v(T) = z,$

where $T > 0$ and $y, z \in R^n$. It is not difficult to see that for every $T > 0$ and every absolutely continuous function $x : [0, T] \to R^n$ the following equalities hold:

$$\int_0^T L(x(t), x'(t))dt$$

$$= \int_0^T [f(x(t), x'(t)) - f(\bar{y}, 0) - \langle (\partial f / \partial z)(\bar{y}, 0), x'(t) \rangle] dt$$

$$= \int_0^T f(x(t), x'(t))dt + Tf(\bar{y}, 0) - \langle (\partial f / \partial z)(\bar{y}, 0), x(T) - x(0) \rangle.$$

The equations above imply that a function $x : [0, T] \to R^n$ is a solution of the problem (P_0) if and only if it is a solution of the problem (P_2). This means that the problems (P_0) and (P_2) are equivalent.

Note that the point $(\bar{y}, 0)$ is the unique solution of the minimization problem

$$L(y, z) \to \min, \quad y, z \in R^n.$$

This optimization problem is also well-posed. More precisely, we claim that the following property holds:

(C) for every sequence $\{(y_i, z_i)\}_{i=1}^\infty \subset R^n \times R^n$ satisfying $\lim_{i \to \infty} L(y_i, z_i) = 0$ we have $\lim_{i \to \infty}(y_i, z_i) = (\bar{y}, 0)$.

Indeed, let a sequence $\{(y_i, z_i)\}_{i=1}^\infty \subset R^n \times R^n$ be such that

$$\lim_{i \to \infty} L(y_i, z_i) = 0.$$

It follows from the growth condition that the sequence $\{(y_i, z_i)\}_{i=1}^\infty$ is bounded. Let (y, z) be its limit point. By the continuity of L,

$$L(y, z) = \lim_{i \to \infty} L(y_i, z_i) = 0$$

and since $(\bar{y}, 0)$ is the unique point of minimum of the function L we have $(y, z) = (\bar{y}, 0)$. Since (y, z) is any limit point of the sequence $\{(y_i, z_i)\}_{i=1}^\infty$ we conclude that $(\bar{y}, 0) = \lim_{i \to \infty}(y_i, z_i)$, as claimed.

Let $y, z \in R^n$, $T > 2$ be a real number and let an absolutely continuous function $\bar{x} : [0, T] \to R^n$ be an optimal solution of the problem (P_0). Evidently, the function $\bar{x}$ is also an optimal solution of the problem (P_2). We claim that

$$\int_0^T L(\bar{x}(t), \bar{x}'(t))dt \leq 2c_0(|y|, |z|)$$

where $c_0(|y|, |z|)$ is a positive constant depending only on $|y|$ and $|z|$.

Set

$$x(t) = y + t(\bar{y} - y), \ t \in [0, 1],$$
$$x(t) = \bar{y}, \ t \in [1, T - 1],$$
$$x(t) = \bar{y} + (t - (T - 1))(z - \bar{y}), \ t \in [T - 1, T].$$

Clearly, $x : [0, T] \to R^n$ is an absolutely continuous function satisfying $x(0) = y$ and $x(T) = z$. Since $\bar{x}$ is a solution of the problem (P2) we have

$$\int_0^T L(\bar{x}(t), \bar{x}'(t))dt \leq \int_0^T L(x(t), x'(t))dt$$
$$= \int_0^1 L(x(t), \bar{y} - y)dt + \int_1^{T-1} L(\bar{y}, 0)dt$$
$$+ \int_{T-1}^T L(x(t), z - \bar{y})dt$$
$$= \int_0^1 L(x(t), \bar{y} - y)dt + \int_{T-1}^T L(x(t), z - \bar{y})dt.$$

It is easy to see that the integrals

$$\int_0^1 L(x(t), \bar{y} - y)dt \text{ and } \int_{T-1}^T L(x(t), z - \bar{y})dt$$

do not exceed a constant $c_0(|y|, |z|) > 0$ which depends only on the norms $|y|, |z|$. Hence

$$\int_0^T L(\bar{x}(t), \bar{x}'(t))dt \leq 2c_0(|y|, |z|).$$

It should be mentioned that here the constant $c_0(|y|, |z|)$ does not depend on T.

Denote by mes(E) the Lebesgue measure of a Lebesgue measurable set $E \subset R^1$.

Now let $\epsilon > 0$ be a real number. Property (C) implies the existence of a real number $\delta > 0$ such that if a point $(y, z) \in R^n \times R^n$ satisfies $L(y, z) \leq \delta$, then $|y - \bar{y}| + |z| \leq \epsilon$. It follows from the choice of δ and the relation

$$\int_0^T L(\bar{x}(t), \bar{x}'(t))dt \leq 2c_0(|y|, |z|)$$

that

$$\text{mes}\{t \in [0, T] : |(\bar{x}(t), \bar{x}'(t)) - (\bar{y}, 0)| > \epsilon\}$$
$$\leq \text{mes}\{t \in [0, T] : L(\bar{x}(t), \bar{x}'(t)) > \delta\}$$

$$\leq \delta^{-1} \int_0^T L(\bar{x}(t), \bar{x}'(t)) dt \leq \delta^{-1} 2c_0(|y|, |z|)$$

and

$$\text{mes}\{t \in [0, T] : |\bar{x}(t) - \bar{y}| > \epsilon\} \leq \delta^{-1} 2c_0(|y|, |z|).$$

In view of the inequality above the minimizer $\bar{x}$ spends most of the time in the ϵ-neighborhood of the point $\bar{y}$. More precisely, the Lebesgue measure of the set of all points t such that $\bar{x}(t)$ does not belong to this ϵ-neighborhood, does not exceed the constant $2\delta^{-1} c_0(|y|, |z|)$ depending only on the norms $|y|, |z|$ and ϵ. Note that it does not depend on the length of the interval T. Following the tradition, the point $\bar{y}$ is called the turnpike. Moreover, it can be shown that the set

$$\{t \in [0, T] : |\bar{x}(t) - \bar{y}| > \epsilon\}$$

is a subset of the union of two intervals $[0, \tau_1] \cup [T - \tau_2, T]$, where $0 < \tau_1, \tau_2 \leq 2\delta^{-1} c_0(|y|, |z|)$.

The main goal of this book is to study the turnpike phenomenon and the structure of approximate solutions in regions close to endpoints for linear optimal control systems.

1.2 Problems with Periodic Convex Integrands

In Chap. 2 of this book we study the structure of approximate optimal trajectories of linear control systems with periodic convex integrands and show that these systems possess a turnpike property. This means that approximate optimal trajectories are determined mainly by the integrand, and are essentially independent of the choice of time interval and data, except in regions close to the endpoints of the time interval. We also study the stability of the turnpike phenomenon under small perturbations of integrands and study the structure of approximate optimal trajectories in regions close to the endpoints of the time intervals.

More precisely, we study the structure of approximate optimal trajectories of linear control systems governed by the equation

$$x'(t) = Ax(t) + Bu(t), \tag{1.1}$$

with periodic convex integrands $f : [0, \infty) \times R^n \times R^m \to R^1$, where A and B are given matrices of dimensions $n \times n$ and $n \times m$, $x(t) \in R^n$, $u(t) \in R^m$ and the admissible controls are Lebesgue measurable functions.

We assume that the integrand f is a Borel measurable function and that the linear system (1.1) is controllable which means that the rank of the matrix $(B, AB, \ldots A^{n-1}B)$ is n.

The performance of the above control system is measured on any finite interval $[T_1, T_2] \subset [0, \infty)$ by the integral functional

$$I(T_1, T_2, x, u) = \int_{T_1}^{T_2} f(t, x(t), u(t))dt.$$

We denote by $| \cdot |$ the Euclidean norm and by $\langle \cdot, \cdot \rangle$ the inner product in the n-dimensional Euclidean space R^n.

Artstein and Leizarowitz [1] analyzed the existence and structure of solutions of the linear system (1.1) with an integrand

$$f(t, x, u) = (x - \Gamma(t))'Q(x - \Gamma(t)) + u'Pu \; (t \in [0, \infty), \; x \in R^n, \; u \in R^m),$$

where P is a given positive definite symmetric matrix, Q is a positive semidefinite symmetric matrix, the pair (A, Q) is observable and $\Gamma : [0, \infty) \rightarrow R^n$ is a measurable function satisfying

$$\Gamma(t + T) = \Gamma(t) \; (t \in [0, \infty))$$

for some constant $T > 0$. Artstein and Leizarowitz [1] showed the existence of a unique solution for the infinite horizon tracking of the periodic trajectory Γ and established a turnpike property for finite time optimizers. Their methods are based on explicit expressions for optimal solutions to tracking on finite intervals. In Chap. 6 of [44] and in [57] we extended the results of [1] to an integrand $f : [0, \infty) \times R^n \times R^n \rightarrow R^1$ which satisfies the following assumptions:

(i) $f(t + \tau, x, u) = f(t, x, u)$ for all $t \in [0, \infty)$, all $x \in R^n$ and all $u \in R^m$ for some constant $\tau > 0$ depending only on f;
(ii)   for any $t \in [0, \infty)$ the function $f(t, \cdot, \cdot) : R^n \times R^m \rightarrow R^1$ is strictly convex;
(iii)   the function f is bounded on any bounded subset of $[0, \infty) \times R^n \times R^m$;
(iv) $f(t, x, u) \rightarrow \infty$ as $|x| \rightarrow \infty$ uniformly in $(t, u) \in [0, \infty) \times R^m$;
(v) $f(t, x, u)|u|^{-1} \rightarrow \infty$ as $|u| \rightarrow \infty$ uniformly in $(t, x) \in [0, \infty) \times R^n$.

In this section we also suppose that the assumptions above hold for the integrand f.

Remark 1.1. It is not difficult to see that if λ is a positive number and if assumptions (i)–(v) hold with $f = f_i$, $i = 1, 2$ and with the same $\tau > 0$, where $f_1, f_2 : [0, \infty) \times R^n \times R^m \rightarrow R^1$ are measurable functions, then assumptions (i)–(v) hold with $f = \lambda f_1$ and $f = f_1 + f_2$.

Example 1.2. Let $\tau > 0$. It is not difficult to see that assumptions (i)–(v) hold with a function $f : [0, \infty) \times R^n \times R^m \rightarrow R^1$ defined by

$$f(t, x, u) = g(t)(x - \Gamma(t))'Q(x - \Gamma(t))$$

$$+ h(t)u'Pu + H(t) \; (t \in [0, \infty), \; x \in R^n, \; u \in R^m),$$

where $\Gamma : [0, \infty) \to R^n$ is a measurable and bounded on $[0, \infty)$ function, P, Q are positive definite symmetric matrices, $H, h, g : [0, \infty) \to R^1$ are measurable bounded functions such that for all $t \geq 0$,

$$\Gamma(t + \tau) = \Gamma(t), \ g(t + \tau) = g(t), \ h(t + \tau) = h(t), \ H(t + \tau) = H(t)$$

and that

$$\inf\{g(t) : t \in [0, \infty)\} > 0, \ \inf\{h(t) : t \in [0, \infty)\} > 0.$$

Example 1.3. Let $\tau > 0$. Assume that $h : [0, \infty) \to R^1$ is a bounded measurable function such that

$$h(t + \tau) = h(t) \text{ for all } t \geq 0,$$

$$\inf\{h(t) : t \in [0, \infty)\} > 0$$

and that a strictly convex function $g = g(x, u) \in C^1(R^{n+m})$ has the following properties:

$$g(x, u) \geq \max\{\psi(|x|), \ \psi(|u|)|u|\},$$

$$\max\{|\partial g/\partial x(x, u)|, \ |\partial g/\partial u(x, u)|\} \leq \psi_0(|x|)(1 + \psi|u|)|u|,$$

$x \in R^n$, $u \in R^m$, where $\psi : [0, \infty) \to (0, \infty)$, $\psi_0 : (0, \infty) \to [0, \infty)$ are monotone increasing functions, $\psi(t) \to \infty$ as $t \to \infty$;
for any $\epsilon > 0$ there exists $\delta(\epsilon) > 0$ such that if $x_1, x_2 \in R^n$ and if $u_1, u_2 \in R^m$ satisfy $|x_1 - x_2| + |u_1 - u_2| \geq \epsilon$, then

$$g(2^{-1}(x_1 + x_2), 2^{-1}(u_1 + u_2)) \leq 2^{-1}[g(x_1, u_1) + g(x_2, u_2)] - \delta(\epsilon).$$

It is not difficult to see that the integrand $f(t, x, u) = h(t)g(x, u)$, $t \in [0, \infty)$, $x \in R^n$, $u \in R^m$ satisfies assumptions (i)–(v) and f is not taken from Remark 1.1 and Example 1.2.

Using Remark 1.1 and Examples 1.2 and 1.3, we can easily construct numerous examples of integrands satisfying assumptions (i)–(v).

Example 1.4. Let $\tau > 0$ and k be a natural number. Assume that $\Gamma_i : [0, \infty) \to R^n$, $i = 1, \ldots, k$ are measurable and bounded on $[0, \infty)$ functions, $P_i, Q_i, i = 1, \ldots, k$ are positive definite symmetric matrices, $h_i, \ i = 1, \ldots, k, g_i, \ i = 1, \ldots, k, H :$ $[0, \infty) \to R^1$ are measurable bounded functions such that for all $t \geq 0$ and all $i = 1, \ldots, k$,

$$\Gamma_i(t + \tau) = \Gamma_i(t), \ g_i(t + \tau) = g_i(t), \ h_i(t + \tau) = h_i(t), \ H(t + \tau) = H(t)$$

and that

$$\inf\{g_i(t) : t \in [0, \infty)\} > 0, \ \inf\{h_i(t) : t \in [0, \infty)\} > 0.$$

In view of Remark 1.1 and Example 1.2, assumptions (i)–(v) hold with a function $f : [0, \infty) \times R^n \times R^m \to R^1$ defined by

$$f(t, x, u) = \sum_{i=1}^{k} g_i(t)(x - \Gamma_i(t))' Q_i (x - \Gamma_i(t))$$

$$+ \sum_{i=1}^{k} h_i(t) u' P_i u + H(t) \ (t \in [0, \infty), \ x \in R^n, \ u \in R^m).$$

We consider the following optimal control problems:

$$I(0, T, x, u) \to \min, \tag{P_1}$$

$x : [0, T] \to R^n, \ u : [0, T] \to R^m$ is a trajectory-control pair such that

$$x(0) = y, \ x(T) = z,$$

$$I(0, T, x, u) \to \min, \tag{P_2}$$

$$x : [0, T] \to R^n, \ u : [0, T] \to R^m$$

is a trajectory-control pair such that $x(0) = y$,

$$I(0, T, x, u) \to \min, \tag{P_3}$$

$x : [0, T] \to R^n, \ u : [0, T] \to R^m$ is a trajectory-control,

where $y, z \in R^n$ and $T > 0$.

The study of these problems is based on the properties of solutions of the corresponding infinite horizon optimal control problem associated with the control system (1.1) and the integrand f.

In [57] (see also Chap. 6 of [44]) we were interested in a turnpike property of the approximate solutions of problems (P_2). It was shown that there exists a trajectory-control pair $x_f : [0, \tau] \to R^n, \ u_f : [0, \tau] \to R^m$ which is the unique solution of the minimization problem

$$I(0, \tau, x, u) \to \min, \ x : [0, \tau] \to R^n, \ u : [0, \tau] \to R^m$$

is a trajectory-control pair such that $x(0) = x(\tau)$.

Put

$$\mu(f) = \tau^{-1} I(0, \tau, x_f, u_f).$$

It was shown in [57] (see also Chap. 6 of [44]) that for any trajectory-control pair $x : [0, \infty) \to R^n$, $u : [0, \infty) \to R^m$ either

$$I(0, T, x, u) - T\mu(f) \to \infty \text{ as } T \to \infty$$

or

$$\sup\{|I(0, T, x, u) - T\mu(f)| : T > 0\} < \infty. \tag{1.2}$$

Moreover, if (1.2) holds, then

$$\sup\{|x(i\tau + t) - x_f(t)| : t \in [0, \tau]\} \to 0 \text{ as } i \to \infty$$

over the integers.

We say that a trajectory-control pair $x : [0, \infty) \to R^n$, $u : [0, \infty) \to R^m$ is good [44, 53] if

$$\sup\{|I(0, T, x, u) - T\mu(f)| : T > 0\} < \infty.$$

We say that a trajectory-control pair $\tilde{x} : [0, \infty) \to R^n$, $\tilde{u} : [0, \infty) \to R^m$ is overtaking optimal [44, 53] if

$$\limsup_{T \to \infty}[I(0, T, \tilde{x}, \tilde{u}) - I(0, T, x, u)] \leq 0$$

for each trajectory-control pair $x : [0, \infty) \to R^n$, $u : [0, \infty) \to R^m$ satisfying $x(0) = \tilde{x}(0)$.

The following existence result was obtained in [57] (see also Chap. 6 of [44]).

Theorem 1.5. *Let $x_0 \in R^n$. Then there exists a unique overtaking optimal trajectory-control pair $\tilde{x} : [0, \infty) \to R^n$, $\tilde{u} : [0, \infty) \to R^m$ satisfying $\tilde{x}(0) = x_0$.*

The next theorem, which was also obtained in [57], establishes the turnpike property for approximate solutions of problems (P_2) with the turnpike $x_f(\cdot)$.

Theorem 1.6. *Let $M, \epsilon > 0$. Then there exist an integer $N \geq 1$ and $\delta > 0$ such that for each $T > 2N\tau$ and each trajectory-control pair $x : [0, T] \to R^n$, $u : [0, T] \to R^m$ which satisfies*

$$|x(0)| \leq M,$$

$$I(0, T, x, u) \leq \inf\{I(0, T, y, v) : y : [0, T] \to R^n, v : [0, T]$$

$$\text{is a trajectory-control pair, } y(0) = x(0)\} + \delta$$

the inequality

$$\sup\{|x(i\tau + t) - x_f(t)| : t \in [0, \tau]\} \leq \epsilon \tag{1.3}$$

holds for all integers $i \in [N, \tau^{-1}T - N]$. *Moreover if* $|x(0) - x_f(0)| \le \delta$, *then inequality* (1.3) *holds for all integers* $i \in [0, \tau^{-1}T - N]$.

In Chap. 2 we establish the turnpike property of the approximate solutions of problems (P_1) and (P_3). We show the stability of the turnpike phenomenon under small perturbations of the integrand f and study the structure of approximate optimal trajectories in regions close to the endpoints of the time intervals.

For the problems (P_2) and (P_3) we show that in regions close to the right endpoint T of the time interval these approximate solutions are determined only by the integrand, and are essentially independent of the choice of interval and the endpoint value y. For the problems (P_3), approximate solutions are determined only by the integrand also in regions close to the left endpoint 0 of the time interval.

The study of these problems is based on the properties of solutions of the corresponding infinite horizon optimal control problem associated with the control system (1.1) and the integrand f.

1.3   Nonconvex Optimal Control Problems

In Chap. 3 we study the existence and structure of optimal trajectories of linear control systems with autonomous integrands. For these control systems we establish the existence of optimal trajectories over an infinite horizon and show that the turnpike phenomenon holds. We also study the structure of approximate optimal trajectories in regions close to the endpoints of the time intervals. It is shown that in these regions optimal trajectories converge to solutions of the corresponding infinite horizon optimal control problem which depend only on the integrand. For this class of optimal control problems the turnpike property holds with the turnpike being a singleton. It should be mentioned that there are many linear optimal control problems for which the turnpike is a singleton. Let us consider a few examples.

We use the notation, definitions, and assumptions introduced in Sect. 1.2. Assume that an integrand $f(t, x, u)$, $(t, x, u) \in [0, \infty) \times R^n \times R^m$ satisfies assumptions (ii)–(v) and does not depend on the variable t. As a matter of fact, we can consider this function as $f : R^n \times R^m \to R^1$. Clearly, f satisfies the assumption (i) with any $\tau > 0$. In view of the results discussed in Sect. 1.2, the turnpike phenomenon holds and the turnpike $x_f(\cdot)$ is a constant.

Remark 1.7. If integrands f_1, f_2 do not depend on t, assumptions (ii)–(v) hold with $f = f_i$, $i = 1, 2$, where $f_1, f_2 : R^n \times R^m \to R^1$ are measurable functions, and if λ is a positive number, then in view of Remark 1.1, assumptions (ii)–(v) hold with $f = \lambda f_1$ and $f = f_1 + f_2$, the turnpike property holds for these integrands and their turnpikes are singletons.

Example 1.8. Assume that a strictly convex function $g = g(x, u) \in C^1(R^{n+m})$ has the following properties:

$$g(x, u) \geq \max\{\psi(|x|), \ \psi(|u|)|u|\},$$

$$\max\{|\partial g/\partial x(x, u)|, \ |\partial g/\partial u(x, u)|\} \leq \psi_0(|x|)(1 + \psi|u|)|u|,$$

$x \in R^n$, $u \in R^m$, where $\psi : [0, \infty) \to (0, \infty)$, $\psi_0 : (0, \infty) \to [0, \infty)$ are monotone increasing functions, $\psi(t) \to \infty$ as $t \to \infty$;

for any $\epsilon > 0$ there exists $\delta(\epsilon) > 0$ such that if $x_1, x_2 \in R^n$ and if $u_1, u_2 \in R^m$ satisfy $|x_1 - x_2| + |u_1 - u_2| \geq \epsilon$, then

$$g(2^{-1}(x_1 + x_2), 2^{-1}(u_1 + u_2)) \leq 2^{-1}[g(x_1, u_1) + g(x_2, u_2)] - \delta(\epsilon).$$

In view of Example 1.3, the function $g(x, u)$, $x \in R^n$, $u \in R^m$ satisfies assumptions (ii)–(v), possesses the turnpike property and the turnpike is a singleton.

Example 1.9. Let k be a natural number. Assume that $\xi_i \in R^n$, $i = 1, \ldots, k$, P_i, Q_i, $i = 1, \ldots, k$ are positive definite symmetric matrices and let

$$f(x, u) = \sum_{i=1}^{k}(x - \xi_i)'Q_i(x - \xi_i) + \sum_{i=1}^{k} u'P_i u, \ x \in R^n, \ u \in R^m.$$

In view of Example 1.4, assumptions (ii)–(v) hold for the function f which possesses the turnpike property and the turnpike is a singleton.

Using Remark 1.7 and Examples 1.8 and 1.9, we can easily construct numerous examples of integrands which do not depend on the variable t and satisfy assumptions (ii)–(v). For these integrands the turnpike property holds and the turnpike is a singleton.

It is clear that in the examples above the integrands are convex functions. In our book we study the existence and structure of optimal trajectories of linear control systems with autonomous integrands which are not necessarily convex.

More precisely, we study the structure of approximate optimal trajectories of linear control systems governed by the equation

$$x'(t) = Ax(t) + Bu(t), \tag{1.4}$$

with integrands $f : R^n \times R^m \to R^1$ satisfying the assumptions below, where n, m are natural numbers, A and B are given matrices of dimensions $n \times n$ and $n \times m$, $x(t) \in R^n$, $u(t) \in R^m$ and the admissible controls are Lebesgue measurable functions.

We assume that the linear system (1.4) is controllable and that the integrand f is a continuous function.

For every $s \in R^1$ set $s_+ = \max\{s, 0\}$. For every nonempty set X and every function $h : X \to R^1 \cup \{\infty\}$ set

$$\inf(h) = \inf\{h(x) : x \in X\}.$$

Let $a_0 > 0$ and $\psi : [0, \infty) \to [0, \infty)$ be an increasing function such that

$$\lim_{t \to \infty} \psi(t) = \infty.$$

Suppose that $f : R^n \times R^m \to R^1$ is a continuous function such that the following assumption holds:

(A1)

(i) for each $(x, u) \in R^n \times R^m$,

$$f(x, u) \geq \max\{\psi(|x|), \ \psi(|u|),$$
$$\psi([|Ax + Bu| - a_0|x|]_+)[|Ax + Bu| - a_0|x|]_+\} - a_0;$$

(ii) for each $x \in R^n$ the function $f(x, \cdot) : R^m \to R^1$ is convex;
(iii) for each $M, \epsilon > 0$ there exist $\Gamma, \delta > 0$ such that

$$|f(x_1, u_1) - f(x_2, u_2)| \leq \epsilon \max\{f(x_1, u_1), f(x_2, u_2)\}$$

for each $u_1, u_2 \in R^m$ and each $x_1, x_2 \in R^n$ which satisfy

$$|x_i| \leq M, \ |u_i| \geq \Gamma, \ i = 1, 2, \quad \max\{|x_1 - x_2|, |u_1 - u_2|\} \leq \delta;$$

(iv) for each $K > 0$ there exists a constant $a_K > 0$ and an increasing function

$$\psi_K : [0, \infty) \to [0, \infty)$$

such that

$$\psi_K(t) \to \infty \text{ as } t \to \infty$$

and

$$f(x, u) \geq \psi_K(|u|)|u| - a_K$$

for each $u \in R^m$ and each $x \in R^n$ satisfying $|x| \leq K$.

Remark 1.10. A function h satisfies (A1) if $h \in C^1(R^n \times R^m)$, (A1)(i), (A1)(ii), (A1)(iv) hold, and for each $K > 0$ there exists an increasing function $\tilde{\psi} : [0, \infty) \to [0, \infty)$ such that for each $x \in R^n$ satisfying $|x| \leq K$ and each $u \in R^m$,

$$\max\{|\partial h/\partial x(x, u)|, \ |\partial h/\partial u(x, u)|\} \leq \tilde{\psi}(|x|)(1 + \psi_K(|u|)|u|).$$

The performance of the above control system is measured on any finite interval $[T_1, T_2] \subset [0, \infty)$ and for any trajectory-control pair $x : [T_1, T_2] \to R^n$, $u : [T_1, T_2] \to R^m$ by the integral functional

$$I(T_1, T_2, x, u) = \int_{T_1}^{T_2} f(x(t), u(t))dt.$$

In Chap. 3 we consider optimal control problems (P_1), (P_2), and (P_3) introduced in Sect. 1.2.

A number

$$\mu(f) := \inf\{\liminf_{T \to \infty} T^{-1} I(0, T, x, u) :$$

$$x : [0, \infty) \to R^n, \ u : [0, \infty) \to R^m \text{ is a trajectory-control pair}\}$$

is called the minimal long-run average cost growth rate of f. By (A1)(i), $-\infty < \mu(f)$.

We say that a trajectory-control pair $\tilde{x} : [0, \infty) \to R^n$, $\tilde{u} : [0, \infty) \to R^m$ is overtaking optimal [44, 53] if

$$\limsup_{T \to \infty} [I(0, T, \tilde{x}, \tilde{u}) - I(0, T, x, u)] \leq 0$$

for each trajectory-control pair $x : [0, \infty) \to R^n$, $u : [0, \infty) \to R^m$ satisfying $x(0) = \tilde{x}(0)$.

Let $(x_f, u_f) \in R^n \times R^m$ satisfy

$$Ax_f + Bu_f = 0.$$

Clearly, $\mu(f) \leq f(x_f, u_f)$.

We suppose that the following assumption holds.

(A2) $\mu(f) = f(x_f, u_f)$ and if $(x, u) \in R^n \times R^m$ satisfies

$$Ax + Bu = 0, \ \mu(f) = f(x, u),$$

then $x = x_f$.

In Chap. 3 we will show that for each trajectory-control pair $x : [0, \infty) \to R^n$, $u : [0, \infty) \to R^m$ either

$$I(0, T, x, u) - T\mu(f) \to \infty \text{ as } T \to \infty$$

or

$$\sup\{|I(0, T, x, u) - T\mu(f)| : \ T > 0\} < \infty.$$

A trajectory-control pair $x : [0, \infty) \to R^n$, $u : [0, \infty) \to R^m$ is called good [44, 53] if

$$\sup\{|I(0, T, x, u) - T\mu(f)| : T > 0\} < \infty.$$

We suppose that the following assumption holds.

(A3)   For each good trajectory-control pair $x : [0, \infty) \to R^n$, $u : [0, \infty) \to R^m$,

$$\lim_{t \to \infty} x(t) = x_f.$$

Let us consider examples of integrands satisfying assumptions (A1)–(A3).

Example 1.11. Assume that a continuous strictly convex function $h : R^n \times R^m \to R^1$ satisfies assumption (A1) (with $f = h$) and

$$h(x, u)/|u| \to \infty \text{ as } |u| \to \infty \text{ uniformly in } x \in R^n.$$

Then the function h satisfies assumptions (A2) and (A3) (with $f = h$). This follows from Corollary 2.11 of Chap. 2.

Let us consider another example of an integrand which satisfies (A1)–(A3).

Example 1.12. Let $c \in R^1$, $a_1 > 0$, $l \in R^n$, $(x_*, u_*) \in R^n \times R^m$ satisfy $Ax_* + Bu_* = 0$ and let $\psi_0 : [0, \infty) \to [0, \infty)$ be an increasing function such that $\lim_{t \to \infty} \psi_0(t) = \infty$. Assume that a continuous function $L : R^n \times R^m \to [0, \infty)$ satisfies for each $(x, u) \in R^n \times R^m$,

$$L(x, u) \geq \max\{\psi_0(|x|), \ \psi_0(|u|)|u|\} - a_1 + |l||Ax + Bu|,$$

$$L(x, u) = 0 \text{ if and only if } x = x_*, \ u = u_*,$$

for each $x \in R^n$, the function $L(x, \cdot) : R^m \to R^1$ is convex and for each $M, \epsilon > 0$ there exist $\Gamma, \delta > 0$ such that

$$|L(x_1, u_1) - L(x_2, u_2)| \leq \epsilon \max\{L(x_1, u_1), \ L(x_2, u_2)\}$$

for each $x_1, x_2 \in R^n$ and each $u_1, u_2 \in R^m$ which satisfy

$$|x_i| \leq M, \ |u_i| \geq \Gamma, \ i = 1, 2, \ |x_1 - x_2|, \ |u_1 - u_2| \leq \delta.$$

For every $(x, u) \in R^n \times R^m$ set

$$h(x, u) = L(x, u) + c + \langle l, Ax + Bu \rangle.$$

It is not difficult to see that for each $(x, u) \in R^n \times R^m$,

$$h(x, u) \geq \max\{\psi_0(|x|), \ \psi_0(|u|)|u|\} - a_1 - |c|$$

and that h satisfies (A1) holds under the appropriate choice of $a_0 > 0$, ψ. In Sect. 3.12 we prove that

$$\mu(h) = h(x_*, u_*) = c,$$

(A2)   holds for $f = h$ and for any good trajectory-control pair $x : [0, \infty) \to R^n$, $u : [0, \infty) \to R^m$,

$$\lim_{t \to \infty} x(t) = x_*$$

(see Proposition 3.6).

In Chap. 3 we study the existence and structure of optimal trajectories of linear control system (1.4) with the integrand f. For these control systems we establish the existence of optimal trajectories over an infinite horizon and show that the turnpike phenomenon holds for approximate solutions of problems (P_1), (P_2) and (P_3). For problems (P_2) and (P_3) we show that, in regions close to the right endpoint T of the time interval, their approximate solutions are determined only by the integrand, and are essentially independent of the choice of interval and the endpoint value y. For problems (P_3), approximate solutions are determined only by the integrand also in regions close to the left endpoint 0 of the time interval. It is shown that in the regions closed to the endpoints optimal closed trajectories converge to solutions of the corresponding infinite horizon optimal control problem which depend only on the integrand.

In Chaps. 4, 5, and 7 we continue to study the class of linear optimal control problems considered in this section. In Chap. 4 we show that the turnpike phenomenon and the convergence, in the regions close to the endpoints of time intervals, are stable under small perturbations of the integrand f. Linear control systems with discounting are studied in Chap. 5. In Chap. 7 we show that for a typical (in the sense of Baire category) integrand the values of approximate solutions at the end points converge to the limit which is a unique solution of the corresponding minimization problem associated with the integrand. In Chap. 6 we study the existence and turnpike properties of approximate solutions for a class of dynamic continuous-time two-player zero-sum games without using convexity-concavity assumptions and with linear control constraints. We describe the structure of approximate solutions which is independent of the length of the interval, for all sufficiently large intervals and show that approximate solutions are determined mainly by the objective function, and are essentially independent of the choice of interval and endpoint conditions.

1.4 Problems of the Calculus of Variations with Extended-Valued Integrands

In Chap. 8 we study the structure of approximate solutions of autonomous variational problems with a lower semicontinuous extended-valued integrand. In our recent research we showed that approximate solutions are determined mainly by the integrand, and are essentially independent of the choice of time interval and data, except in regions close to the endpoints of the time interval. In Chap. 8 our goal is to study the structure of approximate solutions in regions close to the endpoints of the time intervals.

More precisely, in Chap. 8 we consider the following variational problems

$$\int_0^T f(v(t), v'(t))dt \to \min, \tag{P_1}$$

$v : [0, T] \to R^n$ is an absolutely continuous (a.c.) function such that

$$v(0) = x, \ v(T) = y,$$

$$\int_0^T f(v(t), v'(t))dt \to \min, \tag{P_2}$$

$v : [0, T] \to R^n$ is an a. c. function such that $v(0) = x$

and

$$\int_0^T f(v(t), v'(t))dt \to \min, \tag{P_3}$$

$v : [0, T] \to R^n$ is an a. c. function,

where $x, y \in R^n$. Here R^n is the n-dimensional Euclidean space with the Euclidean norm $|\cdot|$ and $f : R^n \times R^n \to R^1 \cup \{\infty\}$ is an extended-valued integrand.

The problems (P_1) and (P_2) were studied in [46, 49, 53], where it was shown, under certain assumptions, that the turnpike property holds and that the turnpike $\bar{x}$ is a unique solution of the minimization problem $f(x, 0) \to \min, x \in R^n$.

In this book we study the structure of approximate solutions of the problems (P_2) and (P_3) in regions close to the endpoints of the time intervals. It is shown that in regions close to the right endpoint T of the time interval these approximate solutions are determined only by the integrand, and are essentially independent of the choice of interval and endpoint value x. For the problems (P_3), approximate solutions are determined only by the integrand also in regions close to the left endpoint 0 of the time interval.

More precisely, we define $\bar{f}(x, y) = f(x, -y)$ for all $x, y \in R^n$ and consider the set $\mathcal{P}(\bar{f})$ of all solutions of a corresponding infinite horizon variational problem associated with the integrand $\bar{f}$. For a given pair of real positive numbers ϵ, τ, we show that if T is large enough and $v : [0, T] \to R^n$ is an approximate solution of the problem (P_2), then $|v(T - t) - y(t)| \leq \epsilon$ for all $t \in [0, \tau]$, where $y(\cdot) \in \mathcal{P}(\bar{f})$.

In this section we describe the class of integrands which is considered in Chap. 8. For each function $f : X \to R^1 \cup \{\infty\}$, where X is a nonempty, set

$$\mathrm{dom}(f) = \{x \in X : f(x) < \infty\}.$$

Let a be a real positive number, $\psi : [0, \infty) \to [0, \infty)$ be an increasing function such that

$$\lim_{t \to \infty} \psi(t) = \infty$$

and let $f : R^n \times R^n \to R^1 \cup \{\infty\}$ be a lower semicontinuous function such that the set

$$\mathrm{dom}(f) = \{(x, y) \in R^n \times R^n : f(x, y) < \infty\}$$

is nonempty, convex, and closed and that

$$f(x, y) \geq \max\{\psi(|x|), \ \psi(|y|)|y|\} - a \text{ for each } x, y \in R^n.$$

We suppose that there exists a point $\bar{x} \in R^n$ such that

$$f(\bar{x}, 0) \leq f(x, 0) \text{ for each } x \in R^n \tag{1.5}$$

and that the following assumptions hold:

(A1) $(\bar{x}, 0)$ is an interior point of the set $\mathrm{dom}(f)$ and the function f is continuous at the point $(\bar{x}, 0)$;

(A2)   for each $M > 0$ there exists $c_M > 0$ such that

$$\int_0^T f(v(t), v'(t))dt \geq Tf(\bar{x}, 0) - c_M$$

for each real number $T > 0$ and each a. c. function $v : [0, T] \to R^n$ satisfying $|v(0)| \leq M$;

(A3)   for each $x \in R^n$ the function $f(x, \cdot) : R^n \to R^1 \cup \{\infty\}$ is convex.

It should be mentioned that inequality (1.5) and assumptions (A1)–(A3) are common in the literature and hold for many infinite horizon optimal control problems. In particular, we need inequality (1.5) and assumption (A2) in the cases when the problems (P_1) and (P_2) possess the turnpike property and the point $\bar{x}$ is its turnpike. Assumption (A2) means that the constant function $\bar{v}(t) = \bar{x}$, $t \in [0, \infty)$

is an approximate solution of the infinite horizon variational problem with the integrand f related to the problems (P_1) and (P_2).

We say that an a. c. function $v : [0, \infty) \to R^n$ is (f)-good [44, 51] if

$$\sup \left\{ \left| \int_0^T f(v(t), v'(t))dt - Tf(\bar{x}, 0) \right| : T \in (0, \infty) \right\} < \infty.$$

The following result was obtained in [46].

Proposition 1.13. *Let $v : [0, \infty) \to R^n$ be an a.c. function. Then either the function v is (f)-good or*

$$\int_0^T f(v(t), v'(t))dt - Tf(\bar{x}, 0) \to \infty \text{ as } T \to \infty.$$

Moreover, if the function v is (f)-good, then $\sup\{|v(t)| : t \in [0, \infty)\} < \infty$.

We suppose that the following assumption holds:

(A4)   (the asymptotic turnpike property) for each (f)-good function $v : [0, \infty) \to R^n$, $\lim_{t\to\infty} |v(t) - \bar{x}| = 0$.

We consider two examples of integrands f which satisfy assumptions (A1)–(A4).

Example 1.14. Let a_0 be a positive number, $\psi_0 : [0, \infty) \to [0, \infty)$ be an increasing function satisfying

$$\lim_{t\to\infty} \psi_0(t) = \infty$$

and let $L : R^n \times R^n \to [0, \infty]$ be a lower semicontinuous function such that

$$\operatorname{dom}(L) := \{(x, y) \in R^n \times R^n : L(x, y) < \infty\}$$

is nonempty, convex, closed set and

$$L(x, y) \geq \max\{\psi_0(|x|), \ \psi_0(|y|)|y|\} - a_0 \text{ for each } x, y \in R^n.$$

Assume that for each point $x \in R^n$ the function $L(x, \cdot) : R^n \to R^1 \cup \{\infty\}$ is convex and that there exists a point $\bar{x} \in R^n$ such that

$$L(x, y) = 0 \text{ if and only if } (x, y) = (\bar{x}, 0),$$

$(\bar{x}, 0)$ is an interior point of $\operatorname{dom}(L)$ and that L is continuous at the point $(\bar{x}, 0)$.

Let $\mu \in R^1$ and $l \in R^n$. Define

$$f(x, y) = L(x, y) + \mu + \langle l, y \rangle, \ x, y \in R^n.$$

It was shown in [46] and in Sect. 3.9 of [53] that all the assumptions introduced in this section hold for f.

Example 1.15. Let a be a positive number, $\psi : [0, \infty) \to [0, \infty)$ be an increasing function such that $\lim_{t \to \infty} \psi(t) = \infty$ and let $f : R^n \times R^n \to R^1 \cup \{\infty\}$ be a convex lower semicontinuous function such that the set dom(f) is nonempty, convex and closed and that

$$f(x, y) \geq \max\{\psi(|x|), \ \psi(|y|)|y|\} - a \text{ for each } x, y \in R^n.$$

We suppose that there exists a point $\bar{x} \in R^n$ such that

$$f(\bar{x}, 0) \leq f(x, 0) \text{ for each } x \in R^n$$

and that $(\bar{x}, 0)$ is an interior point of the set dom(f). It is known that the function f is continuous at the point $(\bar{x}, 0)$. It is a well-known fact of convex analysis [38] that there exists a point $l \in R^n$ such that

$$f(x, y) \geq f(\bar{x}, 0) + \langle l, y \rangle \text{ for each } x, y \in R^n.$$

We assume that for each pair of points (x_1, y_1), $(x_2, y_2) \in \text{dom}(f)$ satisfying $(x_1, y_1) \neq (x_2, y_2)$ and each number $\alpha \in (0, 1)$, we have

$$f(\alpha(x_1, y_1) + (1 - \alpha)(x_2, y_2)) < \alpha f(x_1, y_1) + (1 - \alpha)f(x_2, y_2).$$

Put

$$L(x, y) = f(x, y) - f(\bar{x}, 0) - \langle l, y \rangle \text{ for each } x, y \in R^n.$$

It is not difficult to see that there exist a positive number a_0 and an increasing function $\psi_0 : [0, \infty) \to [0, \infty)$ such that

$$L(x, y) \geq \max\{\psi_0(|x|), \ \psi_0(|y|)|y|\} - a_0 \text{ for all } x, y \in R^n.$$

It is easy to see that L is a convex, lower semicontinuous function and that the equality $L(x, y) = 0$ holds if and only if $(x, y) = (\bar{x}, 0)$. Now it is easy to see that our example is a particular case of Example 1.14 and all the assumptions introduced in this section hold for f.

In the last chapter of the book we study a class of dynamic continuous-time two-player zero-sum unconstrained games with extended-valued integrands. We do not assume convexity-concavity assumptions and establish the existence and the turnpike property of approximate solutions.

The results of Chap. 8 were obtained in [55]. The results which are proved in all other chapters are new.

Chapter 2
Linear Control Systems with Periodic Convex Integrands

We study the structure of approximate optimal trajectories of linear control systems with periodic convex integrands and show that these systems possess a turnpike property. To have this property means, roughly speaking, that the approximate optimal trajectories are determined mainly by the integrand, and are essentially independent of the choice of time interval and data, except in regions close to the endpoints of the time interval. We also show the stability of the turnpike phenomenon under small perturbations of integrands and study the structure of approximate optimal trajectories in regions close to the endpoints of the time intervals.

2.1 Preliminaries and Turnpike Results

In this chapter we study the structure of approximate optimal trajectories of linear control systems

$$x'(t) = Ax(t) + Bu(t), \tag{2.1}$$

$$x(0) = x_0$$

with periodic convex integrands $f : [0, \infty) \times R^n \times R^m \to R^1$, where A and B are given matrices of dimensions $n \times n$ and $n \times m$, $x(t) \in R^n$, $u(t) \in R^m$ and the admissible controls are Lebesgue measurable functions.

We assume that the linear system (2.1) is controllable and that the integrand f is a Borel measurable function.

We denote by $| \cdot |$ the Euclidean norm and by $\langle \cdot, \cdot \rangle$ the inner product in the n-dimensional Euclidean space R^n. Denote by $\mathbf{Z}$ the set of all integers. For every

© Springer International Publishing Switzerland 2015
A.J. Zaslavski, *Turnpike Theory of Continuous-Time Linear Optimal Control Problems*, Springer Optimization and Its Applications 104, DOI 10.1007/978-3-319-19141-6_2

$z \in R^1$ denote by $\lfloor z \rfloor$ the largest integer which does not exceed z: $\lfloor z \rfloor = \max\{i \in \mathbf{Z} : i \leq z\}$.

The performance of the above control system is measured on any finite interval $[T_1, T_2] \subset [0, \infty)$ by the integral functional

$$I^f(T_1, T_2, x, u) = \int_{T_1}^{T_2} f(t, x(t), u(t))dt. \tag{2.2}$$

We suppose that the integrand $f : [0, \infty) \times R^n \times R^m \to R^1$ satisfies the following
   Assumption (A)

(i) $f(t + \tau, x, u) = f(t, x, u)$ for all $t \in [0, \infty)$, all $x \in R^n$ and all $u \in R^m$ for some
   constant $\tau > 0$ depending only on f;
(ii) for any $t \in [0, \infty)$ the function $f(t, \cdot, \cdot) : R^n \times R^m \to R^1$ is strictly convex;
(iii) the function f is bounded on any bounded subset of $[0, \infty) \times R^n \times R^m$;
(iv) $f(t, x, u) \to \infty$ as $|x| \to \infty$ uniformly in $(t, u) \in [0, \infty) \times R^m$;
(v) $f(t, x, u)|u|^{-1} \to \infty$ as $|u| \to \infty$ uniformly in $(t, x) \in [0, \infty) \times R^n$.

Assumption (A) implies that f is bounded below on $[0, \infty) \times R^n \times R^m$.

Let $T_2 > T_1 \geq 0$. A pair of an absolutely continuous (a.c.) function $x : [T_1, T_2] \to R^n$ and a Lebesgue measurable function $u : [T_1, T_2] \to R^m$ is called an (A, B)-trajectory-control pair if for almost every (a. e.) $t \in [T_1, T_2]$ (2.1) holds. Denote by $X(A, B, T_1, T_2)$ the set of all (A, B)-trajectory-control pairs $x : [T_1, T_2] \to R^n$, $u : [T_1, T_2] \to R^m$.

Let $J = [a, \infty)$ be an infinite closed subinterval of $[0, \infty)$. A pair of functions $x : J \to R^n$ and $u : J \to R^m$ is called an (A, B)-trajectory-control pair if it is an (A, B)-trajectory-control pair on any bounded closed subinterval of J. Denote by $X(A, B, a, \infty)$ the set of all (A, B)-trajectory-control pairs $x : J \to R^n$, $u : J \to R^m$.

In this chapter we study the structure of approximate optimal trajectories of the linear control system (2.1) with the integrand f and show that the turnpike property holds. To have this property means, roughly speaking, that the approximate optimal trajectories on sufficiently large intervals are determined mainly by the integrand, and are essentially independent of the choice of time intervals and data, except in regions close to the endpoints of the time intervals. We also show the stability of the turnpike phenomenon under small perturbations of the integrand and study the structure of approximate optimal trajectories in regions close to the endpoints of the time intervals.

More precisely, we consider the following optimal control problems

$$I^f(0, T, x, u) \to \min, \tag{P_1}$$

$$(x, u) \in X(A, B, 0, T) \text{ such that } x(0) = y, \; x(T) = z,$$

$$I^f(0, T, x, u) \to \min, \tag{P_2}$$

$$(x, u) \in X(A, B, 0, T) \text{ such that } x(0) = y,$$

$$I^f(0, T, x, u) \to \min, \qquad\qquad (P_3)$$

$$(x, u) \in X(A, B, 0, T),$$

where $y, z \in R^n$ and $T > 0$. The study of these problems is based on the properties of solutions of the corresponding infinite horizon optimal control problem associated with the control system (2.1) and the integrand f.

In [57] we were interested in a turnpike property of the approximate solutions of problems (P_2). In this chapter we establish the turnpike property of the approximate solutions of problems (P_1) and (P_3). We show the stability of the turnpike phenomenon under small perturbations of the integrand f and study the structure of approximate optimal trajectories in regions close to the endpoints of the time intervals.

For the problems (P_2) and (P_3) we show that in regions close to the right endpoint T of the time interval these approximate solutions are determined only by the integrand, and are essentially independent of the choice of interval and the endpoint value y. For the problems (P_3), approximate solutions are determined only by the integrand also in regions close to the left endpoint 0 of the time interval.

The following result was obtained in [57] (see also Chap. 6 of [44]).

Proposition 2.1. *There exists* $(x_f, u_f) \in X(A, B, 0, \tau)$ *which is the unique solution of the following minimization problem*

$$I^f(0, \tau, x, u) \to \min, \ (x, u) \in X(A, B, 0, \tau) \text{ such that } x(0) = x(\tau).$$

Let a trajectory-control pair $(x_f, u_f) \in X(A, B, 0, \tau)$ be as guaranteed by Proposition 2.1. Put

$$\mu(f) = \tau^{-1} I^f(0, \tau, x_f, u_f). \qquad\qquad (2.3)$$

The following results were obtained in [57] (see also Chap. 6 of [44]).

Theorem 2.2. *For any* $(x, u) \in X(A, B, 0, \infty)$ *either*

$$(i) \ I^f(0, T, x, u) - T\mu(f) \to \infty \text{ as } T \to \infty$$

$$or \ (ii) \ \sup\{|I^f(0, T, x, u) - T\mu(f)| : \ T > 0\} < \infty.$$

Moreover, if relation (ii) holds, then

$$\sup\{|x(i\tau + t) - x_f(t)| : \ t \in [0, \tau]\} \to 0 \text{ as } i \to \infty, \text{ where } i \in \mathbf{Z}.$$

We say that $(x, u) \in X(A, B, 0, \infty)$ is (f, A, B)-good [44, 53] if

$$\sup\{|I^f(0, T, x, u) - T\mu(f)| : \ T > 0\} < \infty.$$

The second statement of Theorem 2.2 describes the asymptotic behavior of (f, A, B)-good trajectory-control pairs, shows that the corresponding infinite horizon optimal control problem has the turnpike property, and that the function x_f is its turnpike.

We say that $(\tilde{x}, \tilde{u}) \in X(A, B, 0, \infty)$ is (f, A, B)-overtaking optimal [44, 53] if for each $(x, u) \in X(A, B, 0, \infty)$ satisfying $x(0) = \tilde{x}(0)$,

$$\limsup_{T \to \infty}[I^f(0, T, \tilde{x}, \tilde{u}) - I^f(0, T, x, u)] \leq 0.$$

Theorem 2.3. *Let* $x_0 \in R^n$. *Then there exists an* (f, A, B)-*overtaking optimal trajectory-control pair* $(\tilde{x}, \tilde{u}) \in X(A, B, 0, \infty)$ *satisfying* $\tilde{x}(0) = x_0$. *Moreover, if* $(x, u) \in X(A, B, 0, \infty) \setminus \{(\tilde{x}, \tilde{u})\}$ *satisfies* $x(0) = x_0$, *then there are* $T_0 > 0$ *and* $\epsilon > 0$ *such that*

$$I^f(0, T, x, u) \geq I^f(0, T, \tilde{x}, \tilde{u}) + \epsilon \ for \ all \ T \geq T_0.$$

The next result describes the limit behavior of overtaking optimal trajectories.

Theorem 2.4. *Let* $M, \epsilon > 0$. *Then there exists a natural number* N *such that for any* (f, A, B)-*overtaking optimal trajectory-control pair* $(x, u) \in X(A, B, 0, \infty)$ *which satisfies* $|x(0)| \leq M$ *the relation*

$$\sup\{|x(i\tau + t) - x_f(t)| : \ t \in [0, \tau]\} \leq \epsilon \tag{2.4}$$

holds for all integers $i \geq N$. *Moreover, there exists* $\delta > 0$ *such that for any* (f, A, B)-*overtaking optimal trajectory-control pair* $(x, u) \in X(A, B, 0, \infty)$ *satisfying* $|x(0) - x_f(0)| \leq \delta$, *the relation* (2.4) *holds for all integers* $i \geq 0$.

Let $T > 0$ and $y, z \in R^n$. Set

$$\sigma(f, y, z, T) = \inf\{I^f(0, T, x, u) :$$

$$(x, u) \in X(A, B, 0, T) \ and \ x(0) = y, \ x(T) = z\}, \tag{2.5}$$

$$\sigma(f, y, T) = \inf\{I^f(0, T, x, u) : \ (x, u) \in X(A, B, 0, T) \ and \ x(0) = y\}, \tag{2.6}$$

$$\hat{\sigma}(f, z, T) = \inf\{I^f(0, T, x, u) : \ (x, u) \in X(A, B, 0, T) \ and \ x(T) = z\}, \tag{2.7}$$

$$\sigma(f, T) = \inf\{I^f(0, T, x, u) : \ (x, u) \in X(A, B, 0, T)\}. \tag{2.8}$$

It follows from assumption (A) and Proposition 2.28 that

$$-\infty < \sigma(f, y, z, T), \sigma(f, y, T), \hat{\sigma}(f, z, T), \sigma(f, T) < \infty.$$

The next theorem establishes the turnpike property for approximate solutions of problems (P_2) with the turnpike $x_f(\cdot)$.

Theorem 2.5. *Let $M, \epsilon > 0$. Then there exist an integer $N \geq 1$ and $\delta > 0$ such that for each $T > 2N\tau$ and each $(x, u) \in X(A, B, 0, T)$ which satisfies*

$$|x(0)| \leq M, \ I^f(0, T, x, u) \leq \sigma(f, x(0), T) + \delta$$

the inequality

$$\sup\{|x(i\tau + t) - x_f(t)| : t \in [0, \tau]\} \leq \epsilon \tag{2.9}$$

holds for all integers $i \in [N, \tau^{-1}T - N]$. Moreover if $|x(0) - x_f(0)| \leq \delta$, then inequality (2.9) holds for all integers $i \in [0, \tau^{-1}T - N]$.

Theorems 2.2–2.5 were obtained in [57] (see also Chap. 6 of [44]). Note that under assumptions of Theorem 2.5, if $|x(\lfloor \tau^{-1}T \rfloor \tau) - x_f(0)| \leq \delta$, then inequality (2.9) holds for all integers $i \in [N, \tau^{-1}T - 1]$.

The next two results establish the turnpike property for approximate solutions of problems (P_1) and (P_3) respectively with the turnpike $x_f(\cdot)$.

Theorem 2.6. *Let $M, \epsilon > 0$. Then there exist an integer $N \geq 1$ and $\delta > 0$ such that for each $T > 2N\tau$ and each $(x, u) \in X(A, B, 0, T)$ which satisfies*

$$|x(0)|, \ |x(T)| \leq M, \ I^f(0, T, x, u) \leq \sigma(f, x(0), x(T), T) + \delta$$

inequality (2.9) holds for all integers $i \in [N, \tau^{-1}T - N]$. Moreover if $|x(0) - x_f(0)| \leq \delta$, then inequality (2.9) holds for all integers $i \in [0, \tau^{-1}T - N]$ and if $|x(\lfloor \tau^{-1}T \rfloor \tau) - x_f(0)| \leq \delta$, then inequality (2.9) holds for all integers $i \in [N, \tau^{-1}T - 1]$.

Theorem 2.7. *Let $\epsilon > 0$. Then there exist an integer $N \geq 1$ and $\delta > 0$ such that for each $T > 2N\tau$ and each $(x, u) \in X(A, B, 0, T)$ which satisfies*

$$I^f(0, T, x, u) \leq \sigma(f, T) + \delta$$

inequality (2.9) holds for all integers $i \in [N, \tau^{-1}T - N]$. Moreover if $|x(0) - x_f(0)| \leq \delta$, then inequality (2.9) holds for all integers $i \in [0, \tau^{-1}T - N]$ and if $|x(\lfloor \tau^{-1}T \rfloor \tau) - x_f(0)| \leq \delta$, then inequality (2.9) holds for all integers $i \in [N, \tau^{-1}T - 1]$.

Theorems 2.5–2.7 are partial cases of Theorem 2.13 stated in Sect. 2.2 which is one of the main results of the chapter. The next theorem establishes a weak version of the turnpike property for approximate solutions of problems (P_1), (P_2), and (P_3) with the turnpike $x_f(\cdot)$.

Theorem 2.8. *Let $\epsilon, M_0, M_1 > 0$. Then there exist natural numbers Q, l such that for each $T > Ql\tau$ and each $(x, u) \in X(A, B, 0, T)$ which satisfies at least one of the following conditions:*

$$|x(0)|, \ |x(T)| \le M_0, \ I^f(0, T, x, u) \le \sigma(f, x(0), x(T), T) + M_1;$$

$$|x(0)| \le M_0, \ I^f(0, T, x, u) \le \sigma(f, x(0), T) + M_1;$$

$$I^f(0, T, x, u) \le \sigma(f, T) + M_1$$

there exist strictly increasing sequences of nonnegative integers

$$\{a_i\}_{i=1}^q, \ \{b_i\}_{i=1}^q \subset [0, \tau^{-1}T]$$

such that $q \le Q$,

$$0 \le b_i - a_i \le l \text{ for all } i = 1, \ldots, q,$$

$b_i \le a_{i+1}$ *for all integers i satisfying $1 \le i < q$ and that for each integer $i \in [0, \tau^{-1}T - 1] \setminus \cup_{j=1}^q [a_j, b_j]$,*

$$|x(i\tau + t) - x_f(t)| \le \epsilon, \ t \in [0, \tau].$$

Theorem 2.8 is a partial case of Theorem 2.14, our stability result (see Sect. 2.2). We say that $(x, u) \in X(A, B, 0, \infty)$ is (f, A, B)-minimal [5, 53] if for each $T > 0$,

$$I^f(0, T, x, u) = \sigma(f, x(0), x(T), T). \tag{2.10}$$

The next result which is proved in Sect. 2.5 shows the equivalence of the optimality criterions introduced above.

Theorem 2.9. *Assume that $(x, u) \in X(A, B, 0, \infty)$. Then the following conditions are equivalent:*

(i) (x, u) is (f, A, B)-overtaking optimal; (ii) (x, u) is (f, A, B)-minimal and (f, A, B)-good; (iii) (x, u) is (f, A, B)-minimal and

$$\max\{|x(i\tau + t) - x_f(t)| : \ t \in [0, \tau]\} \to 0 \text{ as integers } i \to \infty;$$

(iv) (x, u) is (f, A, B)-minimal and $\liminf_{t \to \infty} |x(t)| < \infty$.

The following result is also proved in Sect. 2.5. It shows that if the integrand f does not depend on the variable t, then $x_f(\cdot)$ is a constant function.

Theorem 2.10. *Assume that for each $x \in R^n$, each $u \in R^m$, and each $t_1, t_2 \ge 0$, $f(t_1, x, u) = f(t_2, x, u)$. Then $x_f(t) = x_f(0)$ for all $t \in [0, \tau]$ and $x_f(0)$ does not depend of τ.*

Corollary 2.11. *Assume that for each $x \in R^n$, each $u \in R^m$, and each $t_1, t_2 \ge 0$, $f(t_1, x, u) = f(t_2, x, u)$. Then for all $t \in [0, \tau]$, $x_f(t) = x_*$ and $u_f(t) = u_*$ where $(x_*, u_*) \in R^n \times R^m$ is a unique solution of the minimization problem*

$$f(x, u) \to \min, \ (x, u) \in R^n \times R^m, \ Ax + Bu = 0.$$

2.2 Stability of the Turnpike Phenomenon

In this section we state Theorems 2.12–2.14 which show that the turnpike phenomenon is stable under small perturbations of the integrand f. We use the notation, definitions, and assumptions introduced in Sect. 2.1.

Recall that $f : [0, \infty) \times R^n \times R^m \to R^1$ is a Borel measurable function satisfying assumption (A). Let $a > 0$ and $\psi : [0, \infty) \to [0, \infty)$ be an increasing function such that

$$\lim_{t \to \infty} \psi(t) = \infty. \tag{2.11}$$

We suppose that for all $(t, x, u) \in [0, \infty) \times R^n \times R^m$,

$$f(t, x, u) \geq \max\{\psi(|x|), \ \psi(|u|)|u|\} - a. \tag{2.12}$$

Denote by $\mathcal{M}$ the set of all Borel measurable functions $g : [0, \infty) \times R^n \times R^m \to R^1$ which are bounded on all bounded subsets of $[0, \infty) \times R^n \times R^m$ and such that for all $(t, x, u) \in [0, \infty) \times R^n \times R^m$,

$$g(t, x, u) \geq \max\{\psi(|x|), \ \psi(|u|)|u|\} - a. \tag{2.13}$$

For the set $\mathcal{M}$ we consider the uniformity which is determined by the following base:

$$E(N, \epsilon, \lambda) = \{(g_1, g_2) \in \mathcal{M} \times \mathcal{M} : \ |g_1(t, x, u) - g_2(t, x, u)| \leq \epsilon \text{ for each } t \geq 0,$$

$$\text{each } x \in R^n \text{ satisfying } |x| \leq N \text{ and each } u \in R^m \text{ satisfying } |u| \leq N\}$$

$$\cap \{(g_1, g_2) \in \mathcal{M} \times \mathcal{M} : \ (|g_1(t, x, u)| + 1)(|g_2(t, x, u)| + 1)^{-1} \in [\lambda^{-1}, \lambda]$$

$$\text{for each } t \geq 0, \ \text{each } x \in R^n \text{ satisfying } |x| \leq N \text{ and each } u \in R^m\}, \tag{2.14}$$

where $N > 0, \ \epsilon > 0, \ \lambda > 1$. It is not difficult to see that the space $\mathcal{M}$ with this uniformity is metrizable and complete.

Let $T_2 > T_1 \geq 0, y, z \in R^n$, and $g \in \mathcal{M}$. For each pair of Lebesgue measurable functions $x : [T_1, T_2] \to R^n$, $u : [T_1, T_2] \to R^m$ set

$$I^g(T_1, T_2, x, u) = \int_{T_1}^{T_2} g(t, x(t), u(t)) dt \tag{2.15}$$

and set

$$\sigma(g, y, z, T_1, T_2) = \inf\{I^g(T_1, T_2, x, u) :$$

$$(x, u) \in X(A, B, T_1, T_2) \text{ and } x(T_1) = y, \ x(T_2) = z\}, \tag{2.16}$$

$$\sigma(g, y, T_1, T_2) = \inf\{I^g(T_1, T_2, x, u) :$$

$$(x, u) \in X(A, B, T_1, T_2) \text{ and } x(T_1) = y\}, \tag{2.17}$$

$$\hat{\sigma}(g, z, T_1, T_2) = \inf\{I^g(T_1, T_2, x, u) :$$

$$(x, u) \in X(A, B, T_1, T_2) \text{ and } x(T_2) = z\}, \tag{2.18}$$

$$\sigma(g, T_1, T_2) = \inf\{I^g(T_1, T_2, x, u) : (x, u) \in X(A, B, T_1, T_2)\}. \tag{2.19}$$

Since any $g \in \mathcal{M}$ is bounded on all the bounded subsets of $[0, \infty) \times R^n \times R^m$ it follows from Proposition 2.28 and (2.13) that all the values defined above are finite.

In this chapter we prove the following three stability results.

Theorem 2.12. *Let $\epsilon, M > 0$. Then there exist an integer $L_0 \geq 1$ and $\delta_0 > 0$ such that for each integer $L_1 \geq L_0$ there exists a neighborhood $\mathcal{U}$ of f in $\mathcal{M}$ such that the following assertion holds.*

Assume that $T > 2L_1\tau$, $g \in \mathcal{U}$, $(x, u) \in X(A, B, 0, T)$ and that a finite sequence of integers $\{S_i\}_{i=0}^q$ satisfy

$$S_0 = 0, \ S_{i+1} - S_i \in [L_0, L_1], \ i = 0, \ldots, q-1, \ S_q\tau \in (T - L_1\tau, T], \tag{2.20}$$

$$I^g(S_i\tau, S_{i+1}\tau, x, u) \leq (S_{i+1} - S_i)\tau\mu(f) + M$$

for each integer $i \in [0, q-1]$,

$$I^g(S_i\tau, S_{i+2}\tau, x, u) \leq \sigma(g, x(S_i\tau), x(S_{i+2}\tau), S_i\tau, S_{i+2}\tau) + \delta_0$$

for each nonnegative integer $i \leq q-2$ and

$$I^g(S_{q-2}\tau, T, x, u) \leq \sigma(g, x(S_{q-2}\tau), x(T), S_{q-2}\tau, T) + \delta_0.$$

Then there exist integers $p_1, p_2 \in [0, \tau^{-1}T]$ such that $p_1 \leq p_2$, $p_1 \leq 2L_0$, $p_2 > \tau^{-1}T - 2L_1$ and that for all integers $i = p_1, \ldots, p_2 - 1$,

$$\max\{|x(i\tau + t) - x_f(t)| : t \in [0, \tau]\} \leq \epsilon.$$

Moreover if $|x(0) - x_f(0)| \leq \delta_0$, then $p_1 = 0$ and if $|x(\lfloor \tau^{-1}T \rfloor \tau) - x_f(0)| \leq \delta_0$, then $p_2 = \lceil \tau^{-1}T \rceil$.

Theorem 2.13. *Let $\epsilon \in (0, 1)$, $M_0, M_1 > 0$. Then there exist an integer $L \geq 1$, $\delta \in (0, \epsilon)$ and a neighborhood $\mathcal{U}$ of f in $\mathcal{M}$ such that for each $T > 2L\tau$, each $g \in \mathcal{U}$, and each $(x, u) \in X(A, B, 0, T)$ which satisfies for each $S \in [0, T - L\tau]$,*

$$I^g(S, S + L\tau, x, u) \leq \sigma(g, x(S), x(S + L\tau), S, S + L\tau) + \delta$$

and satisfies at least one of the following conditions:

(a) $|x(0)|$, $|x(T)| \leq M_0$, $I^g(0, T, x, u) \leq \sigma(g, x(0), x(T), 0, T) + M_1$;

(b) $|x(0)| \leq M_0$, $I^g(0, T, x, u) \leq \sigma(g, x(0), 0, T) + M_1$;

(c) $I^g(0, T, x, u) \leq \sigma(g, 0, T) + M_1$

there exist integers $p_1 \in [0, L]$, $p_2 \in [\lfloor \tau^{-1}T \rfloor - L, \tau^{-1}T]$ *such that for all integers* $i = p_1, \ldots, p_2 - 1$,

$$|x(i\tau + t) - x_f(t)| \leq \epsilon \text{ for all } t \in [0, \tau].$$

Moreover if $|x(0) - x_f(0)| \leq \delta$, *then* $p_1 = 0$ *and if* $|x(\lfloor \tau^{-1}T \rfloor \tau) - x_f(0)| \leq \delta$, *then* $p_2 = \lfloor \tau^{-1}T \rfloor$.

Denote by Card(A) the cardinality of the set A.

Theorem 2.14. *Let* $\epsilon \in (0, 1)$, $M_0, M_1 > 0$. *Then there exist an integer* $L \geq 1$ *and a neighborhood* $\mathcal{U}$ *of* f *in* $\mathcal{M}$ *such that for each* $T > L\tau$, *each* $g \in \mathcal{U}$, *and each* $(x, u) \in X(A, B, 0, T)$ *which satisfies at least one of the following conditions:*

(a) $|x(0)|$, $|x(T)| \leq M_0$, $I^g(0, T, x, u) \leq \sigma(g, x(0), x(T), 0, T) + M_1$;

(b) $|x(0)| \leq M_0$, $I^g(0, T, x, u) \leq \sigma(g, x(0), 0, T) + M_1$;

(c) $I^g(0, T, x, u) \leq \sigma(g, 0, T) + M_1$

the following inequality holds:

Card($\{i \in \{0, \ldots, \lfloor \tau^{-1}T \rfloor - 1\} : \max\{|x(i\tau + t) - x_f(t)| : t \in [0, \tau]\} > \epsilon\}) \leq L$.

2.3 Structure of Solutions in the Regions Close to the End Points

In this section we state results which describe the structure of solutions of problems (P_1), (P_2) and (P_3) in the regions close to the end points. Combined with the turnpike results of Sect. 2.2 they provide the full description of the structure of their solutions. We use the notation, definitions, and assumptions introduced in Sects. 2.1 and 2.2.

By Theorem 2.14 for each $z \in R^n$ there exists a unique (f, A, B)-overtaking optimal pair $(\xi^{(z)}, \eta^{(z)}) \in X(A, B, 0, \infty)$ such that $\xi^{(z)}(0) = z$. Let $z \in R^n$. Set

$$\pi^f(z) = \liminf_{T \to \infty, \, T \in Z} [I^f(0, T\tau, \xi^{(z)}, \eta^{(z)}) - T\tau\mu(f)]. \tag{2.21}$$

In view of Theorems 2.2, 2.3, and 2.9, $\pi^f(z)$ is a finite number. Definition (2.21) and the definition of (f, A, B)-overtaking optimal pairs imply the following result.

Proposition 2.15. *1. Let $(x, u) \in X(A, B, 0, \infty)$ be (f, A, B)-good. Then*

$$\pi^f(x(0)) \leq \liminf_{T \to \infty, \, T \in Z} [I^f(0, T\tau, x, u) - T\tau\mu(f)]$$

and for each pair of integers $S > T \geq 0$,

$$\pi^f(x(T\tau)) \leq I^f(T\tau, S\tau, x, u) - (S - T)\tau\mu(f) + \pi^f(x(S\tau)). \qquad (2.22)$$

2. Let $S > T \geq 0$ be integers and $(x, u) \in X(A, B, T\tau, S\tau)$. Then (2.22) holds.

The next result follows from definition (2.21).

Proposition 2.16. *Let $(x, u) \in X(A, B, 0, \infty)$ be (f, A, B)-overtaking optimal. Then for each pair of integers $S > T \geq 0$,*

$$\pi^f(x(T\tau)) = I^f(T\tau, S\tau, x, u) - (S - T)\tau\mu(f) + \pi^f(x(S\tau)).$$

Theorems 2.3–2.5 and (2.21), (2.3) imply the following result.

Proposition 2.17. $\pi^f(x_f(0)) = 0$.

The following result is proved in Sect. 2.11.

Proposition 2.18. *The function π^f is continuous at $x_f(0)$.*

Proposition 2.19. *Let $(x, u) \in X(A, B, 0, \infty)$ be (f, A, B)-overtaking optimal. Then*

$$\pi^f(x(0)) = \lim_{T \to \infty, \, T \in Z} [I^f(0, T\tau, x, u) - T\tau\mu(f)].$$

Proof. It follows from Propositions 2.16–2.18 and Theorems 2.2 and 2.9 that

$$\pi^f(x(0)) = \lim_{T \to \infty, \, T \in Z} (\pi^f(x(0)) - \pi^f(x(T\tau)))$$

$$= \lim_{T \to \infty, \, T \in Z} [I^f(0, T\tau, x, u) - T\tau\mu(f)].$$

Proposition 2.19 is proved.

Proposition 2.20. *The function π^f is strictly convex and continuous.*

Proof. It is sufficient to show that the function π^f is strictly convex. Let $y, z \in R^n$, $y \neq z$ and $\alpha \in (0, 1)$. Consider $(\alpha\xi^{(y)} + (1 - \alpha)\xi^{(z)}, \alpha\eta^{(y)} + (1 - \alpha)\eta^{(z)}) \in X(A, B, 0, \infty)$ which satisfies $(\alpha\xi^{(y)} + (1-\alpha)\xi^{(z)})(0) = \alpha y + (1-\alpha)z$ and in view of (A) and Theorem 2.2, is (f, A, B)-good. By Propositions 2.15 and 2.19, assumption (A) and the relation $y \neq z$,

$$\pi^f(\alpha y + (1-\alpha)z) = \liminf_{T\to\infty,\, T\in Z}[I^f(0, T\tau, \alpha\xi^{(y)} + (1-\alpha)\xi^{(z)}, \alpha\eta^{(y)}$$

$$+ (1-\alpha)\eta^{(z)}) - T\tau\mu(f)]$$

$$< \liminf_{T\to\infty,\, T\in Z}[\alpha(I^f(0, T\tau, \xi^{(y)}, \eta^{(y)}) - T\tau\mu(f))$$

$$+ (1-\alpha)(I^f(0, T\tau, \xi^{(z)}, \eta^{(z)}) - T\tau\mu(f))]$$

$$= \alpha \lim_{T\to\infty,\, T\in Z}[I^f(0, T\tau, \xi^{(y)}, \eta^{(y)}) - T\tau\mu(f)]$$

$$+ (1-\alpha) \lim_{T\to\infty,\, T\in Z}[I^f(0, T\tau, \xi^{(z)}, \eta^{(z)}) - T\tau\mu(f)]$$

$$= \alpha\pi^f(y) + (1-\alpha)\pi^f(z).$$

Proposition 2.20 is proved.

The next result is proved in Sect. 2.11.

Proposition 2.21. *For each $M > 0$ the set $\{x \in R^n : \pi^f(x) \le M\}$ is bounded.*

Set

$$\inf(\pi^f) = \inf\{\pi^f(z) : z \in R^n\}. \tag{2.23}$$

By Propositions 2.20 and 2.21, $\inf(\pi^f)$ is finite and there exists a unique $\theta_f \in R^n$ such that $\pi^f(\theta_f) = \inf(\pi^f)$.

Proposition 2.22. *Let $(x, u) \in X(A, B, 0, \infty)$ be (f, A, B)-good such that for all integers $T > 0$,*

$$I^f(0, T\tau, x, u) - T\tau\mu(f) = \pi^f(x(0)) - \pi^f(x(T\tau)). \tag{2.24}$$

Then $(x, u) \in X(A, B, 0, \infty)$ is (f, A, B)-overtaking optimal.

Proof. Theorem 2.3 implies that there exists an (f, A, B)-overtaking optimal pair $(x_1, u_1) \in X(A, B, 0, \infty)$ such that $x_1(0) = x(0)$. By Proposition 2.16, for each integer $T \ge 1$,

$$I^f(0, T\tau, x_1, u_1) - T\tau\mu(f) = \pi^f(x_1(0)) - \pi^f(x_1(T\tau)).$$

It follows from the equality above, (2.24), Theorems 2.2 and 2.9, and Propositions 2.17 and 2.18 that for all integers $T > 0$,

$$I^f(0, T\tau, x, u) - I^f(0, T\tau, x_1, u_1) = \pi^f(x_1(T\tau)) - \pi^f(x(T\tau)) \to 0 \text{ as } T \to \infty.$$

Thus

$$\lim_{T\to\infty,\ T\in Z}[I^f(0,T\tau,x,u) - I^f(0,T\tau,x_1,u_1)] = 0$$

and in view of Theorem 2.3 this implies the pair (x,u) is (f,A,B)-overtaking optimal. Proposition 2.22 is proved.

Consider a linear control system

$$x'(t) = -Ax(t) - Bu(t), \quad x(0) = x_0$$

which is also controllable. There exists a Borel measurable function $\bar{f} : [0,\infty) \times R^n \times R^m \to R^1$ such that for all $(x,u) \in R^n \times R^m$,

$$\bar{f}(t+\tau,x,u) = \bar{f}(t,x,u) \text{ for all } t \geq 0,$$

$$\bar{f}(t,x,u) = f(\tau - t,x,u) \text{ for all } t \in [0,\tau]. \tag{2.25}$$

Evidently, $\bar{f}$ satisfies assumption (A). For $\bar{f}$ we use all the notation and definitions introduced for f. It is clear that all the results obtained for the triplet (f,A,B) also hold for the triplet $(\bar{f},-A,-B)$.

Assume that integers $S_2 > S_1 \geq 0$ and that $(x,u) \in X(A,B,S_1\tau,S_2\tau)$. For all $t \in [S_1\tau,S_2\tau]$ set

$$\bar{x}(t) = x(S_2\tau - t + S_1\tau), \quad \bar{u}(t) = u(S_2\tau - t + S_1\tau). \tag{2.26}$$

In view of (2.26) for a. e. $t \in [S_1\tau, S_2\tau]$,

$$\bar{x}'(t) = -x'(S_2\tau - t + S_1\tau) = -Ax(S_2\tau - t + S_1\tau) - Bu(S_2\tau - t + S_1\tau)$$
$$= -A\bar{x}(t) - B\bar{u}(t)$$

and $(\bar{x},\bar{u}) \in X(-A,-B,S_1\tau,S_2\tau)$. By (2.25) and (2.26),

$$\int_{S_1\tau}^{S_2\tau} \bar{f}(t,\bar{x}(t),\bar{u}(t))dt = \int_{S_1\tau}^{S_2\tau} \bar{f}(t,x(S_2\tau - t + S_1\tau),u(S_2\tau - t + S_1\tau))dt$$

$$= \int_{S_1\tau}^{S_2\tau} f(S_2\tau - t + S_1\tau, x(S_2\tau - t + S_1\tau),$$
$$u(S_2\tau - t + S_1\tau))dt$$

$$= \int_{S_1\tau}^{S_2\tau} f(t,x(t),u(t))dt. \tag{2.27}$$

For each pair $T_2 > T_1 \geq 0$ and each $(x, u) \in X(-A, -B, T_1, T_2)$ set

$$\bar{I}^f(T_1, T_2, x, u) = \int_{T_1}^{T_2} \bar{f}(t, x(t), u(t))dt.$$

For each $y, z \in R^n$ and each $T > 0$ set

$$\sigma_-(\bar{f}, y, z, T) = \inf\{\bar{I}^f(0, T, x, u) :$$
$$(x, u) \in X(-A, -B, 0, T) \text{ and } x(0) = y, \ x(T) = z\},$$

$$\sigma_-(\bar{f}, y, T) = \inf\{\bar{I}^f(0, T, x, u) : \ (x, u) \in X(-A, -B, 0, T) \text{ and } x(0) = y\},$$

$$\hat{\sigma}_-(\bar{f}, z, T) = \inf\{\bar{I}^f(0, T, x, u) : \ (x, u) \in X(-A, -B, 0, T) \text{ and } x(T) = z\},$$

$$\sigma_-(\bar{f}, T) = \inf\{\bar{I}^f(0, T, x, u) : \ (x, u) \in X(-A, -B, 0, T)\}. \tag{2.28}$$

Relations (2.26) and (2.27) imply the following result.

Proposition 2.23. *Let $S_2 > S_1 \geq 0$ be integers, $M \geq 0$ and that $(x_i, u_i) \in X(A, B, S_1\tau, S_2\tau)$, $i = 1, 2$. Then*

$$I^f(S_1\tau, S_2\tau, x_1, u_1) \geq I^f(S_1\tau, S_2\tau, x_2, u_2) - M$$

if and only if $\bar{I}^f(S_1\tau, S_2\tau, \bar{x}_1, \bar{u}_1) \geq \bar{I}^f(S_1\tau, S_2\tau, \bar{x}_2, \bar{u}_2) - M$.

Proposition 2.23 implies the following result.

Proposition 2.24. *Let $S_2 > S_1 \geq 0$ be integers and*

$$(x, u) \in X(A, B, S_1\tau, S_2\tau).$$

Then the following assertion holds:

$$I^f(S_1\tau, S_2\tau, x, u) \leq \sigma(f, (S_2 - S_1)\tau) + M$$

if and only if $\bar{I}^f(S_1\tau, S_2\tau, \bar{x}, \bar{u}) \leq \sigma_-(\bar{f}, (S_2 - S_1)\tau) + M$;

$$I^f(S_1\tau, S_2\tau, x, u) \leq \sigma(f, x(S_1\tau), x(S_2\tau), (S_2 - S_1)\tau) + M$$

if and only if $\bar{I}^f(S_1\tau, S_2\tau, \bar{x}, \bar{u}) \leq \sigma_-(\bar{f}, \bar{x}(S_1\tau), \bar{x}(S_2\tau), (S_2 - S_1)\tau) + M$;

$$I^f(S_1\tau, S_2\tau, x, u) \leq \sigma(f, x(S_1\tau), (S_2 - S_1)\tau) + M$$

if and only if $\bar{I}^f(S_1\tau, S_2\tau, \bar{x}, \bar{u}) \leq \hat{\sigma}_-(\bar{f}, \bar{x}(S_2\tau), (S_2 - S_1)\tau) + M$;

$$I^f(S_1\tau, S_2\tau, x, u) \leq \hat{\sigma}(f, x(S_2\tau), (S_2 - S_1)\tau) + M$$

if and only if $\bar{I}^f(S_1\tau, S_2\tau, \bar{x}, \bar{u}) \leq \sigma_-(\bar{f}, \bar{x}(S_1\tau), (S_2 - S_1)\tau) + M$.

By Proposition 2.1, $(x_f, u_f) \in X(A, B, 0, \tau)$ is the unique solution of the minimization problem

$$I^f(0, \tau, x, u) \to \min, \ (x, u) \in X(A, B, 0, \tau) \text{ such that } x(0) = x(\tau).$$

Analogously there exists $(x_{\bar{f}}, u_{\bar{f}}) \in X(-A, -B, 0, \tau)$ which is the unique solution of the minimization problem

$$I^{\bar{f}}(0, \tau, x, u) \to \min, \ (x, u) \in X(-A, -B, 0, \tau) \text{ such that } x(0) = x(\tau).$$

In view of Proposition 2.23 and (2.27), for all $t \in [0, \tau]$,

$$x_{\bar{f}}(t) = x_f(\tau - t), \ u_{\bar{f}}(t) = u_f(\tau - t), \ \mu(\bar{f}) = \mu(f). \tag{2.29}$$

For each $z \in R^n$, set

$$\pi^{\bar{f}}(z) = \liminf_{T \to \infty, \ T \in \mathbf{Z}} [I^{\bar{f}}(0, T\tau, x, u) - T\tau\mu(f)], \tag{2.30}$$

where $(x, u) \in X(-A, -B, 0, \infty)$ is the unique $(\bar{f}, -A, -B)$-overtaking optimal pair such that $x(0) = z$. Let $(x_*, u_*) \in X(A, B, 0, \infty)$ be the unique (f, A, B)-overtaking optimal pair such that $\pi^f(x_*(0)) = \inf(\pi^f)$ and

$$(\bar{x}_*, \bar{u}_*) \in X(-A, -B, 0, \infty)$$

be the unique $(\bar{f}, -A, -B)$-overtaking optimal pair such that $\pi^{\bar{f}}(\bar{x}_*(0)) = \inf(\pi^{\bar{f}})$.

The following three theorems describe the structure of solutions of problems (P_1), (P_2), and (P_3) in the regions closed to the end points.

Theorem 2.25. *Let $L_0 > 0$ be an integer, $\epsilon \in (0, 1)$, $M > 0$. Then there exist $\delta > 0$, a neighborhood $\mathcal{U}$ of f in $\mathcal{M}$ and an integer $L_1 > L_0$ such that for each integer $T \geq L_1$, each $g \in \mathcal{U}$ and each $(x, u) \in X(A, B, 0, T\tau)$ which satisfies*

$$|x(0)| \leq M, \ I^g(0, T\tau, x, u) \leq \sigma(g, x(0), 0, T\tau) + \delta$$

the following inequality holds:

$$|x(T\tau - t) - \bar{x}_*(t)| \leq \epsilon \text{ for all } t \in [0, L_0\tau].$$

Theorem 2.26. *Let $L_0 > 0$ be an integer, $\epsilon > 0$. Then there exist $\delta > 0$, a neighborhood $\mathcal{U}$ of f in $\mathcal{M}$ and an integer $L_1 > L_0$ such that for each integer $T \geq L_1$, each $g \in \mathcal{U}$ and each $(x, u) \in X(A, B, 0, T\tau)$ which satisfies*

$$I^g(0, T\tau, x, u) \leq \sigma(g, 0, T\tau) + \delta$$

the following inequalities hold for all $t \in [0, L_0\tau]$:

$$|x(T\tau - t) - \bar{x}_*(t)| \leq \epsilon, \ |x(t) - x_*(t)| \leq \epsilon.$$

Theorem 2.27. *Let $L_0 > 0$ be an integer, $\epsilon > 0$, $M_0 > 0$. Then there exist $\delta > 0$, a neighborhood $\mathcal{U}$ of f in $\mathcal{M}$ and an integer $L_1 > L_0$ such that for each integer $T \geq L_1$, each $g \in \mathcal{U}$ and each $(x, u) \in X(A, B, 0, T\tau)$ which satisfies*

$$|x(0)|, \ |x(T\tau)| \leq M_0, \ I^g(0, T\tau, x, u) \leq \sigma(g, x(0), x(T\tau), 0, T\tau) + \delta$$

the inequalities

$$|x(T\tau - t) - \bar{\xi}(t)| \leq \epsilon, \ |x(t) - \xi(t)| \leq \epsilon$$

hold for all $t \in [0, L_0\tau]$, where $(\xi, \eta) \in X(A, B, 0, \infty)$ is the unique (f, A, B)-overtaking optimal pair such that $\xi(0) = x(0)$ and

$$(\bar{\xi}, \bar{\eta}) \in X(-A, -B, 0, \infty)$$

is the unique $(\bar{f}, -A, -B)$-overtaking optimal pair such that $\bar{\xi}(0) = x(T\tau)$.

2.4 Auxiliary Results

In the sequel we use the following auxiliary results.

Proposition 2.28 (Proposition 6.2.1 of [44]). *For every $\tilde{y}, \tilde{z} \in R^n$ and every $T > 0$ there exists a solution $x(\cdot), y(\cdot)$ of the system*

$$x' = Ax + BB^t y, \ y' = x - A^t y$$

with the boundary conditions $x(0) = \tilde{y}$, $x(T) = \tilde{z}$ (where B^t denotes the transpose of B).

Proposition 2.29 (Proposition 6.2.2 of [44]). *Let $M_1 > 0$ and $0 < \tau_0 < \tau_1$. Then there exists a positive number M_2 such that for each $T \in [\tau_0, \tau_1]$ and each $(x, u) \in X(A, B, 0, T)$ satisfying $I^f(0, T, x, u) \leq M_1$ the inequality $|x(t)| \leq M_2$ holds for all $t \in [0, T]$.*

Proposition 2.30 (Proposition 6.2.4 of [44]). *Let M_1 and T be positive numbers and let $\mathcal{F}$ be the set of all $(x, u) \in X(A, B, 0, T)$ satisfying $I^f(0, T, x, u) \leq M_1$. Then for every sequence $\{(x_i, u_i)\}_{i=1}^{\infty} \subset \mathcal{F}$ there exist a subsequence $\{(x_{i_k}, u_{i_k})\}_{k=1}^{\infty}$ and $(x, u) \in \mathcal{F}$ such that $x_{i_k}(t) \to x(t)$ as $k \to \infty$ uniformly in $[0, T]$, $x'_{i_k} \to x'$ as $k \to \infty$ weakly in $L^1(R^n; (0, T))$, and $u_{i_k} \to u$ as $k \to \infty$ weakly in $L^1(R^m; (0, T))$.*

For each $y, z \in R^n$ define

$$v(y, z) = \inf\{I^f(0, \tau, x, u) : (x, u) \in X(A, B, 0, \tau)$$

$$\text{such that } x(0) = y, \ x(\tau) = z\}. \tag{2.31}$$

It was shown in Sect. 6.2 of [44] that the function v is convex, satisfies

$$-\infty < v(y, z) < \infty \text{ for each } y, z \in R^n,$$

$$v(y, z) \to \infty \text{ as } |y| + |z| \to \infty \tag{2.32}$$

and that there exists $z_f \in R^n$ such that

$$v(z_f, z_f) < v(z, z) \text{ for all } z \in R^n \setminus \{z_f\}, \tag{2.33}$$

$$x_f(0) = z_f, \ \mu(f) = \tau^{-1}v(z_f, z_f). \tag{2.34}$$

Proposition 2.31 (Proposition 6.2.5 of [44]). *There exists $p_f \in R^n$ such that the function $\Theta_f : R^n \times R^n \to R^1$ defined by*

$$\Theta_f(y, z) = v(y, z) - v(z_f, z_f) - \langle p_f, y - z \rangle, \ y, z \in R^n$$

is strictly convex and

$$\Theta_f(z_f, z_f) = 0, \ \Theta_f(y, z) > 0 \text{ for all } (y, z) \in R^n \times R^n \setminus \{(z_f, z_f)\}. \tag{2.35}$$

Define a function $f_0 : R^n \times R^m \to R^1$ by

$$f_0(x, u) = \sup\{f(t, x, u) : t \in [0, \infty)\}, \ (x, u) \in R^n \times R^m. \tag{2.36}$$

In view of assumption (A), the function f_0 is well defined, convex, and bounded on bounded subsets of $R^n \times R^m$. For all $y, z \in R^n$ set

$$v_0(y, z) = \inf\left\{ \int_0^\tau f_0(x(t), u(t))dt : (x, u) \in X(A, B, 0, \tau) \right.$$

$$\left. \text{such that } x(0) = y, \ x(\tau) = z \right\}. \tag{2.37}$$

By (2.37), convexity of f_0, Proposition 2.28 and assumption (A), the function $v_0 : R^n \times R^n \to R^1$ is well defined, convex, and continuous.

Proposition 2.32 (Corollary 6.2.1 of [44]). *Let $x_1, x_2 \in R^n$. Then there is a unique $(x, u) \in X(A, B, 0, \tau)$ such that $x(0) = x_1, \ x(\tau) = x_2$ and $I^f(0, \tau, x, u) = v(x_1, x_2)$.*

Proposition 2.33. *Let $\epsilon > 0$. Then there exist $\delta > 0$ such that for each integer $k \geq 1$ and each $y, z \in R^n$ satisfying $|y - x_f(0)|, |z - x_f(0)| \leq \delta$,*

$$\sigma(f, y, z, k\tau) \leq k\tau\mu(f) + \epsilon.$$

Proof. Since the function v is continuous there exists $\delta \in (0, \epsilon)$ such that for each $y, z \in R^n$ satisfying

$$|y - x_f(0)|, \ |z - x_f(0)| \leq \delta, \tag{2.38}$$

we have

$$|v(y, z) - \tau\mu(f)| = |v(y, z) - v(x_f(0), x_f(0))| \leq \epsilon/4.$$

Let $k \geq 1$ be an integer and $y, z \in R^n$ satisfy (2.38). Assume that $k = 1$. By (2.31) and the choice of δ [see (2.38)],

$$\sigma(f, y, z, \tau) = v(y, z) \leq \tau\mu(f) + \epsilon.$$

Assume that $k > 1$. By Proposition 2.32, there exists $(x, u) \in X(A, B, 0, k\tau)$ such that

$$x(0) = y, \ x(\tau) = x_f(0), \ I^f(0, \tau, x, u) = v(y, x_f(0)),$$

$$x((k - 1)\tau) = x_f(0), \ x(k\tau) = z, \ I^f((k - 1)\tau, k\tau, x, u) = v(x_f(0), z),$$

and that for each integer i satisfying $1 \leq i < k - 1$,

$$x(i\tau + t) = x_f(t), \ u(i\tau + t) = u_f(t), \ t \in [0, \tau].$$

By the definition above, the choice of δ [see (2.38)],

$$\sigma(f, y, z, k\tau) \leq I^f(0, k\tau, x, u) = v(x_f(0), z) + v(y, x_f(0)) + \tau\mu(f)(k - 2)$$

$$\leq 2(\mu(f)\tau + \epsilon/4) + \tau\mu(f)(k - 2) \leq k\tau\mu(f) + \epsilon/2.$$

Proposition 2.33 is proved.

Proposition 2.34. *There exists $M_* > 0$ such that for each $T > 0$ and each $(x, u) \in X(A, B, 0, T)$,*

$$I^f(0, T, x, u) \geq T\mu(f) - M_*.$$

Proof. By (A) there is $c_0 > 0$ such that

$$f(t, x, u) \geq -c_0 \text{ for all } (t, x, u) \in [0, \infty) \times R^n \times R^m. \tag{2.39}$$

In view of Proposition 2.29 there exists $M_0 > 0$ such that for each $T \in [\tau/2, 4\tau]$ and each $(x, u) \in X(A, B, 0, T)$ satisfying

$$I^f(0, T, x, u) \le 4\tau(|\mu(f)| + 1)$$

we have

$$|x(t)| \le M_0 \text{ for all } t \in [0, T]. \tag{2.40}$$

Let $p_f \in R^n$ be as guaranteed by Proposition 2.31. Choose

$$M_* > 4c_0\tau + 4\tau|\mu(f)| + 1 + 4|p_f|M_0. \tag{2.41}$$

Let $T > 0$ and $(x, u) \in X(A, B, 0, T)$. If $T \le 4\tau$, then by (2.39) and (2.41),

$$I^f(0, T, x, u) \ge -4c_0\tau \ge T\mu(f) - 4\tau|\mu(f)| - 4c_0\tau \ge T\mu(f) - M_*$$

and in this case Proposition 2.34 holds.

Assume that

$$T > 4\tau. \tag{2.42}$$

There are two cases:

$$|x(i\tau)| > M_0 \text{ for all integers } i \in [1, \tau^{-1}T - 1]; \tag{2.43}$$

$$\min\{|x(i\tau)| : \text{ an integer } i \in [1, \tau^{-1}T - 1]\} \le M_0. \tag{2.44}$$

Assume that (2.43) holds. Set

$$S_0 = 0, \ S_i = i\tau \text{ for all integers } i \in [0, \lfloor \tau^{-1}T \rfloor - 1], \ S_{\lfloor \tau^{-1}T \rfloor} = T. \tag{2.45}$$

By (2.45),

$$S_{\lfloor \tau^{-1}T \rfloor} - S_{\lfloor \tau^{-1}T \rfloor - 1} = T - \tau(\lfloor \tau^{-1}T \rfloor - 1) \in [\tau, 2\tau]. \tag{2.46}$$

By the choice of M_0 (see (2.40)), (2.42), (2.43), (2.45) and (2.46),

$$I^f(S_i, S_{i+1}, x, u) > 4\tau(|\mu(f)| + 1), \ i = 0, \ldots, \lfloor T/\tau \rfloor - 1.$$

This implies that

$$I^f(0, T, x, u) \ge 4\tau(|\mu(f)| + 1)\lfloor T/\tau \rfloor > 2(|\mu(f)| + 1)T.$$

Thus in this case Proposition 2.34 holds.

Assume that (2.44) holds. Then there exist integers j_1, j_2 such that

$$1 \leq j_1 \leq j_2 \leq \tau^{-1}T - 1, \tag{2.47}$$

$$|x(j_1\tau)|, \ |x(j_2\tau)| \leq M_0, \tag{2.48}$$

$$|x(i\tau)| > M_0 \tag{2.49}$$

for each integer i satisfying $1 \leq i < j_1$ and for each integer i satisfying $j_2 < i \leq T/\tau - 1$. We will estimate $I^f(0, j_1\tau, x, u)$, $I^f(j_2\tau, T, x, u)$ and $I^f(j_1\tau, j_2\tau, x, u)$. By (2.39),

$$I^f(0, \tau, x, u) \geq -c_0\tau. \tag{2.50}$$

In view of (2.49) and the choice of j_1 and M_0 [see (2.40)], for each integer i satisfying $1 \leq i < j_1$,

$$I^f(i\tau, (i+1)\tau, x, u) > 4\tau(|\mu(f)| + 1).$$

Together with (2.50) this implies that

$$I^f(0, j_1\tau, x, u) - j_1\tau\mu(f) \geq -c_0\tau - \tau|\mu(f)|. \tag{2.51}$$

By (2.39),

$$I^f(\tau\lfloor\tau^{-1}T\rfloor - \tau, T, x, u) \geq -2c_0\tau. \tag{2.52}$$

It follows from (2.49) and the choice of j_2 and M_0 [see (2.40)] that for each integer i satisfying $j_2 \leq i < T/\tau - 1$,

$$I^f(i\tau, (i+1)\tau, x, u) \geq 4\tau(|\mu(f)| + 1).$$

Together with (2.52) this implies that

$$I^f(j_2\tau, T, x, u) - (T - j_2\tau)\mu(f) \geq -2c_0\tau - 2\tau|\mu(f)|. \tag{2.53}$$

If $j_1 = j_2$, then (2.51) and (2.53) imply that

$$I^f(0, T, x, u) - T\mu(f) \geq -3c_0\tau - 3\tau|\mu(f)| > -M_*.$$

Therefore we may assume without loss of generality that $j_1 < j_2$. We estimate $I^f(j_1\tau, j_2\tau, x, u)$. By (2.31), the choice of p_f, Proposition 2.31, (2.34), (2.42), and (2.48),

$$I^f(j_1\tau, j_2\tau, x, u) - (j_2 - j_1)\tau\mu(f) \geq \sum_{i=j_1}^{j_2-1} v(x(i\tau), x((i+1)\tau)) - (j_2 - j_1)v(z_f, z_f)$$

$$\geq \langle p_f, x(j_1\tau) - x(j_2\tau) \rangle \geq -2|p_f|M_0. \qquad (2.54)$$

It follows from (2.41), (2.54), (2.51), and (2.53) that

$$I^f(0, T, x, u) - T\mu(f) = I^f(0, j_1\tau, x, u) - j_1\tau\mu(f) + I^f(j_1\tau, j_2\tau, x, u)$$

$$- (j_2 - j_1)\tau\mu(f) + I^f(j_2\tau, T, x, u) - (T - j_2\tau)\mu(f)$$

$$\geq -c_0\tau - \tau|\mu(f)| - 2|p_f|M_0 - 2c_0\tau - 2\tau|\mu(f)|$$

$$\geq -3c_0\tau - 3\tau|\mu(f)| - 2|p_f|M_0 > -M_*.$$

Proposition 2.34 is proved.

Proposition 2.35. *Let $M_0 > 0$. Then there exists $M > 0$ such that for each $T \geq 3\tau$ and each $y, z \in R^n$ satisfying $|y|, |z| \leq M_0$,*

$$\sigma(f, y, z, T) \leq T\mu(f) + M.$$

Proof. We may assume without loss of generality that

$$M_0 \geq |x_f(t)| \text{ for all } t \in [0, \tau]. \qquad (2.55)$$

Since the function v_0 is continuous there exists $M_1 > 0$ such that

$$|v_0(y, z)| \leq M_1 \text{ for all } y, z \in R^n \text{ satisfying } |y|, |z| \leq M_0. \qquad (2.56)$$

In view of assumption (A), there exists $c_0 > 0$ such that

$$f(t, x, u) \geq -c_0 \text{ for all } (t, x, u) \in [0, \infty) \times R^n \times R^m. \qquad (2.57)$$

Choose

$$M > 2M_1 + 2c_0\tau + 2 + 2\tau|\mu(f)|. \qquad (2.58)$$

Assume that

$$T \geq 3\tau, \ y, z \in R^n, \ |y|, |z| \leq M_0. \qquad (2.59)$$

There exists

$$\xi \in [0, \tau) \qquad (2.60)$$

such that

$$(T - \xi)\tau^{-1} \in \mathbf{Z}. \tag{2.61}$$

By (2.37), there exists $(x_1, u_1) \in X(A, B, 0, \tau)$ such that

$$x_1(0) = y, \ x_1(\tau) = x_f(0), \tag{2.62}$$

$$\int_0^\tau f_0(x_1(t), u_1(t))dt \leq v_0(y, x_f(0)) + 1. \tag{2.63}$$

By (2.37), there exists $(x_2, u_2) \in X(A, B, T - \tau, T)$ such that

$$x_2(T - \tau) = x_f(\xi), \ x_2(T) = z, \tag{2.64}$$

$$\int_{T-\tau}^T f_0(x_2(t), u_2(t))dt \leq v_0(x_f(\xi), z) + 1. \tag{2.65}$$

It follows from (2.36), (2.55), (2.56), (2.59), and (2.62)–(2.65) that

$$\int_0^\tau f(t, x_1(t), u_1(t))dt \leq \int_0^\tau f_0(x_1(t), u_1(t))dt \leq M_1 + 1,$$

$$\int_{T-\tau}^T f(t, x_2(t), u_2(t))dt \leq \int_{T-\tau}^T f_0(x_2(t), u_2(t))dt \leq M_1 + 1. \tag{2.66}$$

Define

$$x(t) = x_1(t), \ u(t) = u_1(t), \ t \in [0, \tau],$$

$$x(t) = x_f(t - \tau\lfloor \tau^{-1}t \rfloor), \ u(t) = u_f(t - \tau\lfloor \tau^{-1}t \rfloor), \ t \in (\tau, T - \tau],$$

$$x(t) = x_2(t), \ u(t) = u_2(t), \ t \in (T - \tau, T]. \tag{2.67}$$

By (2.61), (2.62), (2.64), (2.67),

$$(x, u) \in X(A, B, 0, T), \ x(0) = y, \ x(T) = z.$$

In view of (2.3), (2.47), (2.58), (2.66), and (2.67),

$$I^f(0, T, x, u) = I^f(0, \tau, x_1, u_1) + I^f(T - \tau, T, x_2, u_2) + I^f(\tau, T - \tau, x, u)$$

$$\leq 2M_1 + 2 + (\lfloor \tau^{-1}T \rfloor - 1)\tau\mu(f) + c_0\tau < T\mu(f) + M.$$

Proposition 2.35 is proved.

Proposition 2.36. *Let $M, \epsilon > 0$. Then there exists a natural number L such that for each $(x, u) \in X(A, B, 0, L\tau)$ satisfying*

$$I^f(0, L\tau, x, u) \leq L\tau\mu(f) + M$$

there exists an integer $i \in [0, L - 1]$ such that

$$\sup\{|x(\tau i + t) - x_f(t)| \,:\, t \in [0, \tau]\} \leq \epsilon.$$

Proof. Assume that the proposition does not hold. Then there exist a strictly increasing sequence of natural numbers $\{L_k\}_{k=1}^{\infty}$ such that $L_k \geq k$ for all integers $k \geq 1$ and a sequence $(x_k, u_k) \in X(A, B, 0, L_k\tau)$, $k = 1, 2, \ldots$ such that for each integer $k \geq 1$ and each integer $i \in [0, L_k - 1]$,

$$I^f(0, L_k\tau, x_k, u_k) \leq L_k\tau\mu(f) + M, \tag{2.68}$$

$$\sup\{|x_k(\tau i + t) - x_f(t)| \,:\, t \in [0, \tau]\} > \epsilon. \tag{2.69}$$

By Proposition 2.34, there exists $M_* > 0$ such that for each $T > 0$ and each $(x, u) \in X(A, B, 0, T)$,

$$I^f(0, T, x, u) \geq T\mu(f) - M_*. \tag{2.70}$$

Let $p \geq 1$ be an integer. It follows from (2.68) and (2.70) that for each integer $k > p$,

$$\begin{aligned}
I^f(0, p\tau, x_k, u_k) &= I^f(0, L_k\tau, x_k, u_k) - I^f(p\tau, L_k\tau, x_k, u_k) \\
&\leq L_k\tau\mu(f) + M - (L_k\tau - p\tau)\mu(f) + M_* \\
&\leq p\tau\mu(f) + M + M_*. \tag{2.71}
\end{aligned}$$

By (2.71) and Proposition 2.30, extracting a subsequence and re-indexing if necessary, we may assume without loss of generality that there exists $(x, u) \in X(A, B, 0, \infty)$ such that for each integer $p \geq 1$,

$$x_k(t) \to x(t) \text{ as } k \to \infty \text{ uniformly on } [0, p\tau], \tag{2.72}$$

$$x'_k \to x' \text{ as } k \to \infty \text{ weakly in } L^1(R^n; (0, p\tau)),$$

$$u_k \to u \text{ as } k \to \infty \text{ weakly in } L^1(R^m; (0, p\tau)),$$

$$I^f(0, p\tau, x, u) \leq p\tau\mu(f) + M + M_*. \tag{2.73}$$

In view of (2.73) and Theorem 2.2, $(x, u) \in X(A, B, 0, \infty)$ is (f, A, B)-good and there exists an integer $i_0 \geq 1$ such that for each integer $i \geq i_0$,

$$\sup\{|x(\tau i + t) - x_f(t)| \,:\, t \in [0, \tau]\} \leq \epsilon/4. \tag{2.74}$$

Relation (2.72) implies that there exists an integer $k_0 > i_0 + 4$ such that for each integer $k \geq k_0$,

$$|x_k(t) - x(t)| \leq \epsilon/4 \text{ for all } t \in [0, (i_0 + 4)\tau]. \tag{2.75}$$

By (2.74) and (2.75), for each integer $k \geq k_0$ and each $t \in [0, \tau]$,

$$|x_f(t) - x_k(i_0\tau + t)| \leq |x_f(t) - x(i_0\tau + t)| + |x(i_0\tau + t) - x_k(i_0\tau + t)|$$
$$\leq \epsilon/2.$$

This contradicts (2.69). The contradiction we have reached proves Proposition 2.36.

Proposition 2.37. *Any (f, A, B)-overtaking optimal $(x, u) \in X(A, B, 0, \infty)$ is (f, A, B)-good.*

Proof. Let $(x, u) \in X(A, B, 0, \infty)$ be (f, A, B)-overtaking optimal. By Theorem 2.4, there is a number M_0 such that $M_0 > |x(t)|$ for all $t \geq 0$. Together with Proposition 2.35 this implies the existence of $M_1 > 0$ such that for each $T \geq 3\tau$, $I^f(0, T, x, u) \leq T\mu(f) + M_1$. In view of Theorem 2.2, the pair (x, u) is (f, A, B)-good.

Proposition 2.38. *Let $M, \epsilon > 0$. Then there exists a natural number L such that for each $T \geq L\tau$, each $(x, u) \in X(A, B, 0, T)$ satisfying*

$$I^f(0, T, x, u) \leq T\mu(f) + M \tag{2.76}$$

and each integer S satisfying

$$[S\tau, (S + L)\tau] \subset [0, T] \tag{2.77}$$

there exists an integer $i \in [S, S + L - 1]$ such that

$$|x(\tau i + t) - x_f(t)| \leq \epsilon \text{ for all } t \in [0, \tau]. \tag{2.78}$$

Proof. By Proposition 2.34, there exists $M_* > 0$ such that for each $T > 0$ and each $(x, u) \in X(A, B, 0, T)$,

$$I^f(0, T, x, u) \geq T\mu(f) - M_*. \tag{2.79}$$

By Proposition 2.36, there exists a natural number L such that the following property holds:

(i) for each $(x, u) \in X(A, B, 0, L\tau)$ satisfying

$$I^f(0, L\tau, x, u) \leq L\tau\mu(f) + M + 2M_*$$

there exists an integer $i \in [0, L-1]$ such that $|x(\tau i + t) - x_f(t)| \le \epsilon$ for all $t \in [0, \tau]$.

Assume that $T \ge L\tau$, an $(x, u) \in X(A, B, 0, T)$ satisfies (2.76) and an integer S satisfies (2.77). By the choice of M_* [see (2.79)],

$$I^f(0, S\tau, x, u) \ge S\tau\mu(f) - M_*, \quad I^f(\tau(S+L), T, x, u) \ge (T - \tau(S+L))\mu(f) - M_*.$$

Together with (2.76) this implies that

$$\begin{aligned}
I^f(\tau S, \tau(S+L), x, u) &= I^f(0, T, x, u) - I^f(0, S\tau, x, u) - I^f((S+L)\tau, T, x, u) \\
&\le T\mu(f) + M - S\mu(f) + M_* - (T - (S+L)\tau)\mu(f) \\
&\quad + M_* \le L\tau\mu(f) + 2M_*.
\end{aligned}$$

By the inequality above and property (i), there exists an integer i such that $[i\tau, (i+1)\tau] \subset [S\tau, (S+L)\tau]$ and (2.78) holds. Proposition 2.38 is proved.

Proposition 2.39. *Let* $\epsilon \in (0, 1)$. *Then there exists* $\delta > 0$ *such that for each integer* $p \ge 1$, *each* $(x, u) \in X(A, B, 0, p\tau)$ *satisfying*

$$|x(0) - x_f(0)|, \ |x(p\tau) - x_f(0)| \le \delta, \ I^f(0, p\tau, x, u) \le \sigma(f, x(0), x(p\tau), p\tau) + \delta$$

and each integer $i \in [0, p-1]$, *the inequality* $|x(i\tau + t) - x_f(t)| \le \epsilon$ *holds for all* $t \in [0, \tau]$.

Proof. By the continuity of v, (2.34) and Proposition 2.33, for each integer $k \ge 1$, there is

$$\delta_k \in (0, 4^{-k}\epsilon) \tag{2.80}$$

such that the following properties hold:

(ii) for each $y, z \in R^n$ satisfying $|y - x_f(0)|, |z - x_f(0)| \le \delta_k$ we have $|v(y, z) - \mu(f)\tau| \le 4^{-k}$;

(iii) for each integer $p \ge 1$ and each $y, z \in R^n$ satisfying $|y - x_f(0)|, |z - x_f(0)| \le \delta_k$ we have $\sigma(f, y, z, p\tau) \le p\tau\mu(f) + 4^{-k}$.

We may assume without loss of generality that the sequence $\{\delta_k\}_{k=1}^{\infty}$ is decreasing. Assume that the proposition does not hold. Then for each natural number k there exist an integer $p_k \ge 1$ and $(x_k, u_k) \in X(A, B, 0, p_k\tau)$ satisfying

$$|x_k(0) - x_f(0)| \le \delta_k, \ |x_k(p_k\tau) - x_f(0)| \le \delta_k, \tag{2.81}$$

$$I^f(0, p_k\tau, x_k, u_k) \le \sigma(f, x_k(0), x_k(p_k\tau), p_k\tau) + \delta_k, \tag{2.82}$$

$$\sup\{\sup\{|x_k(i\tau + t) - x_f(t)| : t \in [0, \tau]\} : i = 0, \ldots, p_k - 1\} > \epsilon. \tag{2.83}$$

By property (iii) and (2.80)–(2.82), for each integer $k \geq 1$,

$$I^f(0, p_k\tau, x_k, u_k) \leq p_k\tau\mu(f) + 2 \cdot 4^{-k}. \tag{2.84}$$

In view of Proposition 2.32 there exists $(x, u) \in X(A, B, 0, \infty)$ such that

$$x(t) = x_1(t), \ u(t) = u_1(t), \ t \in [0, p_1\tau], \ x((p_1 + 1)\tau) = x_2(0), \tag{2.85}$$

$$I^f(\tau p_1, (p_1 + 1)\tau, x, u) = v(x_1(p_1\tau), x_2(0)) \tag{2.86}$$

and for each integer $k \geq 1$,

$$x\left(\sum_{i=1}^{k}(p_i + 1)\tau + t \right) = x_{k+1}(t), \tag{2.87}$$

$$u\left(\sum_{i=1}^{k}(p_i + 1)\tau + t \right) = u_{k+1}(t), \tag{2.88}$$

$$x\left(\sum_{i=1}^{k+1}(p_i + 1)\tau \right) = x_{k+2}(0), \tag{2.89}$$

$$I^f\left(\left(\sum_{i=1}^{k+1}(p_i + 1) - 1 \right)\tau, \sum_{i=1}^{k+1}(p_i + 1)\tau, x, u \right) = v(x_{k+1}(p_{k+1}\tau), x_{k+2}(0)).$$

$$\tag{2.90}$$

By (2.81), (2.84)–(2.90) and property (ii), for each integer $k \geq 2$,

$$I^f\left(0, \left(\sum_{i=1}^{k}(p_i + 1) \right)\tau, x, u \right) = \sum_{i=1}^{k}(I^f(0, p_i\tau, x_i, u_i) + v(x_i(p_i\tau), x_{i+1}(0)))$$

$$\leq \sum_{i=1}^{k}[p_i\tau\mu(f) + 2 \cdot 4^{-i} + \mu(f)\tau + 4^{-i}]$$

$$\leq \mu(f) \sum_{i=1}^{k}(p_i + 1)\tau + 6.$$

Since the relation above holds for any integer $k \geq 2$ it follows from Theorem 2.2 that the pair (x, u) is (f, A, B)-good and

$$\sup\{|x(i\tau + t) - x_f(t)| : \ t \in [0, \tau]\} \to 0 \text{ as } i \to \infty.$$

Thus there exists an integer $i_0 \geq 1$ such that for each integer $i \geq i_0$,

$$\epsilon \geq |x(i\tau + t) - x_f(t)|, \quad t \in [0, \tau].$$

In view of (2.85)–(2.90) this contradicts (2.83). The contradiction we have reached proves Proposition 2.39.

2.5   Proofs of Theorems 2.9 and 2.10

Proof of Theorem 2.9. In view of Proposition 2.37, (i) implies (ii). By Theorem 2.2, (ii) implies (iii). Clearly, (iv) follows from (iii). Let us show that (iv) implies (i).
    Assume that $(x, u) \in X(A, B, 0, \infty)$ is (f, A, B)-minimal and that

$$\liminf_{t \to \infty} |x(t)| < \infty. \tag{2.91}$$

We show that (x, u) is (f, A, B)-overtaking optimal. Since (x, u) is (f, A, B)-minimal it follows from (2.91), Proposition 2.35, and Theorem 2.2 that (x, u) is (f, A, B)-good and that

$$\lim_{i \to \infty, \, i \in Z} \max\{|x(i\tau + t) - x_f(t)| : t \in [0, \tau]\} = 0. \tag{2.92}$$

Assume that (x, u) is not (f, A, B)-overtaking optimal. By Theorem 2.3, there exist an (f, A, B)-overtaking optimal $(\tilde{x}, \tilde{u}) \in X(A, B, 0, \infty)$, $\epsilon > 0$ and $T_0 > 0$ such that

$$\tilde{x}(0) = x(0), \tag{2.93}$$

$$I^f(0, T, x, u) \geq I^f(0, T, \tilde{x}, \tilde{u}) + \epsilon \text{ for all } T \geq T_0. \tag{2.94}$$

In view of Theorem 2.2 and Proposition 2.37,

$$\lim_{i \to \infty, \, i \in Z} \max\{|\tilde{x}(i\tau + t) - x_f(t)| : t \in [0, \tau]\} = 0. \tag{2.95}$$

Since the function v is continuous there exists $\delta > 0$ such that

$$|v(z_1, z_2) - v(x_f(0), x_f(0))| \leq \epsilon/4 \text{ for all } z_1, z_2 \in R^n$$

$$\text{satisfying } |z_i - x_f(0)| \leq \delta, \, i = 1, 2. \tag{2.96}$$

By (2.92) and (2.95), there exists a natural number $p > T_0/\tau$ such that for all integers $i \geq p$,

$$|\tilde{x}(i\tau) - x_f(0)|, \, |x(i\tau) - x_f(0)| \leq \delta. \tag{2.97}$$

Proposition 2.32 implies that there exists $(x_1, u_1) \in X(A, B, 0, (p+1)\tau)$ such that

$$x_1(t) = \tilde{x}(t), \ u_1(t) = \tilde{u}(t), \ t \in [0, p\tau], \ x_1((p+1)\tau) = x((p+1)\tau),$$

$$I^f(\tau p, \tau(p+1), x_1, u_1) = v(\tilde{x}(p\tau), x((p+1)\tau)). \qquad (2.98)$$

By (2.94), (2.96), (2.97), (2.98), and (f, A, B)-minimality of (x, u),

$$I^f(0, (p+1)\tau, x_1, u_1) - I^f(0, (p+1)\tau, x, u) = I^f(0, \tau p, \tilde{x}, \tilde{u}) - I^f(0, \tau p, x, u)$$
$$+ I^f(\tau p, \tau(p+1), x_1, u_1)$$
$$- I^f(\tau p, \tau(p+1)x, u)$$
$$\leq -\epsilon + v(\tilde{x}(p\tau), x((p+1)\tau))$$
$$- v(x(p\tau), x((p+1)\tau))$$
$$\leq -\epsilon + \epsilon/2.$$

This contradicts (f, A, B)-minimality of (x, u). The contradiction we have reached completes the proof of Theorem 2.9. $\qquad \square$

Proof of Theorem 2.10. Since f satisfies assumption (A) with any $\tau > 0$ all our results hold for all $\tau > 0$. By Theorems 2.2, 2.3, and 2.9,

$$\mu(f) = \lim_{T \to \infty} T^{-1} I^f(0, T, x, u),$$

for any (f, A, B)-overtaking optimal $(x, u) \in X(A, B, 0, \infty)$. In view of Proposition 2.1, for any $\tau > 0$, there exists $(x_\tau, u_\tau) \in X(A, B, 0, \tau)$ which is a unique solution of the minimization problem

$$I^f(0, \tau, x, u) \to \min, \ (x, u) \in X(A, B, 0, \tau), \ x(0) = x(\tau), \qquad (2.99)$$

$$\tau \mu(f) = I^f(0, \tau, x_\tau, u_\tau). \qquad (2.100)$$

By (2.99) and (2.100), for any $\tau > 0$, $x_\tau(t) = x_{\tau/2}(t)$, $t \in [0, \tau/2]$ and $x_\tau(\tau/2+t) = x_\tau(t)$, $t \in [0, \tau/2]$. Since the relation above holds for any $\tau > 0$ we conclude that for any $\tau > 0$, $x_\tau(\cdot)$ is a constant function. In view of Theorem 2.2, $x_\tau(0)$ does not depend on τ. Theorem 2.10 is proved. $\qquad \square$

2.6 Auxiliary Results for Theorem 2.12

Proposition 2.40. *Let $M_1 > 0$, $0 < \tau_0 < \tau_1$. Then there exists $M_2 > 0$ such that for each $g \in \mathcal{M}$, each $T_2 > T_1 \geq 0$ satisfying*

$$T_2 - T_1 \in [\tau_0, \tau_1] \qquad (2.101)$$

and each $(x, u) \in X(A, B, T_1, T_2)$ satisfying

$$I^g(T_1, T_2, x, u) \leq M_1 \tag{2.102}$$

the following inequality holds:

$$|x(t)| \leq M_2 \text{ for all } t \in [T_1, T_2]. \tag{2.103}$$

Proof. Fix

$$\delta \in (0, \min\{8^{-1}\tau_0, (2\|A\| + 2)^{-1}\}). \tag{2.104}$$

By (2.11) and (2.13), there exists $c_0 > 1$ such that

$$g(t, x, u) \geq 8|u|(\|B\| + 1) \tag{2.105}$$

for each $g \in \mathcal{M}$ and each $(t, x, u) \in [0, \infty) \times R^n \times R^m$ satisfying $|u| \geq c_0$ and $h_0 > 0$ such that

$$g(t, x, u) \geq 4M_1(\min\{1, \tau_0\})^{-1}\delta^{-1} + 2a\tau_1\delta^{-1} \tag{2.106}$$

for each $g \in \mathcal{M}$ and each $(t, x, u) \in [0, \infty) \times R^n \times R^m$ satisfying $|x| \geq h_0$. Fix

$$M_2 > 2 + 2M_1 + 2a\tau_1 + 2c_0(1 + \tau_1)\|B\| + 2h_0. \tag{2.107}$$

Let $g \in \mathcal{M}$, $T_2 > T_1 \geq 0$ satisfy (2.101) and let $(x, u) \in X(A, B, T_1, T_2)$ satisfy (2.102). We show that (2.103) holds.

Assume the contrary. Then there exists $t_0 \in [T_1, T_2]$ such that

$$|x(t_0)| > M_2. \tag{2.108}$$

By the choice of h_0 [see (2.106)], (2.13), (2.102) and (2.104), there exists $t_1 \in [T_1, T_2]$ satisfying

$$|x(t_1)| \leq h_0, \ |t_1 - t_0| \leq \delta. \tag{2.109}$$

There exists a number t_2 such that

$$\min\{t_0, t_1\} \leq t_2 \leq \max\{t_0, t_1\},$$

$$|x(t_2)| \geq |x(t)|, \ t \in [\min\{t_0, t_1\}, \max\{t_0, t_1\}]. \tag{2.110}$$

It follows from (2.1), (2.109), and (2.110) that

$$|x(t_1) - x(t_2)| = \left| \int_{t_1}^{t_2} x'(t)dt \right| \leq \|A\| \left| \int_{t_1}^{t_2} |x(t)|dt \right| + \|B\| \left| \int_{t_1}^{t_2} |u(t)|dt \right|$$

$$\leq \|A\| |x(t_2)| \delta + \|B\| \int_{t_1}^{t_2} |u(t)|dt|. \tag{2.111}$$

By the choice of c_0 (see (2.105)), (2.13), (2.101), (2.102) and (2.109),

$$\left| \int_{t_1}^{t_2} |u(t)|dt \right| \leq \left| \int_{t_1}^{t_2} [8^{-1}g(t, x(t), u(t))(\|B\| + 1)^{-1} + c_0]dt \right|$$

$$\leq c_0|t_1 - t_2| + 8^{-1}a\tau_1(\|B\| + 1)^{-1} + 8^{-1}(\|B\| + 1)^{-1}I^g(T_1, T_2, x, u)$$

$$\leq c_0\delta + a\tau_1 8^{-1}(\|B\| + 1)^{-1} + 8^{-1}(\|B\| + 1)^{-1}M_1.$$

By this relation, (2.104) and (2.111),

$$|x(t_1) - x(t_2)| \leq 2^{-1}|x(t_2)| + \|B\|c_0\delta + a\tau_1 + M_1.$$

Combined with (2.108) and (2.109) this implies that $2^{-1}M_2 - h_0 \leq \|B\|c_0\delta + a\tau_1 + M_1$. This contradicts (2.107). The contradiction we have reached proves Proposition 2.40.

Proposition 2.41. *Let* $0 < c_1 < c_2$, $D, \epsilon > 0$. *Then there exists a neighborhood V of f in* $\mathcal{M}$ *such that for each* $g \in V$, *each* $T_2 > T_1 \geq 0$ *satisfying* $T_2 - T_1 \in [c_1, c_2]$ *and each* $(x, u) \in X(A, B, T_1, T_2)$ *satisfying*

$$\min\{I^f(T_1, T_2, x, u), I^g(T_1, T_2, x, u)\} \leq D \tag{2.112}$$

the inequality $|I^f(T_1, T_2, x, u) - I^g(T_1, T_2, x, u)| \leq \epsilon$ *holds.*

Proof. By Proposition 2.40, there exists $S > 0$ such that

$$|x(t)| \leq S \text{ for all } t \in [T_1, T_2] \tag{2.113}$$

for each $g \in \mathcal{M}$, each $T_2 > T_1 \geq 0$ satisfying $T_2 - T_1 \in [c_1, c_2]$ and each $(x, u) \in X(A, B, T_1, T_2)$ satisfying $I^g(T_1, T_2, x, u) \leq D + 1$.

Choose $\delta \in (0, 1)$, $N > S$, $\Gamma > 1$ such that

$$\delta(c_2 + 1) \leq 4^{-1}\epsilon, \quad \psi(N)N > 4a, \quad (\Gamma - 1)(c_2 + D + a(c_2 + 1)) \leq \epsilon/4 \tag{2.114}$$

and set

$$V = \{g \in \mathcal{M} : (f, g) \in E(N, \delta, \Gamma)\}. \tag{2.115}$$

Assume that

$$g \in V, \ T_2 > T_1 \geq 0, \ T_2 - T_1 \in [c_1, c_2] \tag{2.116}$$

and that $(x, u) \in X(A, B, T_1, T_2)$ satisfies (2.112). By the choice of S, (2.112) and (2.116), (2.113) is true. Set

$$E_1 = \{t \in [T_1, T_2] : |u(t)| \leq N\}, \ E_2 = [T_1, T_2] \setminus E_1. \tag{2.117}$$

By (2.113), (2.115), (2.116), (2.117), and the inequality $N > S$,

$$|f(t, x(t), u(t)) - g(t, x(t), u(t))| \leq \delta, \ t \in E_1. \tag{2.118}$$

Set

$$h(t) = \min\{f(t, x(t), u(t)), \ g(t, x(t), u(t))\}, \ t \in [T_1, T_2]. \tag{2.119}$$

In view of (2.13), (2.113), (2.114), (2.115), (2.116), (2.117), (2.119), and the inequality $N > S$, for all $t \in E_2$,

$$(f(t, x(t), u(t)) + 1)(g(t, x(t), u(t)) + 1)^{-1} \in [\Gamma^{-1}, \Gamma]$$

and

$$|f(t, x(t), u(t)) - g(t, x(t), u(t))| \leq (\Gamma - 1)(h(t) + 1). \tag{2.120}$$

Relations (2.13), (2.112), (2.114), and (2.116)–(2.120) imply that

$$\begin{aligned}
|I^f(T_1, T_2, x, u) - I^g(T_1, T_2, x, u)| &\leq \int_{E_1} |f(t, x(t), u(t)) - g(t, x(t), u(t))| dt \\
&\quad + \int_{E_2} |f(t, x(t), u(t)) - g(t, x(t), u(t))| dt \\
&\leq \delta c_2 + (\Gamma - 1) \int_{E_2} (h(t) + 1) dt \leq \delta c_2 \\
&\quad + (\Gamma - 1)c_2 + (\Gamma - 1)(D + ac_2) \leq \epsilon.
\end{aligned}$$

Proposition 2.41 is proved.

2.7 Proof of Theorem 2.12

By Proposition 2.39, there exists $\delta_0 \in (0, 1/8)$ such that the following property holds:

(P1)   for each integer $p \geq 1$, each $(x, u) \in X(A, B, 0, p\tau)$ satisfying

$$|x(0) - x_f(0)|, \ |x(p\tau) - x_f(0)| \leq 4\delta_0,$$
$$I^f(0, p\tau, x, u) \leq \sigma(f, x(0), x(p\tau), p\tau) + 4\delta_0$$

and each integer $i \in [0, p-1]$, the inequality $|x(i\tau + t) - x_f(t)| \leq \epsilon$ holds for all $t \in [0, \tau]$.

By Proposition 2.38, there exists an integer $L_0 \geq 5$ such that the following property holds:

(P2)   for each $T \geq (L_0 - 4)\tau$, each $(x, u) \in X(A, B, 0, T)$ satisfying

$$I^f(0, T, x, u) \leq T\mu(f) + M + 4$$

and each integer S satisfying $[S\tau, (S + L_0 - 4)\tau] \subset [0, T]$ there exists an integer $i \in [S, S + L_0 - 5]$ such that $|x(\tau i + t) - x_f(t)| \leq \delta_0$ for all $t \in [0, \tau]$.

Let an integer $L_1 \geq L_0$. By Proposition 2.35, there exists a number $M_0 > 0$ such that for each $S \geq 3\tau$ and each $y, z \in R^n$ satisfying $|y|, |z| \leq \max\{|x_f(t)| : t \in [0, \tau]\} + 4$,

$$\sigma(f, y, z, S) \leq S\mu(f) + M_0. \tag{2.121}$$

By Proposition 2.41, there exists a neighborhood $\mathcal{U}$ of f in $\mathcal{M}$ such that the following property holds:

(P3)   for each $g \in \mathcal{U}$, each $T_1 \geq 0$, each $T_2 \in [T_1 + 1, T_1 + 4L_1]$ and each $(x, u) \in X(A, B, T_1\tau, T_2\tau)$ satisfying

$$\min\{I^f(T_1\tau, T_2\tau, x, u), I^g(T_1\tau, T_2\tau, x, u)\} \leq (4L_1|\mu(f)| + M + M_0 + 1)(\tau + 1),$$
$$|I^f(T_1\tau, T_2\tau, x, u) - I^g(T_1\tau, T_2\tau, x, u)| \leq \delta_0.$$

Assume that $T > 2L_1\tau$, $g \in \mathcal{U}$, $(x, u) \in X(A, B, 0, T)$ and that a finite sequence of integers $\{S_i\}_{i=0}^q$ satisfy

$$S_0 = 0, \ S_{i+1} - S_i \in [L_0, L_1], \ i = 0, \ldots, q-1, \ S_q\tau \in (T - L_1\tau, T], \tag{2.122}$$
$$I^g(S_i\tau, S_{i+1}\tau, x, u) \leq (S_{i+1} - S_i)\tau\mu(f) + M \tag{2.123}$$

for each integer $i \in [0, q-1]$,

$$I^g(S_i\tau, S_{i+2}\tau, x, u) \leq \sigma(g, x(S_i\tau), x(S_{i+2}\tau), S_i\tau, S_{i+2}\tau) + \delta_0 \tag{2.124}$$

for each nonnegative integer $i \leq q - 2$ and

$$I^g(S_{q-2}\tau, T, x, u) \leq \sigma(g, x(S_{q-2}\tau), x(T), S_{q-2}\tau, T) + \delta_0. \tag{2.125}$$

Let $i \in [0, q-1]$ be an integer. By (2.122), (2.123) and the choice of $\mathcal{U}$ (see property (P3)),

$$I^f(S_i\tau, S_{i+1}\tau, x, u) \leq I^g(S_i\tau, S_{i+1}\tau, x, u) + \delta_0 \leq (S_{i+1} - S_i)\tau\mu(f) + M + 1.$$

The inequality above, (2.122) and property (P2) imply that there exists an integer p_i such that

$$p_i \in [S_i + 3, S_i + L_0], \quad |x(p_i\tau) - x_f(0)| \leq \delta_0. \tag{2.126}$$

Let an integer $i \in [0, q-2]$. In view of (2.122) and (2.126),

$$p_i, \ p_{i+1} \in [S_i + 3, S_{i+2}], \ 3 \leq p_{i+1} - p_i \leq 2L_1. \tag{2.127}$$

It follows from (2.124) and (2.127) that

$$I^g(p_i\tau, p_{i+1}\tau, x, u) \leq \sigma(g, x(p_i\tau), x(p_{i+1}\tau), p_i\tau, p_{i+1}\tau) + \delta_0 \tag{2.128}$$

Thus we have shown that there exists a strictly increasing sequence of nonnegative integers $\{p_i\}_{i=0}^k$ where k is a natural number such that

$$p_0 \leq L_0, \ p_k\tau > T - 2\tau L_1, \ |x(p_i\tau) - x_f(0)| \leq \delta_0, \ i = 0, \ldots, k,$$

$$3 \leq p_{i+1} - p_i \leq 2L_1, \ i = 0, \ldots, k-1 \tag{2.129}$$

and (2.128) holds for all $i = 0, \ldots, k-1$. It is not difficult to see that if $|x(0) - x_f(0)| \leq \delta_0$, then we may assume that $p_0 = 0$ and if $|x(\lfloor T/\tau \rfloor \tau) - x_f(0)| \leq \delta_0$, then we may assume that $p_k = \lfloor T/\tau \rfloor$.

Let $i \in \{0, \ldots, k-1\}$. By (2.129) and the choice of M_0 (see (2.121)),

$$\sigma(f, x(p_i\tau), x(p_{i+1}\tau), (p_{i+1} - p_i)\tau) \leq \mu(f)(p_{i+1} - p_i)\tau + M_0. \tag{2.130}$$

Combined with (2.129) and the choice of $\mathcal{U}$ (see property (P3)) this implies that

$$|\sigma(f, x(p_i\tau), x(p_{i+1}\tau), (p_{i+1} - p_i)\tau) - \sigma(g, x(p_i\tau), x(p_{i+1}\tau), p_i\tau, p_{i+1}\tau)| \leq \delta_0.$$

The inequality above and (2.128) imply that

$$I^g(p_i\tau, p_{i+1}\tau, x, u) \leq \sigma(f, x(p_i\tau), x(p_{i+1}\tau), (p_{i+1} - p_i)\tau) + 2\delta_0$$

$$\leq \mu(f)(p_{i+1} - p_i)\tau + M_0 + 1.$$

Together with (2.129) and the choice of $\mathcal{U}$ (see property (P3)) this implies that

$$I^f(p_i\tau, p_{i+1}\tau, x, u) \leq I^g(p_i\tau, p_{i+1}\tau, x, u) + \delta_0$$

$$\leq \sigma(f, x(p_i\tau), x(p_{i+1}\tau), (p_{i+1} - p_i)\tau) + 3\delta_0.$$

By the relation above, (2.129) and property (P1), for all $j \in \{p_i, \ldots, p_{i+1} - 1\}$,

$$|x(\tau j + t) - x_f(t)| \leq \epsilon, \ t \in [0, \tau].$$

Thus the relation above holds for all $j \in \{p_0, \ldots, p_k - 1\}$ and Theorem 2.12 is proved.   □

2.8   Basic Lemma for Theorem 2.13

Lemma 2.42. *Let $\epsilon \in (0, 1)$, $M_0, M_1 > 0$. Then there exist an integer $L \geq 1$ and a neighborhood $\mathcal{U}$ off in $\mathcal{M}$ such that the following assertion holds.*
  Assume that $T > L\tau$, $g \in \mathcal{U}$, integers S_1, S_2 satisfy

$$0 \leq S_1 \leq S_2 - L, \ [S_1\tau, S_2\tau] \subset [0, T] \tag{2.131}$$

and $(x, u) \in X(A, B, 0, T)$ satisfies at least one of the following conditions:

(a) $|x(0)|, \ |x(T)| \leq M_0, \ I^g(0, T, x, u) \leq \sigma(g, x(0), x(T), 0, T) + M_1$;
(b) $|x(0)| \leq M_0, \ I^g(0, T, x, u) \leq \sigma(g, x(0), 0, T) + M_1$;
(c) $I^g(0, T, x, u) \leq \sigma(g, 0, T) + M_1$.

Then

$$\min\{|x(i\tau) - x_f(0)| : \ i = S_1, \ldots, S_2\} \leq \epsilon. \tag{2.132}$$

Proof. By Proposition 2.38 there exists a natural number L_0 such that the following property holds:
(P4)   for each $T \geq L_0\tau$, each $(x, u) \in X(A, B, 0, T)$ satisfying

$$I^f(0, T, x, u) \leq T\mu(f) + 16(1 + a)(\tau + 1)$$

and each integer S satisfying $[S\tau, (S + L_0)\tau] \subset [0, T]$,

$$\min\{|x(\tau i) - x_f(0)| : \ i = S, \ldots, S + L_0 - 1\} \leq \epsilon.$$

We may assume without loss of generality that

$$M_0 > \sup\{|x_f(t)| : \ t \in [0, \tau]\} + 4. \tag{2.133}$$

We use the functions f_0 and v_0 introduced in Sect. 2.4 [see (2.36) and (2.37)]. We may assume without loss of generality that

$$M_1 > \sup\{|v_0(z_1, z_2)| : z_1, z_2 \in R^n, |z_1|, |z_2| \le M_0\}. \tag{2.134}$$

By Proposition 2.35, there exists a number $M_2 > M_1 + M_0$ such that for each $S \ge 3\tau$ and each $y, z \in R^n$ satisfying $|y|, |z| \le M_0$,

$$\sigma(f, y, z, S) \le S\mu(f) + M_2. \tag{2.135}$$

Choose a natural number l such that

$$\min\{1, \tau\}l > 4 + M_1 + 4(\tau + 1)\min\{1, \tau\}^{-1} + (\tau + 1)|\mu(f)|(2L_0 + M_2 + 4)$$
$$+ (2L_0 + 18)(1 + a)(1 + \tau) + 2M_2 + 18$$
$$+ a(L_0\tau + \tau + 1) + \tau + a + 1 + \tau a \tag{2.136}$$

and set

$$L = 2(L_0 + 1)l. \tag{2.137}$$

By Proposition 2.41, there exists a neighborhood $\mathcal{U}$ of f in $\mathcal{M}$ such that the following property holds:

(P5)　for each $g \in \mathcal{U}$, each $T_1 \ge 0$, each $T_2 \in [T_1 + \min\{1, \tau\}, T_1 + 4L \max\{1, \tau\}]$ and each $(x, u) \in X(A, B, T_1, T_2)$ satisfying

$$\min\{I^f(T_1, T_2, x, u), I^g(T_1, T_2, x, u)\}$$
$$\le ((4L + 2)|\mu(f)| + 4M_2 + 4 + 16(1 + a))(\tau + 1),$$
$$|I^f(T_1, T_2, x, u) - I^g(T_1, T_2, x, u)| \le (4L)^{-1}(\tau + 1)^{-1}.$$

Assume that

$$T > L\tau, \ g \in \mathcal{U}, \tag{2.138}$$

integers S_1, S_2 satisfy (2.131) and $(x, u) \in X(A, B, 0, T)$ satisfies at least one of the conditions (a), (b), (c). We show that (2.132) holds. Assume the contrary. Then

$$|x(i\tau) - x_f(0)| > \epsilon, \ i = S_1, \dots, S_2. \tag{2.139}$$

We may assume without loss of generality that at least one of the following conditions hold:

$$S_1 = 0, \ S_2 = \lfloor T\tau^{-1}\rfloor; \tag{2.140}$$

$$S_1 \ge 1, \ |x((S_1 - 1)\tau) - x_f(0)| \le \epsilon, \ S_2 = \lfloor T\tau^{-1}\rfloor; \tag{2.141}$$

$$S_1 = 0, \ S_2 < \lfloor T\tau^{-1} \rfloor, \ |x((S_2 + 1)\tau) - x_f(0)| \le \epsilon \tag{2.142}$$

$$S_1 \ge 1, \ S_2 < \lfloor T\tau^{-1} \rfloor, \ |x(j\tau) - x_f(0)| \le \epsilon, j = S_1 - 1, S_2 + 1. \tag{2.143}$$

In view of (2.131) and (2.137),

$$\lfloor (S_2 - S_1)L_0^{-1} \rfloor \ge \lfloor LL_0^{-1} \rfloor \ge 2l. \tag{2.144}$$

It follows from (2.13) that

$$
\begin{aligned}
I^g(S_1\tau, S_2\tau, x, u) &= I^g(S_1\tau, S_1\tau + \lfloor (S_2 - S_1)L_0^{-1} \rfloor L_0\tau, x, u) \\
&\quad + I^g(S_1\tau + \lfloor (S_2 - S_1)L_0^{-1} \rfloor L_0\tau, S_2\tau, x, u) \\
&\ge \sum_{i=0}^{\lfloor (S_2-S_1)L_0^{-1} \rfloor - 1} I^g((S_1 + iL_0)\tau, (S_1 + (i + 1)L_0)\tau, x, u) - aL_0\tau.
\end{aligned}
\tag{2.145}
$$

Let

$$j \in \{0, \dots, \lfloor (S_2 - S_1)L_0^{-1} \rfloor - 1\}. \tag{2.146}$$

By (2.139), (2.146) and property (P4),

$$I^f((S_1 + jL_0)\tau, (S_1 + (j + 1)L_0)\tau, x, u) > L_0\tau\mu(f) + 16(1 + a)(\tau + 1). \tag{2.147}$$

We show that

$$I^g((S_1 + jL_0)\tau, (S_1 + (j + 1)L_0)\tau, x, u) \ge L_0\tau\mu(f) + 16(1 + a)(\tau + 1) - 1. \tag{2.148}$$

Assume the contrary. Then

$$I^g((S_1 + jL_0)\tau, (S_1 + (j + 1)L_0)\tau, x, u) < L_0\tau\mu(f) + 16(1 + a)(\tau + 1) - 1.$$

Combined with the choice of $\mathcal{U}$ [see property (P5)], (2.137) and (2.138) this implies that

$$
\begin{aligned}
&I^f((S_1 + jL_0)\tau, (S_1 + (j + 1)L_0)\tau, x, u) \\
&\le 1 + I^g((S_1 + jL_0)\tau, (S_1 + (j + 1)L_0)\tau, x, u) \le L_0\tau\mu(f) + 16(1 + a)(\tau + 1).
\end{aligned}
$$

This contradicts (2.147). The contradiction we have reached proves (2.148). Thus (2.148) holds for all $j \in \{0, \dots, \lfloor (S_2 - S_1)L_0^{-1} \rfloor - 1\}$. Set

$$z_0 = x(0) \text{ if } |x(0)| \le M_0, \ z_0 = 0 \text{ if } |x(0)| > M_0,$$
$$z_1 = x(T) \text{ if } |x(T)| \le M_0, \ z_1 = 0 \text{ if } |x(T)| > M_0. \tag{2.149}$$

It is not difficult to see that there exists $(x_1, u_1) \in X(A, B, 0, T)$ such that:
if (2.140) holds, then

$$x_1(0) = z_0, \ x_1(\tau) = x_f(0), \ f^f(0, \tau, x_1, u_1) \le v_0(z_0, x_f(0)) + 1,$$

$$x_1(T - \tau) = x_f(T - \lfloor \tau^{-1}T \rfloor \tau), \ x_1(T) = z_1,$$

$$f^f(T - \tau, T, x_1, u_1) \le v_0(x_f(T - \lfloor \tau^{-1}T \rfloor \tau), z_1) + 1,$$

$$x_1(t) = x_f(t - \lfloor \tau^{-1}t \rfloor \tau), \ u_1(t) = u_f(t - \lfloor \tau^{-1}t \rfloor \tau), \ t \in [\tau, T - \tau];$$

if (2.141) holds, then

$$x_1(t) = x(t), \ u_1(t) = u(t), \ t \in [0, \tau(S_1 - 1)], \ x_1(T - \tau) = x_f(T - \lfloor \tau^{-1}T \rfloor \tau),$$

$$x_1(T) = z_1, \ f^f(T - \tau, T, x_1, u_1) \le v_0(x_f(T - \lfloor \tau^{-1}T \rfloor \tau), z_1) + 1,$$

$$x_1(t) = x_f(t - \lfloor \tau^{-1}t \rfloor \tau), \ u_1(t) = u_f(t - \lfloor \tau^{-1}t \rfloor \tau), \ t \in [\tau S_1, T - \tau];$$

$$f^f(\tau(S_1 - 1), \tau S_1, x_1, u_1) \le v_0(x_1(\tau(S_1 - 1)), x_1(\tau S_1)) + 1;$$

if (2.142) holds, then

$$x_1(0) = z_0, \ x_1(\tau) = x_f(0), \ f^f(0, \tau, x_1, u_1) \le v_0(z_0, x_f(0)) + 1,$$

$$x_1(t) = x(t), \ u_1(t) = u(t), \ t \in [\tau(S_2 + 1), T],$$

$$x_1(t) = x_f(t - \lfloor \tau^{-1}t \rfloor \tau), \ u_1(t) = u_f(t - \lfloor \tau^{-1}t \rfloor \tau), \ t \in [\tau, S_2\tau],$$

$$f^f(\tau S_2, \tau(S_2 + 1), x_1, u_1) \le v_0(x_1(\tau S_2), x_1(\tau(S_2 + 1))) + 1;$$

if (2.143) holds, then

$$x_1(t) = x(t), \ u_1(t) = u(t), \ t \in [0, \tau(S_1 - 1)] \cup [\tau(S_2 + 1), T],$$

$$x_1(t) = x_f(t - \lfloor \tau^{-1}t \rfloor \tau), \ u_1(t) = u_f(t - \lfloor \tau^{-1}t \rfloor \tau), \ t \in [\tau(S_1 + 1), \tau S_2],$$

$$f^f((S_1 - 1)\tau, S_1\tau, x_1, u_1) \le v_0(x_1(S_1 - 1)\tau, x_1(S_1\tau)) + 1,$$

$$f^f(\tau S_2, \tau(S_2 + 1), x_1, u_1) \le v_0(x_1(\tau S_2), x_1(\tau(S_2 + 1))) + 1.$$

In view of (2.149), conditions (a), (b), (c) and the choice of (x_1, u_1),

$$I^g(0, T, x, u) \le I^g(0, T, x_1, u_1) + M. \tag{2.150}$$

We consider the cases (2.140)–(2.143) separately and obtain a lower bound for $I^g(0, T, x, u) - I^g(0, T, x_1, u_1)$. Assume that (2.140) holds. By (2.13), (2.140) and (2.148),

$$I^g(0, T, x, u) \geq I^g(0, \lfloor T\tau^{-1} \rfloor \tau, x, u) - \tau a$$

$$\geq I^g(0, \lfloor \lfloor T\tau^{-1} \rfloor L_0^{-1} \rfloor L_0 \tau, x, u) - \tau a(L_0 + 1)$$

$$= \sum_{j=0}^{\lfloor \lfloor T\tau^{-1} \rfloor L_0^{-1} \rfloor - 1} I^g((jL_0)\tau, (j+1)L_0\tau, x, u) - (L_0 + 1)a\tau$$

$$\geq \lfloor L_0 \tau \mu(f) + 16(1 + a)(1 + \tau) - 1 \rfloor \lfloor \lfloor T\tau^{-1} \rfloor L_0^{-1} \rfloor - (L_0 + 1)\tau a$$

$$\geq \lfloor L_0 \tau \mu(f) + 16(1 + a)(1 + \tau) - 1 \rfloor \lfloor T\tau^{-1} \rfloor L_0^{-1}$$

$$-L_0 \tau \mu(f) - 16(1 + a)(1 + \tau) - (L_0 + 1)\tau a$$

$$\geq \lfloor L_0 \tau \mu(f) + 16(1 + a)(1 + \tau) - 1 \rfloor T\tau^{-1} L_0^{-1}$$

$$-2(L_0 \tau \mu(f) + 16(1 + a)(1 + \tau) + (L_0 + 1)\tau a)$$

$$\geq T\mu(f) + TL_0^{-1} 8(1 + a) - 2(L_0 \tau \mu(f) + (L_0 + 17)(1 + a)(1 + \tau)).$$

$$(2.151)$$

Clearly,

$$I^g(0, T, x_1, u_1) = I^g(0, \tau, x_1, u_1) + I^g(\tau, T - \tau, x_1, u_1) + I^g(T - \tau, T, x_1, u_1).$$

$$(2.152)$$

By (2.133), (2.134), (2.140), (2.149), and the choice of (x_1, u_1)

$$I^f(0, \tau, x_1, u_1) \leq v_0(z_0, x_f(0)) + 1 \leq M_1 + 1,$$

$$I^f(T - \tau, T, x_1, u_1) \leq v_0(x_f(T - \lfloor \tau^{-1} T \rfloor \tau), z_1) + 1 \leq M_1 + 1.$$

In view of these inequalities, (2.138) and property (P5),

$$I^g(0, \tau, x_1, u_1), \ I^g(T - \tau, T, x_1, u_1) \leq M_1 + 5/4. \tag{2.153}$$

It follows from (2.3), (2.138), (2.140), and the choice of (x_1, u_1) that

$$I^g(\tau, T - \tau, x_1, u_1) = I^g(\tau, (\lfloor T\tau^{-1} \rfloor - 1)\tau, x_1, u_1)$$

$$+ I^g((\lfloor T\tau^{-1} \rfloor - 1)\tau, T - \tau, x_1, u_1)$$

$$= \sum_{i=1}^{\lfloor T\tau^{-1} \rfloor - 2} I^g(i\tau, (i+1)\tau, x_1, u_1)$$

$$+ I^g((\lfloor T\tau^{-1} \rfloor - 1)\tau, T - \tau, x_1, u_1), \tag{2.154}$$

for all integers $i = 1, \dots, \lfloor \tau^{-1} T \rfloor - 2$,

$$I^f(i\tau, (i+1)\tau, x_1, u_1) = I^f(0, \tau, x_f, u_f) = \tau \mu(f)$$

and in view of property (P5),

$$I^g(i\tau, (i+1)\tau, x_1, u_1) \leq \tau\mu(f) + (\tau+1)^{-1}(4L)^{-1}.$$

This implies that

$$\sum_{i=1}^{\lfloor T\tau^{-1}\rfloor-2} I^g(i\tau, (i+1)\tau, x_1, u_1) \leq T\mu(f) + (\tau+1)^{-1}(4L)^{-1}(T/\tau) + 3\tau|\mu(f)|.$$

$$(2.155)$$

By (2.13), (2.138), (2.140), property (P5), and the choice of (x_1, u_1),

$$I^g((\lfloor T\tau^{-1}\rfloor - 1)\tau, T - \tau, x_1, u_1)$$

$$= \int_{(\lfloor T\tau^{-1}\rfloor-1)\tau}^{T-\tau} g(t, x_f(t - (\lfloor \tau^{-1}T\rfloor - 1)\tau), u_f(t - (\lfloor \tau^{-1}T\rfloor - 1)\tau))dt$$

$$\leq \int_{(\lfloor T\tau^{-1}\rfloor-1)\tau}^{\lfloor T/\tau\rfloor\tau} g(t, x_f(t - (\lfloor \tau^{-1}T\rfloor - 1)\tau), u_f(t - (\lfloor \tau^{-1}T\rfloor - 1)\tau))dt + \tau a$$

$$\leq I^f(0, \tau, x_f, u_f) + 1 + \tau a.$$

Combined with (2.152)–(2.155) this implies that

$$I^g(0, T, x_1, u_1) \leq 2(M_2 + 1) + I^g(\tau, T - \tau, x_1, u_1)$$

$$\leq 2(M_2 + 2) + T\mu(f) + (\tau+1)^{-1}(4L)^{-1}(T/\tau)$$

$$+ 3\tau|\mu(f)| + \tau\mu(f) + \tau a + 1.$$

The relation above, (2.136)–(2.138), (2.150), and (2.151) imply that

$$M_1 \geq I^g(0, T, x, u) - I^g(0, T, x_1, u_1)$$

$$\geq T\mu(f) + 8TL_0^{-1}(1 + a) - 2[L_0\tau\mu(f) + (L_0 + 17)(1 + a)(\tau + 1)]$$

$$- T\mu(f) - 2M_2 - 4 - (\tau + 1)^{-1}(4L)^{-1}T\tau^{-1} - 4\tau|\mu(f)| - 1 - \tau a$$

$$= T[L_0^{-1}8(1 + a) - (4L)^{-1}(\tau + 1)^{-1}\tau^{-1}]$$

$$- 2L_0\tau|\mu(f)| - 2(L_0 + 17)(1 + a)(\tau + 1) - 2M_2 - 4 - 4\tau|\mu(f)| - 1 - \tau a$$

$$\geq 4TL_0^{-1}(1 + a) - \tau|\mu(f)|(2L_0 + 4)$$

$$- 2(L_0 + 17)(1 + a)(\tau + 1) - 2M_2 - 5 - \tau a$$

$$\geq 4l\tau - \tau|\mu(f)|(2L_0 + 4) - 2(L_0 + 17)(1 + a)(\tau + 1) - 2M_2 - 5 - \tau a.$$

This contradicts (2.137). Thus if (2.140) holds we have reached a contradiction.

Assume that (2.141) holds. By (2.141) and the choice of (x_1, u_1),

$$I^g(0, T, x, u) - I^g(0, T, x_1, u_1) = I^g(\tau(S_1 - 1), T, x, u) - I^g(\tau(S_1 - 1), T, x_1, u_1).$$
(2.156)

It follows from (2.13), (2.136)–(2.138), (2.141), and (2.148) that

$$I^g(\tau(S_1 - 1), T, x, u) \geq -a + I^g(\tau S_1, \lfloor T\tau^{-1} \rfloor \tau, x, u) - \tau a$$

$$\geq I^g(\tau S_1, \tau S_1 + \lfloor \lfloor T\tau^{-1} - S_1 \rfloor L_0^{-1} \rfloor L_0 \tau, x, u)$$
$$- L_0 \tau a - a(\tau + 1) = -a(L_0 \tau + \tau + 1)$$

$$+ \sum_{j=0}^{\lfloor \lfloor T\tau^{-1} - S_1 \rfloor L_0^{-1} \rfloor - 1} I^g(S_1 \tau + jL_0 \tau, S_1 \tau + (j + 1)L_0 \tau, x, u)$$

$$\geq -a(L_0 \tau + \tau + 1) + \lfloor \lfloor T\tau^{-1} - S_1 \rfloor L_0^{-1} \rfloor (L_0 \tau \mu(f))$$
$$+ 16(1 + a)(1 + \tau) - 1) \geq -a(L_0 \tau + \tau + 1)$$
$$- (L_0 |\mu(f)| + 16(1 + a)(1 + \tau)) + (\lfloor T\tau^{-1} \rfloor - S_1)\tau \mu(f)$$
$$+ (\lfloor T\tau^{-1} \rfloor - S_1)L_0^{-1}(16(1 + a)(1 + \tau) - 1).$$
(2.157)

It is clear that

$$I^g((S_1 - 1)\tau, T, x_1, u_1) = I^g(\tau(S_1 - 1), T - \tau, x_1, u_1) + I^g(T - \tau, T, x_1, u_1).$$
(2.158)

By the choice of (x_1, u_1), (2.133), (2.134), (2.141), and (2.149),

$$I^f(T - \tau, T, x_1, u_1) \leq v_0(x_f(T - \lfloor \tau^{-1}T \rfloor \tau), z_1) + 1 \leq M_1 + 1,$$
$$I^f(\tau(S_1 - 1), \tau S_1, x_1, u_1) \leq v_0(x_1(\tau(S_1 - 1)), x_1(\tau S_1)) + 1 \leq M_1 + 1.$$

Together with (2.138) and property (P5) this implies that

$$I^g(T - \tau, T, x_1, u_1), \ I^g(\tau(S_1 - 1), \tau S_1, x_1, u_1) \leq M_1 + 2.$$
(2.159)

Clearly,

$$I^g((S_1 - 1)\tau, T - \tau, x_1, u_1) = I^g(S_1 \tau, \lfloor T\tau^{-1} \rfloor \tau - \tau, x_1, u_1)$$
$$+ I^g((S_1 - 1)\tau, S_1 \tau, x_1, u_1)$$
$$+ I^g(\lfloor T\tau^{-1} \rfloor \tau - \tau, T - \tau, x_1, u_1).$$
(2.160)

It follows from (2.141) and the choice of (x_1, u_1) that for each

$$j \in \{S_1, \ldots, \lfloor \tau^{-1} T \rfloor - 2\},$$
$$I^f(j\tau, (j+1)\tau, x_1, u_1) = I^f(0, \tau, x_f, u_f) = \tau \mu(f)$$

and in view of property (P5),

$$I^g(j\tau, (j+1)\tau, x_1, u_1) \leq \tau \mu(f) + (\tau + 1)^{-1}(4L)^{-1}. \qquad (2.161)$$

By (2.13), (2.138), (2.141), property (P5), and the choice of (x_1, u_1),

$$I^g((\lfloor T\tau^{-1} \rfloor - 1)\tau, T - \tau, x_1, u_1)$$
$$= \int_{(\lfloor T\tau^{-1} \rfloor - 1)\tau}^{T-\tau} g(t, x_f(t - (\lfloor \tau^{-1} T \rfloor - 1)\tau), u_f(t - (\lfloor \tau^{-1} T \rfloor - 1)\tau)) dt$$
$$\leq \int_{(\lfloor T\tau^{-1} \rfloor - 1)\tau}^{\lfloor T/\tau \rfloor \tau} g(t, x_f(t - (\lfloor \tau^{-1} T \rfloor - 1)\tau), u_f(t - (\lfloor \tau^{-1} T \rfloor - 1)\tau)) dt + \tau a$$
$$\leq I^f(0, \tau, x_f, u_f) + 1 + \tau a.$$

In view of the relation above and (2.159)–(2.161),

$$I^g((S_1 - 1)\tau, T - \tau, x_1, u_1) \leq M_2 + 2 + \tau \mu(f) + 1$$
$$+ \tau a + I^g(S_1 \tau, \lfloor T\tau^{-1} \rfloor \tau - \tau, x_1, u_1)$$
$$\leq M_2 + 4 + \mu(f)\tau + 1 + \tau a + (\lfloor T\tau^{-1} \rfloor - S_1)(\tau \mu(f)$$
$$+ (\tau + 1)^{-1}(4L)^{-1}) + 2\tau |\mu(f)|.$$

By the relation above, (2.131), (2.137), (2.141), (2.150), and (2.156)–(2.159),

$$M_1 \geq I^g(0, T, x, u) - I^g(0, T, x_1, u_1)$$
$$= I^g(\tau(S_1 - 1), T, x, u) - I^g(\tau(S_1 - 1), T, x_1, u_1)$$
$$\geq -a(L_0\tau + \tau + 1) - L_0|\mu(f)| - 16(a+1)(\tau + 1) + (\lfloor T\tau^{-1} \rfloor - S_1)\tau \mu(f)$$
$$+ (\lfloor T\tau^{-1} \rfloor - S_1)L_0^{-1}(16(1 + a)(1 + \tau) - 1) - 2M_2 - 8 - 3|\mu(f)|\tau - 1$$
$$- \tau a - (\lfloor T\tau^{-1} \rfloor - S_1)\tau \mu(f) - (\lfloor T\tau^{-1} \rfloor - S_1)(\tau + 1)^{-1}(4L)^{-1}$$
$$\geq -a(L_0\tau + \tau + 1) - L_0|\mu(f)| - 16(a+1)(\tau + 1) - 2M_2 - 8$$
$$- 3|\mu(f)|\tau - 1 - \tau a + \lfloor \lfloor T\tau^{-1} \rfloor - S_1 \rfloor(L_0^{-1}(16(1 + a)(1 + \tau) - 1)$$
$$- (4L)^{-1}(1 + \tau)^{-1}) \geq -a(L_0\tau + \tau + 1) - L_0|\mu(f)| - 16(a+1)(\tau + 1)$$

$$- 2M_2 - 8 - 3|\mu(f)|\tau - 1 - \tau a + \lfloor\lfloor T\tau^{-1}\rfloor - S_1\rfloor 4L_0^{-1}(1+a)$$
$$\geq l - a(L_0\tau + \tau + 1) - L_0|\mu(f)| - 16(a+1)(\tau+1)$$
$$- 2M_2 - 8 - 3|\mu(f)|\tau - 1 - \tau a.$$

This contradicts (2.136). Thus if (2.141) holds we have reached a contradiction.
Assume that (2.142) holds. By (2.142) and the choice of (x_1, u_1),

$$I^g(0, T, x, u) - I^g(0, T, x_1, u_1) = I^g(0, \tau(S_2+1), x, u) - I^g(0, \tau(S_2+1), x_1, u_1).$$
(2.162)

By (2.13),

$$I^g(0, \tau(S_2+1), x, u) \geq I^g(0, S_2\tau, x, u) - \tau a$$

$$= \sum_{j=0}^{\lfloor S_2 L_0^{-1}\rfloor - 1} I^g(jL_0\tau, (j+1)L_0\tau, x, u) - a(L_0\tau + \tau)$$

$$\geq \lfloor S_2 L_0^{-1}\rfloor(L_0\tau\mu(f) + 16(1+a)(\tau+1) - 1) - a\tau(L_0+1).$$
(2.163)

By the choice of (x_1, u_1), (8.133), (2.134), (2.138), (2.142), (2.149), and property
(P5),

$$I^f(0, \tau, x_1, u_1) \leq M_1 + 1, \; I^g(0, \tau, x_1, u_1) \leq M_1 + 2,$$

$$I^f(\tau S_2, \tau(S_2+1), x_1, u_1) \leq M_1 + 1, \; I^g(\tau S_2, \tau(S_2+1), x_1, u_1) \leq M_1 + 2.$$
(2.164)

By (2.164),

$$I^g(0, (S_2+1)\tau, x_1, u_1) = I^g(\tau, (S_2+1)\tau, x_1, u_1) + I^g(0, \tau, x_1, u_1)$$

$$\leq M_2 + 2 + I^g(\tau_1, (S_2+1)\tau, x_1, u_1).$$
(2.165)

It follows from (2.142) and the choice of (x_1, u_1), (2.138) and property (P5) that for
each $j \in \{1, \ldots, S_2 - 1\}$,

$$I^f(j\tau, (j+1)\tau, x_1, u_1) = I^f(0, \tau, x_f, u_f) = \tau\mu(f),$$
$$I^g(j\tau, (j+1)\tau, x_1, u_1) \leq \tau\mu(f) + (\tau+1)^{-1}(4L)^{-1}.$$

These relations imply that

$$I^g(\tau, S_2\tau, x_1, u_1) \leq S_2\tau\mu(f) + S_2(4L)^{-1})(\tau+1)^{-1}.$$
(2.166)

By (2.131), (2.137), (2.142), (2.150), and (2.162)–(2.166),

$$M_1 \geq I^g(0, T, x, u) - I^g(0, T, x_1, u_1)$$

$$\geq \lfloor S_2 L_0^{-1} \rfloor (L_0 \tau \mu(f) + 16(1 + a)(\tau + 1) - 1) - a\tau(L_0 + 1)$$

$$- 2M_2 - 4 - S_2 \tau \mu(f) - S_2(4L)^{-1}(\tau + 1)^{-1}$$

$$\geq S_2(2L_0)^{-1}(16(1 + a)(\tau + 1) - 1) - a\tau(L_0 + 1)$$

$$- 2M_2 - 4 - S_2(4L)^{-1}(\tau + 1)^{-1} - L_0 \tau |\mu(f)|$$

$$\geq 4S_2 L_0^{-1} - a\tau(L_0 + 1) - 2M_2 - 4 - L_0 \tau |\mu(f)|$$

$$\geq 4l - a\tau(L_0 + 1) - L_0 |\mu(f)| \tau - 2M_2 - 4.$$

This contradicts (2.137). Thus if (2.142) holds we have reached a contradiction. Assume that (2.143) holds. By (2.143) and the choice of (x_1, u_1),

$$I^g(0, T, x, u) - I^g(0, T, x_1, u_1) = I^g(\tau(S_1 - 1), \tau(S_2 + 1), x, u)$$

$$- I^g(\tau(S_1 - 1), \tau(S_2 + 1), x_1, u_1). \qquad (2.167)$$

By (2.13) and (2.148),

$$I^g(\tau(S_1 - 1), \tau(S_2 + 1), x, u) \geq I^g(\tau S_1, S_2 \tau, x, u) - 2\tau a$$

$$\geq -2a\tau + I^g(\tau S_1, S_1 \tau$$

$$+ \lfloor (S_2 - S_1) L_0^{-1} \rfloor L_0 \tau, x, u) - L_0 \tau a$$

$$\geq -\tau a(L_0 + 2) + \lfloor (S_2 - S_1) L_0^{-1} \rfloor (L_0 \tau \mu(f)$$

$$+ 16(1 + a)(\tau + 1) - 1)). \qquad (2.168)$$

By the choice of (x_1, u_1), (2.133), (2.134), (2.143), and property (P5),

$$I^f(\tau(S_1 - 1), \tau S_1, x_1, u_1) \leq M_1 + 1, \quad I^f(\tau S_2, \tau(S_2 + 1), x_1, u_1) \leq M_1 + 1,$$

$$I^g(\tau(S_1 - 1), \tau S_1, x_1, u_1) \leq M_1 + 2, \quad I^g(\tau S_2, \tau(S_2 + 1), x_1, u_1) \leq M_1 + 2.$$
$$(2.169)$$

It follows from (2.141) and the choice of (x_1, u_1), (2.138) and property (P5) that for each $j \in \{S_1, \ldots, S_2 - 1\}$,

$$I^f(j\tau, (j + 1)\tau, x_1, u_1) = I^f(0, \tau, x_f, u_f) = \tau \mu(f),$$

$$I^g(j\tau, (j + 1)\tau, x_1, u_1) \leq \tau \mu(f) + (\tau + 1)^{-1}(4L)^{-1}.$$

These relations and (2.169) imply that

$$I^g(\tau(S_1 - 1), \tau(S_2 + 1), x_1, u_1) \le (S_2 - S_1)\tau\mu(f)$$
$$+ (S_2 - S_1)(4L)^{-1})(\tau + 1)^{-1} + 2M_1 + 4. \tag{2.170}$$

By (2.131), (2.137), (2.150), (2.167), (2.168), and (2.170),

$$
\begin{aligned}
M_1 &\ge I^g(\tau(S_1 - 1), \tau(S_2 + 1), x, u) - I^g(\tau(S_1 - 1), \tau(S_2 + 1), x_1, u_1) \\
&\ge -\tau a(2 + L_0) + \lfloor (S_2 - S_1)L_0^{-1}\rfloor L_0\tau\mu(f) \\
&\quad + \lfloor (S_2 - S_1)L_0^{-1}\rfloor(16(1 + a)(\tau + 1) - 1) \\
&\quad - (S_2 - S_1)\tau\mu(f) - (S_2 - S_1)(4L)^{-1}(\tau + 1)^{-1} - 2M_2 - 4 \\
&\ge -(L_0 + 2)|\mu(f)|\tau - \tau a(2 + L_0) - 2M_2 - 4 - 16(1 + a)(\tau + 1) \\
&\quad + (S_2 - S_1)(L_0^{-1}(16(1 + a)(1 + \tau) - 1) - (4L)^{-1}) \\
&\ge -(L_0 + 2)\tau|\mu(f)| - \tau a(2 + L_0) - 2M_2 - 16(1 + a)(1 + \tau) - 6 + 4l.
\end{aligned}
$$

This contradicts (2.137). Thus in all the cases we have reached a contradiction which proves (2.132) and Lemma 2.42 itself.

2.9 Proof of Theorem 2.13

By Proposition 2.39, there exists $\delta_0 \in (0, \epsilon)$ such that the following property holds:

(P6)   for each integer $p \ge 1$, each $(x, u) \in X(A, B, 0, p\tau)$ satisfying

$$|x(0) - x_f(0)|, \ |x(p\tau) - x_f(0)| \le \delta_0, \ I^f(0, p\tau, x, u) \le \sigma(f, x(0), x(p\tau), p\tau) + \delta_0$$

and each integer $i \in [0, p - 1]$, the inequality $|x(i\tau + t) - x_f(t)| \le \epsilon$ holds for all $t \in [0, \tau]$.

By Lemma 2.42, there exist an integer $L_0 \ge 1$ and a neighborhood $\mathcal{U}_0$ of f in $\mathcal{M}$ such that the following property holds:

(P7)   for each $T > L_0\tau$, each $g \in \mathcal{U}_0$, each pair of integers S_1, S_2 satisfying $0 \le S_1 \le S_2 - L_0$, $S_2\tau \le T$ and each $(x, u) \in X(A, B, 0, T)$ for which at least one of the conditions (a), (b), (c) holds,

$$\min\{|x(i\tau) - x_f(0)| : i = S_1, \ldots, S_2\} \le \delta_0.$$

Fix an integer $L \ge 4(L_0 + 1)$ and $\delta \in (0, 4^{-1}\delta_0)$. By assumption (A), Proposition 2.35 and the boundedness of v_0 on bounded sets there is $M_2 > 0$ such that the following property holds:

(P8)   for each $i \in \{1, \ldots, L\}$ and each $y, z \in R^n$ satisfying $|y|, |z| < 1 + \sup\{|x_f(t)| :$
$t \in [0, \tau]\}$ we have $|\sigma(f, y, z, i\tau)| \leq M_2$.

By Proposition 2.41, there exists a neighborhood $\mathcal{U}$ of f in $\mathcal{M}$ such that $\mathcal{U} \subset \mathcal{U}_0$
and that the following property holds:

(P9)   for each $g \in \mathcal{U}$, each $T_1 \geq 0$, each $T_2 \in [T_1 + \min\{1, \tau\}, \ T_1 + 4L \max\{1, \tau\}]$
and each $(x, u) \in X(A, B, T_1, T_2)$ satisfying

$$\min\{I^f(T_1, T_2, x, u), I^g(T_1, T_2, x, u)\} \leq M_2 + 4$$

the inequality $|I^f(T_1, T_2, x, u) - I^g(T_1, T_2, x, u)| \leq \delta$ holds.

Assume that

$$T > 2L\tau, \ g \in \mathcal{U}, \tag{2.171}$$

$(x, u) \in X(A, B, 0, T)$ satisfies for each $S \in [0, T - L\tau]$,

$$I^g(S, S + L\tau, x, u) \leq \sigma(g, x(S), x(S + L\tau), S, S + L\tau) + \delta \tag{2.172}$$

and satisfies at least one of the following conditions:

(a) $|x(0)|, \ |x(T)| \leq M_0, \ I^g(0, T, x, u) \leq \sigma(g, x(0), x(T), 0, T) + M_1;$

(b) $|x(0)| \leq M_0, \ I^g(0, T, x, u) \leq \sigma(g, x(0), 0, T) + M_1;$

(c) $I^g(0, T, x, u) \leq \sigma(g, 0, T) + M_1.$

By conditions (a)–(c) and property (P7), there exists a finite strictly increasing
sequence of integers $S_i, \ i = 1, \ldots, q$ such that

$$0 \leq S_1 < L_0, \ T \geq S_q \tau \geq T - \tau(1 + L_0), \ S_{i+1} - S_i \leq L_0 + 1, \ i = 1, \ldots, q - 1, \tag{2.173}$$

$$|x(S_i \tau) - x_f(0)| \leq \delta_0, \ i = 1, \ldots, q. \tag{2.174}$$

We may assume without loss of generality that

$$\text{if } |x(0) - x_f(0)| \leq \delta, \text{ then } S_1 = 0$$

$$\text{and if } |x(\lfloor T\tau^{-1}\rfloor\tau) - x_f(0)| \leq \delta, \text{ then } S_q = \lfloor T\tau^{-1}\rfloor. \tag{2.175}$$

Assume that an integer $i \in [S_1, S_q)$. Then there exists a natural number $k \in \{1, \ldots, q - 1\}$ such that

$$S_k \leq i < S_{k+1}. \tag{2.176}$$

In view of (2.171), (2.173), and the inequality $L \geq 4(L_0+1)$, there is $S \in [0, T-L\tau]$ such that

$$[S_k\tau, S_{k+1}\tau] \subset [S, S + L\tau]. \tag{2.177}$$

It follows from (2.172) and (2.177) that

$$I^g(S_k\tau, S_{k+1}\tau, x, u) \leq \sigma(g, x(S_k\tau), x(S_{k+1}\tau), S_k\tau, S_{k+1}\tau) + \delta. \tag{2.178}$$

By (2.173), (2.174), and property (P8),

$$\sigma(f, x(S_k\tau), x(S_{k+1}\tau), (S_{k+1} - S_k)\tau) < M_2. \tag{2.179}$$

By (2.171), (2.173), (2.178), (2.179), and property (P9),

$$\sigma(g, x(S_k\tau), x(S_{k+1}\tau), S_k\tau, S_{k+1}\tau)$$
$$\leq \sigma(f, x(S_k\tau), x(S_{k+1}\tau), (S_{k+1} - S_k)\tau) + \delta \tag{2.180}$$

and

$$I^g(S_k\tau, S_{k+1}\tau, x, u) \leq M_2 + 2.$$

In view of the relation above, (2.171), (2.173), (2.178), (2.180), and property (P9),

$$I^f(S_k\tau, S_{k+1}\tau, x, u) \leq I^g(S_k\tau, S_{k+1}\tau, x, u) + \delta$$
$$\leq \sigma(f, x(S_k\tau), x(S_{k+1}\tau), (S_{k+1} - S_k)\tau) + 3\delta.$$

Together with (2.174), (2.176), and property (P6), $|x(i\tau + t) - x_f(t)| \leq \epsilon$ for all $t \in [0, \tau]$. Theorem 2.13 is proved. $\square$

2.10   Proof of Theorem 2.14

By Proposition 2.39, there exists $\delta_0 \in (0, \epsilon)$ such that the following property holds:

(P10)   for each integer $p \geq 1$, each $(x, u) \in X(A, B, 0, p\tau)$ satisfying

$$|x(0) - x_f(0)|, \ |x(p\tau) - x_f(0)| \leq 4\delta_0,$$
$$I^f(0, p\tau, x, u) \leq \sigma(f, x(0), x(p\tau), p\tau) + 4\delta_0$$

and each integer $i \in [0, p-1]$, the inequality $|x(i\tau + t) - x_f(t)| \leq \epsilon$ holds for all $t \in [0, \tau]$.

By Lemma 2.42, there exist an integer $L_0 \geq 1$ and a neighborhood $\mathcal{U}_0$ of f in $\mathcal{M}$ such that the following property holds:

(P11)   for each $T > L_0 \tau$, each $g \in \mathcal{U}_0$, each pair of integers S_1, S_2 satisfying $0 \leq S_1 \leq S_2 - L_0$, $S_2 \tau \leq T$ and each $(x, u) \in X(A, B, 0, T)$ which satisfies at least one of the conditions (a), (b) and (c),

$$\min\{|x(i\tau) - x_f(0)| : i = S_1, \ldots, S_2\} \leq \delta_0.$$

Fix an integer

$$L \geq 4(L_0 + 1)(9 + 2\delta_0^{-1} M_1), \tag{2.181}$$

$$\delta \in (0, 4^{-1}\delta_0). \tag{2.182}$$

By Proposition 2.35 and the boundedness of v_0 on bounded sets there is $M_2 > 0$ such that for each $i \in \{1, \ldots, L\}$ and each $y, z \in R^n$ satisfying $|y|, |z| \leq 1 + \sup\{|x_f(t)| : t \in [0, \tau]\}$ we have

$$|\sigma(f, y, z, i\tau)| \leq M_2. \tag{2.183}$$

By Proposition 2.41, there exists a neighborhood $\mathcal{U}$ of f in $\mathcal{M}$ such that $\mathcal{U} \subset \mathcal{U}_0$ and that the following property holds:

(P12)   for each $g \in \mathcal{U}$, each $T_1 \geq 0$, each $T_2 \in [T_1 + \min\{1, \tau\}, T_1 + 4L \max\{1, \tau\}]$ and each $(x, u) \in X(A, B, T_1, T_2)$ satisfying

$$\min\{I^f(T_1, T_2, x, u), I^g(T_1, T_2, x, u)\} \leq M_2 + 4$$

the inequality $|I^f(T_1, T_2, x, u) - I^g(T_1, T_2, x, u)| \leq \delta$ holds.

Assume that

$$T > L\tau, \quad g \in \mathcal{U} \tag{2.184}$$

and that $(x, u) \in X(A, B, 0, T)$ satisfies at least one of the following conditions:

(a) $|x(0)|, |x(T)| \leq M_0$, $I^g(0, T, x, u) \leq \sigma(g, x(0), x(T), 0, T) + M_1$;

(b) $|x(0)| \leq M_0$, $I^g(0, T, x, u) \leq \sigma(g, x(0), 0, T) + M_1$;

(c) $I^g(0, T, x, u) \leq \sigma(g, 0, T) + M_1$.

By conditions (a)–(c), (2.184), and property (P11), there exists a finite strictly increasing sequence of integers S_i, $i = 1, \ldots, q$ such that

$$0 \leq S_1 \leq L_0, \ T \geq S_q \tau \geq T - \tau(1 + L_0), \ S_{i+1} - S_i \leq L_0 + 1, \ i = 1, \ldots, q-1, \tag{2.185}$$

$$|x(S_i\tau) - x_f(0)| \leq \delta_0, \ i = 1, \ldots, q. \tag{2.186}$$

Define by induction a finite strictly increasing sequence of natural numbers $i_1, \ldots, i_k \in \{1, \ldots, q\}$. Set

$$i_1 = 1. \tag{2.187}$$

Assume that $p \geq 1$ is an integer and that we defined integers $i_1 < \ldots < i_p$ belonging to $\{1, \ldots, q\}$ such that for each natural number $m < p$ the following properties hold:

(i)

$$I^g(S_{i_m}\tau, S_{i_{m+1}}\tau, x, u) > \sigma(g, x(S_{i_m}\tau), x(S_{i_{m+1}}\tau), S_{i_m}\tau, S_{i_{m+1}}\tau) + \delta_0; \tag{2.188}$$

(ii) if $i_{m+1} > i_m + 1$, then

$$I^g(S_{i_m}\tau, S_{i_{m+1}-1}\tau, x, u) \leq \sigma(g, x(S_{i_m}\tau), x(S_{i_{m+1}-1}\tau), S_{i_m}\tau, S_{i_{m+1}-1}\tau) + \delta_0. \tag{2.189}$$

(Note that by (2.187) our assumption holds for $p = 1$.) Let us define i_{p+1}. If $i_p = q$, then our construction is completed, $k = p$, $i_k = q$ and for each natural number $m < p = k$, properties (i) and (ii) hold.

Assume that $i_p < q$. There are two cases:

$$I^g(S_{i_p}\tau, S_q\tau, x, u) \leq \sigma(g, x(S_{i_p}\tau), x(S_q\tau), S_{i_p}\tau, S_q\tau) + \delta_0; \tag{2.190}$$

$$I^g(S_{i_p}\tau, S_q\tau, x, u) > \sigma(g, x(S_{i_p}\tau), x(S_q\tau), S_{i_p}\tau, S_q\tau) + \delta_0. \tag{2.191}$$

Assume that (2.190) holds. Then we set $k = p + 1$, $i_k = q$ the construction is completed, for each natural number $m < k - 1$, (2.188) is true and for each natural number $m < k$, property (ii) holds.

Assume that (2.191) holds. Then we set

$$i_{p+1} = \min\{j > S_{i_p} : \ j \text{ is an integer and}$$
$$I^g(S_{i_p}\tau, S_j\tau, x, u) > \sigma(g, x(S_{i_p}\tau), x(S_j\tau), S_{i_p}\tau, S_j\tau) + \delta_0\}. \tag{2.192}$$

It is easy to see that the assumption made for p also holds for $p + 1$. As a result we obtain a finite strictly increasing sequence of integers $i_1, \ldots, i_k \in \{1, \ldots, q\}$ such that $i_k = q$, for all integers m satisfying $1 \leq m < k - 1$, (2.188) holds and for each integer m satisfying $1 \leq m < k$, (ii) holds. By conditions (a)–(c) and (2.188),

$$M_1 \geq I^g(0, T, x, u) - \sigma(g, x(0), x(T), 0, T)$$

$$\geq \sum \{I^g(S_{i_j}\tau, S_{i_{j+1}}\tau, x, u) - \sigma(g, x(S_{i_j}\tau), x(S_{i_{j+1}}\tau), S_{i_j}\tau, S_{i_{j+1}}\tau) :$$

j is an integer satisfying $1 \le j < k - 1\} \ge \delta_0(k - 2)$,

$$k \le \delta_0^{-1} M_1 + 2. \tag{2.193}$$

Set

$$A = \{j \in \{1, \ldots, k\} : j < k \text{ and } S_{i_{j+1}} - S_{i_j} \ge 4(L_0 + 1)\}. \tag{2.194}$$

Let

$$j \in A. \tag{2.195}$$

By (2.185), (2.189), (2.194), (2.195), and property (ii),

$$I^g(S_{i_j}\tau, S_{i_{j+1}-1}\tau, x, u) \le \sigma(g, x(S_{i_j}\tau), x(S_{i_{j+1}-1}\tau), S_{i_j}\tau, S_{i_{j+1}-1}\tau)$$
$$+ \delta_0, \ i_{j+1} > i_j + 3. \tag{2.196}$$

Let

$$p \in \{i_j, \ldots, i_{j+1} - 2\}.$$

This implies that $\{S_p, S_{p+1}\} \subset \{S_{i_j}, \ldots, S_{i_{j+1}-1}\}$ and in view of (2.196),

$$I^g(S_p\tau, S_{p+1}\tau, x, u) \le \sigma(g, x(S_p\tau), x(S_{p+1}\tau), S_p\tau, S_{p+1}\tau) + \delta_0. \tag{2.197}$$

By the choice of M_2 [see (2.183)], (2.185), and (2.186),

$$\sigma(f, x(S_p\tau), x(S_{p+1}\tau), S_{p+1}\tau - S_p\tau) \le M_2. \tag{2.198}$$

Together with (2.181), (2.184), (2.185), and property (P12) this implies that

$$\sigma(g, x(S_p\tau), x(S_{p+1}\tau), S_p\tau, S_{p+1}\tau)$$
$$\le \sigma(f, x(S_p\tau), x(S_{p+1}\tau), (S_{p+1} - S_p)\tau) + \delta_0.$$

By (2.181), (2.184), (2.185), (2.197), (2.198), the inequality above and property (P12),

$$I^f(S_p\tau, S_{p+1}\tau, x, u) \le \sigma(f, x(S_p\tau), x(S_{p+1}\tau), (S_{p+1} - S_p)\tau) + 3\delta_0. \tag{2.199}$$

It follows from (2.186), (2.199), and property (P10) that for all integers $m \in \{S_p, \ldots, S_{p+1} - 1\}$,

$$|x(m\tau + t) - x_f(t)| \le \epsilon, \ t \in [0, \tau]. \tag{2.200}$$

Thus (2.200) holds for all integers $m \in \{S_{i_j}, \ldots, S_{i_{j+1}-1} - 1\}$. Since j is any integer belonging to A we conclude that (2.200) holds for all

$$m \in \cup \{\{S_{i_j}, \ldots, S_{i_{j+1}-1} - 1\} : j \in A\}$$

and that

$$\{m \in \{0, \ldots, \lfloor T\tau^{-1} \rfloor - 1\} : (2.200) \text{ does not hold}\}$$
$$\subset \{0, \ldots, S_1\} \cup \{S_q, \ldots, \lfloor T\tau^{-1} \rfloor\} \cup \{\{S_{i_j}, \ldots, S_{i_{j+1}}\} : j \in \{1, \ldots, k\} \setminus A\}$$
$$\cup \{\{S_{i_{j+1}-1}, \ldots, S_{i_{j+1}}\} : j \in A\}$$

and in view of (2.181), (2.185), (2.193), and (2.194), the cardinality of the right-hand side of the inclusion above does not exceed

$$4(L_0 + 1)(2 + 2k) \leq 4(L_0 + 1)(8 + 2\delta_0^{-1}M_1) < L.$$

Theorem 2.14 is proved.   □

2.11 Proofs of Propositions 2.18 and 2.21

Proof of Proposition 2.18. Let $\epsilon > 0$. Since the function v is continuous there exists $\delta > 0$ such that

$$|v(z_1, z_2) - \mu(f)\tau| \leq \epsilon/4 \tag{2.201}$$

for all $z_1, z_2 \in R^n$ satisfying $|z_i - x_f(0)| \leq \delta$, $i = 1, 2$.

Let

$$z_1, z_2 \in R^n, \ |z_i - x_f(0)| \leq \delta, \ i = 1, 2. \tag{2.202}$$

By Proposition 2.32, there exists $(x, u) \in X(A, B, 0, \infty)$

$$x(0) = z_2, \ x(\tau) = z_1, \ I^f(0, \tau, x, u) = v(z_2, z_1), \tag{2.203}$$

$$x(t) = \xi^{(z_1)}(t - \tau), \ u(t) = \eta^{(z_1)}(t - \tau) \text{ for all } t \geq \tau. \tag{2.204}$$

By (2.21), (2.202)–(2.204), and Proposition 2.15,

$$\pi^f(z_2) = \pi^f(x(0)) \leq \liminf_{T \in Z, T \to \infty} [I^f(0, \tau, x, u) - T\mu(f)]$$

$$= I^f(0, \tau, x, u) - \tau\mu(f) + \liminf_{T \in Z, T \to \infty} [I^f(0, T, \xi^{(z_1)}, \eta^{(z_1)}) - T\mu(f)]$$

$$\leq v(z_2, z_1) + \pi^f(z_1) - \mu(f)\tau \leq \pi^f(z_1) + \epsilon/2.$$

Proposition 2.18 is proved.   □

Proof of Proposition 2.21. Let $M > 0$. by Proposition 2.18, there exists $\delta > 0$ such that for each $x \in R^n$ satisfying $|x - x_f(0)| \leq \delta$,

$$|\pi^f(x)| \leq 1. \tag{2.205}$$

By Proposition 2.38, there exists a natural number L_0 such that the following property holds:

(i) for each $T \geq L_0\tau$, each $(x, u) \in X(A, B, 0, T)$ satisfying

$$I^f(0, T, x, u) \leq T\mu(f) + M + 1$$

and each integer S satisfying $[S\tau, (S + L_0)\tau] \subset [0, T]$ there exists an integer $i \in [S, S + L_0 - 1]$ such that

$$|x(\tau i + t) - x_f(t)| \leq \delta \text{ for all } t \in [0, \tau].$$

By Proposition 2.40, there exists $M_1 > 0$ such that the following property holds:

(ii) for each $T \in [\tau, (L_0 + 1)\tau]$ and each $(x, u) \in X(A, B, 0, T)$ satisfying

$$I^f(0, T, x, u) \leq \tau|\mu(f)|(L_0 + 1) + M + 1$$

we have

$$|x(t)| \leq M_1 \text{ for all } t \in [0, T].$$

Assume that $x \in R^n$ satisfies

$$\pi^f(x) \leq M. \tag{2.206}$$

By (2.21) and (2.206),

$$\pi^f(x) = \liminf_{T \in Z, T \to \infty} [I^f(0, T, \xi^{(x)}, \eta^{(x)}) - T\mu(f)] \leq M. \tag{2.207}$$

In view of (2.207) and property (i), there exists an integer $t_0 \in [1, L_0 + 1]$ such that

$$|\xi^{(x)}(t_0\tau) - x_f(0)| \leq \delta. \tag{2.208}$$

It follows from (2.208) and the choice of δ [see (2.205)] that

$$|\pi^f(\xi^{(x)}(t_0\tau))| \leq 1. \tag{2.209}$$

Proposition 2.16, (2.206) and (2.209) imply that

$$I^f(0, t_0\tau, \xi^{(x)}, \eta^{(x)}) - t_0\tau\mu(f) = \pi^f(x) - \pi^f(\xi^{(x)}(t_0\tau)) \leq M + 1. \tag{2.210}$$

By (2.210) and property (ii), $|\xi^{(x)}(t)| \leq M_1$ for all $t \in [0, t_0\tau]$. Thus $|x| \leq M_1$. Proposition 2.21 is proved. $\qquad\square$

2.12 Auxiliary Results for Theorem 2.25

We continue to use the notation, definitions, and assumptions introduced in Sects. 2.1–2.3.

Assume that $S_2 > S_1 \geq 0$ are integers and $g \in \mathcal{M}$. There exists a unique function $\mathcal{L}_{S_1,S_2}(g)(t, x, u)$, $(t, x, u) \in [0, \infty) \times R^n \times R^m$ such that for each $x \in R^n$ and each $u \in R^m$,

$$\mathcal{L}_{S_1,S_2}(g)(t, x, u) = g(S_2\tau - t + S_1\tau, x, u) \text{ for each } t \in [S_1\tau, S_2\tau], \tag{2.211}$$

$$\mathcal{L}_{S_1,S_2}(g)(t + (S_2 - S_1)\tau, x, u) = \mathcal{L}_{S_1,S_2}(g)(t, x, u) \text{ for each } t \geq 0. \tag{2.212}$$

Clearly, $\mathcal{L}_{S_1,S_2}(g) \in \mathcal{M}$ and $\mathcal{L}_{S_1,S_2}$ is a self-mapping of $\mathcal{M}$. It is easy to see that the following proposition holds.

Proposition 2.43. *1. Let V be a neighborhood of $\bar{f}$ in $\mathcal{M}$. Then there exists a neighborhood U of f in $\mathcal{M}$ such that $\mathcal{L}_{S_1,S_2}(g) \in V$ for all $g \in U$ and all integers $S_2 > S_1 \geq 0$.*
2. Let V be a neighborhood of f in $\mathcal{M}$. Then there exists a neighborhood U of $\bar{f}$ in $\mathcal{M}$ such that $\mathcal{L}_{S_1,S_2}(g) \in V$ for all $g \in U$ and all integers $S_2 > S_1 \geq 0$.

Let $S_2 > S_1 \geq 0$ be integers, $g \in \mathcal{M}$ and $(x, u) \in X(A, B, S_1\tau, S_2\tau)$ ($X(-A, -B, S_1\tau, S_2\tau)$ respectively). Then in view of (2.26) and (2.211),

$$\int_{S_1\tau}^{S_2\tau} \mathcal{L}_{S_1,S_2}(g)(t, \bar{x}(t), \bar{u}(t))dt$$

$$= \int_{S_1\tau}^{S_2\tau} g(S_2\tau - t + S_1\tau, x(S_2\tau - t + S_1\tau), u(S_2\tau - t + S_1\tau))dt$$

$$= \int_{S_1\tau}^{S_2\tau} g(t, x(t), u(t))dt. \tag{2.213}$$

Let $T_2 > T_1 \geq 0$, $y, z \in R^n$ and $g \in \mathcal{M}$. For each $(x, u) \in X(-A, -B, T_1, T_2)$, put

$$I^g(T_1, T_2, x, u) = \int_{T_1}^{T_2} g(t, x(t), u(t))dt \tag{2.214}$$

and set

$$\sigma_-(g, y, z, T_1, T_2) = \inf\{I^g(T_1, T_2, x, u) :$$

$$(x, u) \in X(-A, -B, T_1, T_2) \text{ and } x(T_1) = y, \; x(T_2) = z\}, \tag{2.215}$$

$$\sigma_-(g, y, T_1, T_2) = \inf\{I^g(T_1, T_2, x, u) :$$

$$(x, u) \in X(-A, -B, T_1, T_2) \text{ and } x(T_1) = y\}, \tag{2.216}$$

$$\hat{\sigma}_-(g, z, T_1, T_2) = \inf\{I^g(T_1, T_2, x, u) :$$

$$(x, u) \in X(-A, -B, T_1, T_2) \text{ and } x(T_2) = z\}, \tag{2.217}$$

$$\sigma_-(g, T_1, T_2) = \inf\{I^g(T_1, T_2, x, u) : (x, u) \in X(-A, -B, T_1, T_2)\}. \tag{2.218}$$

Relation (2.213) implies the following result.

Proposition 2.44. *Let* $S_2 > S_1 \geq 0$ *be integers,* $g \in \mathcal{M}$ *and* $(x_i, u_i) \in X(A, B, S_1\tau, S_2\tau)$, $i = 1, 2$. *Then*

$$I^g(S_1\tau, S_2\tau, x_1, u_1) \geq I^g(S_1\tau, S_2\tau, x_2, u_2) - M \tag{2.219}$$

if and only if

$$I^{\bar{g}}(S_1\tau, S_2\tau, \bar{x}_1, \bar{u}_1) \geq I^{\bar{g}}(S_1\tau, S_2\tau, \bar{x}_2, \bar{u}_2) - M, \tag{2.220}$$

where $\bar{g} = \mathcal{L}_{S_1, S_2}(g)$.

Proposition 2.44 [see (2.219) and (2.220)] implies the following result.

Proposition 2.45. *Let* $S_2 > S_1 \geq 0$ *be integers,* $M \geq 0$, $g \in \mathcal{M}$, $\bar{g} = \mathcal{L}_{S_1, S_2}(g)$ *and* $(x, u) \in X(A, B, S_1\tau, S_2\tau)$. *Then the following assertions are equivalent:*

$$I^g(S_1\tau, S_2\tau, x, u) \leq \sigma(g, S_1\tau, S_2\tau) + M$$

if and only if

$$I^{\bar{g}}(S_1\tau, S_2\tau, \bar{x}, \bar{u}) \leq \sigma_-(\bar{g}, S_1\tau, S_2\tau) + M;$$

$$I^g(S_1\tau, S_2\tau, x, u) \leq \sigma(g, x(S_1\tau), x(S_2\tau), S_1\tau, S_2\tau) + M$$

if and only if

$$I^{\bar{g}}(S_1\tau, S_2\tau, \bar{x}, \bar{u}) \le \sigma_-(\bar{g}, \bar{x}(S_1\tau), \bar{x}(S_2\tau), S_1\tau, S_2\tau) + M;$$
$$I^{g}(S_1\tau, S_2\tau, x, u) \le \hat{\sigma}(g, x(S_2\tau), S_1\tau, S_2\tau) + M$$

if and only if

$$I^{\bar{g}}(S_1\tau, S_2\tau, \bar{x}, \bar{u}) \le \sigma_-(\bar{g}, \bar{x}(S_1\tau), S_1\tau, S_2\tau) + M;$$
$$I^{g}(S_1\tau, S_2\tau, x, u) \le \sigma(g, x(S_1\tau), S_1\tau, S_2\tau) + M$$

if and only if

$$I^{\bar{g}}(S_1\tau, S_2\tau, \bar{x}, \bar{u}) \le \hat{\sigma}_-(\bar{g}, \bar{x}(S_2\tau), S_1\tau, S_2\tau) + M.$$

2.13   The Basic Lemma for Theorem 2.25

Let $\theta_f \in R^n$ satisfy

$$\pi^f(\theta_f) = \inf(\pi^f). \tag{2.221}$$

Lemma 2.46. *Let $S_0 \ge 1$ be an integer, $\epsilon \in (0, 1)$ and*

$$(x_*, u_*) \in X(A, B, 0, \infty)$$

be an (f, A, B)-overtaking optimal pair satisfying

$$x_*(0) = \theta_f. \tag{2.222}$$

Then there exists $\delta \in (0, \epsilon)$ such that for each $(x, u) \in X(A, B, 0, S_0\tau)$ which satisfies

$$\pi^f(x(0)) \le \inf(\pi^f) + \delta,$$
$$I^f(0, S_0\tau, x, u) - S_0\tau\mu(f) - \pi^f(x(0)) + \pi^f(x(S_0\tau)) \le \delta$$

the inequality $|x(t) - x_*(t)| \le \epsilon$ holds for all $t \in [0, S_0\tau]$.

Proof. Assume that the lemma does not hold. Then there exist a sequence $\{\delta_k\}_{k=1}^{\infty} \subset (0, 1]$ and a sequence $\{(x_k, u_k)\}_{k=1}^{\infty} \subset X(A, B, 0, S_0\tau)$ such that

$$\lim_{k \to \infty} \delta_k = 0$$

and that for all integers $k \geq 1$,

$$\pi^f(x_k(0)) \leq \inf(\pi^f) + \delta_k, \tag{2.223}$$

$$I^f(0, S_0\tau, x_k, u_k) - S_0\tau\mu(f) - \pi^f(x_k(0)) + \pi^f(x_k(S_0\tau)) \leq \delta_k, \tag{2.224}$$

$$\sup\{|x_k(t) - x_*(t)| : t \in [0, S_0\tau]\} > \epsilon. \tag{2.225}$$

In view of (2.223) and Proposition 2.21, the sequence $\{x_k(0)\}_{k=1}^{\infty}$ is bounded. By (2.224), the continuity and the boundedness from below of the function π^f (see Proposition 2.22) and boundedness of the sequence $\{x_k(0)\}_{k=1}^{\infty}$, the sequence $\{I^f(0, S_0\tau, x_k, u_k)\}_{k=1}^{\infty}$ is bounded. By Proposition 2.30, we may assume without loss of generality that there exists $(x, u) \in X(A, B, 0, S_0\tau)$ such that

$$x_k(t) \to x(t) \text{ as } k \to \infty \text{ uniformly on } [0, S_0\tau], \tag{2.226}$$

$$I^f(0, S_0\tau, x, u) \leq \liminf_{k \to \infty} I^f(0, S_0\tau, x_k, u_k), \tag{2.227}$$

$$u_k \to u \text{ as } k \to \infty \text{ weakly in } L^1(R^m; (0, S_0\tau)).$$

It follows from (2.221), (2.223), (2.226) and the continuity and strict convexity of π^f (see Proposition 2.20) that

$$\pi^f(x(0)) = \lim_{k \to \infty} \pi^f(x_k(0)) = \inf(\pi^f), \ x(0) = \theta_f.$$

By (2.224), (2.226), (2.227), and the continuity of π^f (see Proposition 2.20),

$$\pi^f(x(S_0\tau)) = \lim_{k \to \infty} \pi^f(x_k(S_0\tau)),$$

$$I^f(0, S_0\tau, x, u) - S_0\tau\mu(f) - \pi^f(x(0)) + \pi^f(x(S_0\tau))$$

$$\leq \liminf_{k \to \infty}[I^f(0, S_0\tau, x_k, u_k) - S_0\tau\mu(f) - \pi^f(x_k(0)) + \pi^f(x_k(S_0\tau))] \leq 0.$$

In view of the inequality above and Proposition 2.15,

$$I^f(0, S_0\tau, x, u) - S_0\tau\mu(f) - \pi^f(x(0)) + \pi^f(x(S_0\tau)) = 0. \tag{2.228}$$

Theorem 2.3 implies that there exists an (f, A, B)-overtaking optimal pair $(\tilde{x}, \tilde{u}) \in X(A, B, 0, \infty)$ such that

$$\tilde{x}(0) = x(S_0\tau). \tag{2.229}$$

For all $t > S_0\tau$ set

$$x(t) = \tilde{x}(t - S_0\tau), \ u(t) = \tilde{u}(t - S_0\tau). \tag{2.230}$$

It is not difficult to see that the pair $(x, u) \in X(A, B, 0, \infty)$ is an (f, A, B)-good pair. By its definition, (2.228)–(2.230) and Propositions 2.15 and 2.16,

$$I^f(0, S\tau, x, u) - S\tau\mu(f) - \pi^f(x(0)) + \pi^f(x(S\tau)) = 0 \text{ for all integers } S \geq 1.$$

Combined with Proposition 2.23 this implies that (x, u) is an (f, A, B)-overtaking optimal pair satisfying $x(0) = \theta_f$. By Theorem 2.3 and (2.222), $x(t) = x_*(t)$ and $u(t) = u_*(t)$ for all $t \geq 0$. Together with (2.226) this implies that for all sufficiently large natural numbers k,

$$|x_k(t) - x_*(t)| \leq \epsilon/2 \text{ for all } t \in [0, S_0\tau].$$

This contradicts (2.225). The contradiction we have reached proves Lemma 2.46.

Note that Lemma 2.46 can also be applied for the triplet $(\bar{f}, -A, -B)$.

2.14   Proof of Theorem 2.25

By Lemma 2.46 applied to the triplet $(\bar{f}, -A - B)$ there exist

$$\delta_1 \in (0, \epsilon/4)$$

such that the following property holds:

(P13)   for each $(x, u) \in X(-A, -B, 0, L_0\tau)$ which satisfies

$$\pi^{\bar{f}}(x(0)) \leq \inf(\pi^{\bar{f}}) + \delta_1,$$

$$I^{\bar{f}}(0, L_0\tau, x, u) - L_0\tau\mu(f) - \pi^{\bar{f}}(x(0)) + \pi^{\bar{f}}(x(L_0\tau)) \leq \delta_1$$

we have

$$|x(t) - \bar{x}_*(t)| \leq \epsilon \text{ holds for all } t \in [0, L_0\tau].$$

In view of the continuity of π^f, Proposition 2.17, and (2.34), there exists $\delta_2 \in (0, \delta_1)$ such that for each $z \in R^n$ satisfying $|z - x_f(0)| \leq 2\delta_2$,

$$|\pi^{\bar{f}}(z)| = |\pi^{\bar{f}}(z) - \pi^{\bar{f}}(x_f(0))| \leq \delta_1/8; \tag{2.231}$$

for each $y, z \in R^n$ satisfying $|y - x_f(0)| \leq 2\delta_2$, $|z - x_f(0)| \leq 2\delta_2$,

$$|v(y, z) - \tau\mu(f)| \leq \delta_1/8. \tag{2.232}$$

By Theorem 2.13, there exist an integer $l_0 \geq 1$, $\delta_3 \in (0, \delta_2/8)$ and a neighborhood $\mathcal{U}_1$ of f in $\mathcal{M}$ such that the following property holds:

(P14)   for each integer $T > 2l_0$, each $g \in \mathcal{U}_1$ and each

$$(x, u) \in X(A, B, 0, T\tau)$$

such that

$$|x(0)| \leq M, \ I^g(0, T\tau, x, u) \leq \sigma(g, x(0), 0, T\tau) + \delta_3$$

we have

$$|x(i\tau) - x_f(0)| \leq \delta_2 \text{ for all } i = l_0, \ldots, T - l_0. \tag{2.233}$$

Since the pair $(\bar{x}_*, \bar{u}_*) \in X(-A, -B, 0, \infty)$ is $(\bar{f}, -A, -B)$-good it follows from Theorem 2.2 and (2.29) that there exists an integer $l_1 \geq 1$ such that

$$|\bar{x}_*(i\tau) - x_f(0)| \leq \delta_2 \text{ for all integers } i \geq l_1. \tag{2.234}$$

By Proposition 2.41, there exists a neighborhood $\mathcal{U} \subset \mathcal{U}_1$ of f in $\mathcal{M}$ such that the following property holds:

(P15)   for each $g \in \mathcal{U}$, each integer $j \in \{1, \ldots, 2L_0 + 2l_0 + 2l_1 + 4\}$ and each $(x, u) \in X(A, B, 0, j\tau)$ satisfying

$$\min\{I^f(0, j\tau, x, u), I^g(0, j\tau, x, u)\}$$

$$\leq (|\mu(f)| + 2)(2L_0 + 2l_0 + 2l_1 + 4)\tau + |\pi^{\bar{f}}(\bar{x}_*(0))| + 2$$

we have $|I^f(0, j\tau, x, u) - I^g(0, j\tau, x, u)| \leq \delta_3/8$.

Choose $\delta > 0$ and an integer L_1 such that

$$\delta \leq \delta_3/4, \tag{2.235}$$

$$L_1 > 2L_0 + 2l_0 + 2l_1 + 4. \tag{2.236}$$

Assume that an integer

$$T \geq L_1, \ g \in \mathcal{U} \tag{2.237}$$

and $(x, u) \in X(A, B, 0, T\tau)$ satisfies

$$|x(0)| \leq M, \ I^g(0, T\tau, x, u) \leq \sigma(g, x(0), 0, T\tau) + \delta. \tag{2.238}$$

By property (P14) and (2.235)–(2.238), (2.233) holds. In view of (2.236) and (2.237),

$$[T - l_0 - l_1 - L_0 - 4, T - l_0 - l_1 - L_0] \subset [l_0, T - l_0 - l_1 - L_0]. \tag{2.239}$$

In view of (2.233) and (2.239),

$$|x(i\tau) - x_f(0)| \leq \delta_2 \text{ for all } i \in \{T - l_0 - l_1 - L_0 - 4, \ldots, T - l_0 - l_1 - L_0\}. \tag{2.240}$$

Proposition 2.32 implies that there exists $(x_1, u_1) \in X(A, B, 0, T\tau)$ such that

$$x_1(t) = x(t), \ u_1(t) = u(t), \ t \in [0, \tau(T - l_0 - l_1 - L_0 - 4)], \tag{2.241}$$

$$x_1(t) = \bar{x}_*(T\tau - t), \ u_1(t) = \bar{u}_*(T\tau - t), \ t \in [\tau(T - l_0 - l_1 - L_0 - 3), \tau T], \tag{2.242}$$

$$I^f(\tau(T - l_0 - l_1 - L_0 - 4), \tau(T - l_0 - l_1 - L_0 - 3), x_1, u_1)$$
$$= v(x(\tau(T - l_0 - l_1 - L_0 - 4)), \bar{x}_*(\tau(l_0 + l_1 + L_0 + 3))). \tag{2.243}$$

By (2.238) and (2.241),

$$-\delta \leq I^g(0, T\tau, x_1, u_1) - I^g(0, T\tau, x, u)$$
$$= I^g(\tau(T - l_0 - l_1 - L_0 - 4), \tau(T - l_0 - l_1 - L_0 - 3), x_1, u_1)$$
$$+ I^g(\tau(T - l_0 - l_1 - L_0 - 3), \tau T, x_1, u_1)$$
$$- I^g(\tau(T - l_0 - l_1 - L_0 - 4), \tau(T - l_0 - l_1 - L_0 - 3), x, u)$$
$$- I^g(\tau(T - l_0 - l_1 - L_0 - 3), \tau T, x, u). \tag{2.244}$$

We show that

$$I^g(\tau(T - l_0 - l_1 - L_0 - 4), \tau(T - l_0 - l_1 - L_0 - 3), x_1, u_1)$$
$$- I^g(\tau(T - l_0 - l_1 - L_0 - 4), \tau(T - l_0 - l_1 - L_0 - 3), x, u)$$
$$\leq \delta_1/8 + \delta_3/8 + \delta_1/2. \tag{2.245}$$

In view of (2.234), (2.240), (2.243), and the choice of δ_2 [see (2.232)],

$$I^f(\tau(T - l_0 - l_1 - L_0 - 4), \tau(T - l_0 - l_1 - L_0 - 3), x_1, u_1) \leq \tau\mu(f) + \delta_1/8.$$

Combined with (2.237) and property (P15) this implies that

$$I^g(\tau(T - l_0 - l_1 - L_0 - 4), \tau(T - l_0 - l_1 - L_0 - 3), x_1, u_1) \leq \tau\mu(f) + \delta_1/8 + \delta_3/8. \tag{2.246}$$

It follows from (2.240) and the choice of δ_2 [see (2.232)] that

$$I^f(\tau(T-l_0-l_1-L_0-4),\tau(T-l_0-l_1-L_0-3),x,u) \geq \tau\mu(f)-\delta_1/8. \qquad (2.247)$$

If

$$I^g(\tau(T-l_0-l_1-L_0-4),\tau(T-l_0-l_1-L_0-3),x,u) < \tau\mu(f)-\delta_1/2,$$

then by property (P15) and (2.237),

$$I^f(\tau(T-l_0-l_1-L_0-4),\tau(T-l_0-l_1-L_0-3),x,u)$$
$$< \tau\mu(f)-\delta_1/2+\delta_3/8 < \tau\mu(f)-3\delta_1/8$$

and this contradicts (2.247). Thus

$$I^g(\tau(T-l_0-l_1-L_0-4),\tau(T-l_0-l_1-L_0-3),x,u) \geq \tau\mu(f)-\delta_1/2. \qquad (2.248)$$

It follows from (2.246) and (2.248) that (2.245) holds. By (2.244) and (2.245),

$$I^g(\tau(T-l_0-l_1-L_0-3),\tau T,x_1,u_1) - I^g(\tau(T-l_0-l_1-L_0-3),\tau T,x,u)$$
$$\geq -\delta-\delta_1/8-\delta_3/8-\delta_1/2. \qquad (2.249)$$

Since $(\bar{x}_*,\bar{u}_*)$ is an $(\bar{f},-A,-B)$-overtaking optimal pair it follows from (2.27), (2.242), and Proposition 2.16 that

$$I^f(\tau(T-l_0-l_1-L_0-3),\tau T,x_1,u_1) = I^{\bar{f}}(0,\tau(l_0+l_1+L_0+3),\bar{x}_*,\bar{u}_*)$$
$$= \mu(f)\tau(l_0+l_1+L_0+3) + \pi^{\bar{f}}(\bar{x}_*(0))$$
$$- \pi^{\bar{f}}(\bar{x}_*(\tau(l_0+l_1+L_0+3))). \qquad (2.250)$$

By (2.234) and the choice of δ_2 (see (2.231)),

$$|\pi^{\bar{f}}(\bar{x}_*(\tau(l_0+l_1+L_0+3)))| \leq \delta_1/8.$$

Together with (2.250) this implies that

$$I^f(\tau(T-l_0-l_1-L_0-3),\tau T,x_1,u_1) \leq \pi^{\bar{f}}(\bar{x}_*(0))+\mu(f)\tau(l_0+l_1+L_0+3)+\delta_1/8. \qquad (2.251)$$

Property (P15), (2.237), and (2.251) imply that

$$I^g(\tau(T-l_0-l_1-L_0-3),\tau T,x_1,u_1)$$
$$\leq \pi^{\bar{f}}(\bar{x}_*(0)) + \mu(f)\tau(l_0+l_1+L_0+3)+\delta_1/8+\delta_3/8. \qquad (2.252)$$

In view of (2.249) and (2.252),

$$I^g(\tau(T - l_0 - l_1 - L_0 - 3), \tau T, x, u)$$
$$\leq \pi^{\bar{f}}(\bar{x}_*(0)) + \mu(f)\tau(l_0 + l_1 + L_0 + 3) + \delta + 3\delta_1/4 + \delta_3/4. \qquad (2.253)$$

Property (P15) and (2.253) imply that

$$I^f(\tau(T - l_0 - l_1 - L_0 - 3), \tau T, x, u)$$
$$\leq \pi^{\bar{f}}(\bar{x}_*(0)) + \mu(f)\tau(l_0 + l_1 + L_0 + 3) + \delta + 3\delta_1/4 + 3\delta_3/8. \qquad (2.254)$$

Set

$$\tilde{x}(t) = x(T\tau - t), \ \tilde{u}(t) = u(T\tau - t), \ t \in [0, \tau T]. \qquad (2.255)$$

Clearly, $(\tilde{x}, \tilde{u}) \in X(-A, -B, 0, T\tau)$ and in view of (2.27), (2.254), and (2.255),

$$I^{\bar{f}}(0, \tau(l_0 + l_1 + L_0 + 3), \tilde{x}, \tilde{u}) = I^f(\tau(T - l_0 - l_1 - L_0 - 3), \tau T, x, u)$$
$$\leq \pi^{\bar{f}}(\bar{x}_*(0)) + \mu(f)\tau(l_0 + l_1 + L_0 + 3) + \delta + 3\delta_1/4 + 3\delta_3/8. \qquad (2.256)$$

It follows from (2.240) and (2.255) that

$$|\tilde{x}(\tau(l_0 + l_1 + L_0 + 3)) - x_f(0)| \leq \delta_2. \qquad (2.257)$$

By (2.257) and the choice of δ_2 (see (2.231)),

$$|\pi^{\bar{f}}(\tilde{x}(\tau(l_0 + l_1 + L_0 + 3)))| \leq \delta_1/8. \qquad (2.258)$$

By (2.238), (2.256), and Proposition 2.15,

$$\pi^{\bar{f}}(\tilde{x}(0)) - \pi^{\bar{f}}(\bar{x}_*(0)) + I^{\bar{f}}(0, L_0\tau, \tilde{x}, \tilde{u}) - L_0\tau\mu(f) - \pi^{\bar{f}}(\tilde{x}(0)) + \pi^{\bar{f}}(\tilde{x}(L_0\tau))$$
$$\leq \pi^{\bar{f}}(\tilde{x}(0)) - \pi^{\bar{f}}(\bar{x}_*(0)) + I^{\bar{f}}(0, \tau(l_0 + l_1 + L_0 + 3), \tilde{x}, \tilde{u})$$
$$\quad - \mu(f)\tau(l_0 + l_1 + L_0 + 3) - \pi^{\bar{f}}(\tilde{x}(0)) + \pi^{\bar{f}}(\tilde{x}(\tau(l_0 + l_1 + L_0 + 3)))$$
$$\leq \pi^{\bar{f}}(\tilde{x}(0)) - \pi^{\bar{f}}(\bar{x}_*(0)) + \mu(f)\tau(l_0 + l_1 + L_0 + 3) + \delta + 3\delta_3/8 + 3\delta_1/4$$
$$\quad + \pi^{\bar{f}}(\bar{x}_*(0)) - \mu(f)\tau(l_0 + l_1 + L_0 + 3) - \pi^{\bar{f}}(\tilde{x}(0)) + \delta_1/8$$
$$\leq \delta + 3\delta_3/8 + 3\delta_1/4 + \delta_1/8 \leq \delta_1.$$

By the relation above, Proposition 2.15 and the relation $\pi^{\bar{f}}(\bar{x}_*(0)) = \inf(\pi^{\bar{f}})$,

$$\pi^{\bar{f}}(\tilde{x}(0)) \le \pi^{\bar{f}}(\bar{x}_*(0)) + \delta_1, \tag{2.259}$$

$$I^{\bar{f}}(0, L_0\tau, \tilde{x}, \tilde{u}) - L_0\tau\mu(f) + \pi^{\bar{f}}(\tilde{x}(0)) + \pi^{\bar{f}}(\tilde{x}(L_0\tau)) \le \delta_1. \tag{2.260}$$

It follows from (2.259), (2.260), and property (P13) that

$$|\tilde{x}(t) - \bar{x}_*(t)| \le \epsilon \text{ holds for all } t \in [0, L_0\tau].$$

Together with (2.245) this implies that

$$|x(T\tau - t) - \bar{x}_*(t)| \le \epsilon \text{ holds for all } t \in [0, L_0\tau].$$

Theorem 2.25 is proved.   □

2.15   Proof of Theorem 2.26

Theorems 2.13 and 2.25 imply the following result.

Theorem 2.47. *Let $L_0 > 0$ be an integer, $\epsilon > 0$. Then there exist $\delta > 0$, a neighborhood $\mathcal{U}$ of f in $\mathcal{M}$ and an integer $L_1 > L_0$ such that for each integer $T \ge L_1$, each $g \in \mathcal{U}$ and each $(x, u) \in X(A, B, 0, T\tau)$ which satisfies*

$$I^g(0, T\tau, x, u) \le \sigma(g, 0, T\tau) + \delta$$

the following inequality holds for all $t \in [0, L_0\tau]$:

$$|x(T\tau - t) - \bar{x}_*(t)| \le \epsilon.$$

Theorem 2.47 and Propositions 2.43 and 2.45 imply Theorem 2.26.

2.16   Proof of Theorem 2.27

Theorem 2.27 follows from Propositions 2.43 and 2.45 and the next result.

Theorem 2.48. *Let $L_0 > 0$ be an integer, $\epsilon > 0$, $M_0 > 0$. Then there exist $\delta > 0$, a neighborhood $\mathcal{U}$ of f in $\mathcal{M}$ and an integer $L_1 > L_0$ such that for each integer $T \ge L_1$, each $g \in \mathcal{U}$ and each $(x, u) \in X(A, B, 0, T\tau)$ which satisfies*

$$|x(0)|, \ |x(T\tau)| \le M_0, \ I^g(0, T\tau, x, u) \le \sigma(g, x(0), x(T\tau), 0, T\tau) + \delta$$

for all $t \in [0, L_0 \tau]$,

$$|x(t) - \xi(t)| \leq \epsilon,$$

where $(\xi, \eta) \in X(A, B, 0, \infty)$ is the unique (f, A, B)-overtaking optimal pair such that $\xi(0) = x(0)$.

Proof. Denote by d the metric of the space $\mathcal{M}$. Assume that Theorem 2.48 does not hold. Then there exist a sequence $\{\delta_k\}_{k=1}^{\infty} \subset (0, 1)$ such that

$$\delta_k < 4^{-k}, \; k = 1, 2, \ldots, \tag{2.261}$$

a sequence of integers

$$T_k \geq L_0 + 2k, \; k = 1, 2, \ldots \tag{2.262}$$

a sequence $\{g_k\}_{k=1}^{\infty} \subset \mathcal{M}$ such that

$$d(g_k, f) \leq k^{-1}, \; k = 1, 2, \ldots \tag{2.263}$$

and a sequence $(x_k, u_k) \in X(A, B, 0, T_k\tau)$, $k = 1, 2, \ldots$ such that for each natural number k,

$$|x_k(0)|, \; |x_k(T_k\tau)| \leq M_0, \tag{2.264}$$

$$I^{g_k}(0, T_k\tau, x_k, u_k) \leq \sigma(g_k, x_k(0), x_k(T_k\tau), 0, T_k\tau) + \delta_k, \tag{2.265}$$

$$\max\{|x_k(t) - \xi_k(t)| : t \in [0, L_0\tau]\} > \epsilon, \tag{2.266}$$

where the pair $(\xi_k, \eta_k) \in X(A, B, 0, \infty)$ is (f, A, B)-overtaking optimal and

$$x_k(0) = \xi_k(0). \tag{2.267}$$

In view of (2.264), (2.267) and Theorem 2.4, the following property holds:

(P16) for each $\gamma > 0$ there exists a natural number $m(\gamma)$ such that for each integer $k \geq 1$ and each integer $p \geq m(\gamma)$,

$$|\xi_k(p\tau) - x_f(0)| \leq \gamma.$$

Proposition 2.16 implies that for each pair of natural number p, k,

$$I^f(0, p\tau, \xi_k, \eta_k) - p\tau\mu(f) + \pi^f(\xi_k(0)) + \pi^f(\xi_k(p\tau)) = 0. \tag{2.268}$$

It follows from (2.264), (2.267), (2.268), and the continuity and boundedness from below of the function π^f that for each integer $p \geq 1$ the sequence $\{I^f(0, p\tau, \xi_k, \eta_k)\}_{k=1}^{\infty}$ is bounded. By Proposition 2.30, extracting subsequences and re-indexing we may assume without loss of generality that there exists $(\xi, \eta) \in X(A, B, 0, \infty)$ such that for each integer $p \geq 1$,

$$\xi_k(t) \to \xi(t) \text{ as } k \to \infty \text{ uniformly on } [0, p\tau], \tag{2.269}$$

$$I^f(0, p\tau, \xi, \eta) \leq \liminf_{k \to \infty} I^f(0, p\tau, \xi_k, \eta_k). \tag{2.270}$$

In view of (2.268)–(2.270) and the continuity of π^f, for each integer $p \geq 1$,

$$I^f(0, p\tau, \xi, \eta) - p\tau\mu(f) + \pi^f(\xi(0)) + \pi^f(\xi(p\tau)) \leq 0.$$

Together with Proposition 2.15 this implies that for each integer $p \geq 1$,

$$I^f(0, p\tau, \xi, \eta) - p\tau\mu(f) - \pi^f(\xi(0)) + \pi^f(\xi(p\tau)) = 0. \tag{2.271}$$

By (2.271), the boundedness from below of π^f and Theorem 2.2, the pair $(\xi, \eta) \in X(A, B, 0, \infty)$ is (f, A, B)-good. Together with (2.271) and Proposition 2.22 this implies that the pair (ξ, η) is (f, A, B)-overtaking optimal. In view of (2.267) and (2.269),

$$\xi(0) = \lim_{k \to \infty} \xi_k(0) = \lim_{k \to \infty} x_k(0). \tag{2.272}$$

Let $\Delta > 0$. Proposition 2.17, (2.34), and the continuity of π^f imply that there exists $\delta > 0$ such that for each $y, z \in R^n$ satisfying $|y - x_f(0)|, |z - x_f(0)| \leq \delta$,

$$|\pi^f(y)| \leq \Delta/8, \ |v(y, z) - \tau\mu(f)| \leq \Delta/8. \tag{2.273}$$

By (2.262)–(2.265), (2.267), and Theorems 2.13 and 2.4, there exists an integer $k(\Delta) \geq 1$ such that for each integer $k \geq 4k(\Delta) + 8$,

$$|x_k(i\tau) - x_f(0)| \leq \delta, \ i = k(\Delta) + 1, \ldots, T_k - k(\Delta), \tag{2.274}$$

$$|\xi_k(i\tau) - x_f(0)| \leq \delta \text{ for all integers } i \geq k(\Delta) + 1. \tag{2.275}$$

Let $S \geq 1$ and $k \geq 4k(\Delta) + S + 8$ be integers. By Proposition 2.32 and (2.262), there exists $(\tilde{x}_k, \tilde{u}_k) \in X(A, B, 0, (S + k(\Delta) + 2)\tau)$ such that

$$\tilde{x}_k(t) = \xi_k(t), \ \tilde{u}_k(t) = \eta_k(t), \ t \in [0, (S + k(\Delta) + 1)\tau], \tag{2.276}$$

$$\tilde{x}_k((S + k(\Delta) + 2)\tau) = x_k((S + k(\Delta) + 2)\tau), \tag{2.277}$$

$$I^f((S + k(\Delta) + 1)\tau, (S + k(\Delta) + 2)\tau, \tilde{x}_k, \tilde{u}_k)$$
$$= v(\xi_k((S + k(\Delta) + 1)\tau), x_k((S + k(\Delta) + 2)\tau)). \tag{2.278}$$

In view of (2.267), (2.268), and (2.273)–(2.278), for each integer $k \geq 4k(\Delta) + S + 8$,

$$
\begin{aligned}
I^f(0, (S + k(\Delta) + 2)\tau, \tilde{x}_k, \tilde{u}_k) &= I^f(0, (S + k(\Delta) + 1)\tau, \xi_k, \eta_k) \\
&\quad + v(\xi_k((S + k(\Delta) + 1)\tau), x_k((S + k(\Delta) + 2)\tau)) \\
&\leq \mu(f)\tau(S + k(\Delta) + 1) + \pi^f(x_k(0)) \\
&\quad - \pi^f(\xi_k((S + k(\Delta) + 1)\tau)) + \tau\mu(f) + \Delta/8 \\
&\leq \mu(f)\tau(S + k(\Delta) + 2) + \pi^f(x_k(0)) + \Delta/4.
\end{aligned}
$$
(2.279)

Proposition 2.41 and (2.263) imply that there exists an integer $k_1 \geq 4k(\Delta) + S + 8$ such that for all integers $k \geq k_1$,

$$
I^{g_k}(0, (S + k(\Delta) + 2)\tau, \tilde{x}_k, \tilde{u}_k) \leq \mu(f)\tau(S + k(\Delta) + 2) + \pi^f(x_k(0)) + 3\Delta/8.
$$
(2.280)

By (2.261), (2.276), and (2.267), there exists an integer $k_2 \geq k_1$ such that for all integers $k \geq k_2$,

$$
I^{g_k}(0, (S+k(\Delta)+2)\tau, x_k, u_k) \leq \mu(f)\tau(S+k(\Delta)+2) + \pi^f(x_k(0)) + \Delta/2.
$$
(2.281)

Proposition 2.41, (2.263), and (2.281) imply that there exists an integer $k_3 \geq k_2$ such that for all integers $k \geq k_3$,

$$
I^f(0, (S+k(\Delta)+2)\tau, x_k, u_k) \leq \mu(f)\tau(S+k(\Delta)+2) + \pi^f(x_k(0)) + 5\Delta/8.
$$
(2.282)

Since S is any natural number we conclude that for any integer $p \geq 1$, the sequence $\{I^f(0, p\tau, x_k, u_k)\}_{k=p}^{\infty}$ is bounded. By Proposition 2.30, extracting subsequences and re-indexing we may assume without loss of generality that there exists $(x, u) \in X(A, B, 0, \infty)$ such that for each integer $p \geq 1$,

$$
x_k(t) \to x(t) \text{ as } k \to \infty \text{ uniformly on } [0, p\tau],
$$
(2.283)

$$
I^f(0, p\tau, x, u) \leq \liminf_{k \to \infty} I^f(0, p\tau, x_k, u_k).
$$
(2.284)

In view of (2.274), (2.282)–(2.284), and the continuity of π^f,

$$
I^f(0, (S + k(\Delta) + 2)\tau, x, u) \leq \mu(f)(S + k(\Delta) + 2)\tau + \pi^f(x(0)) + 5\Delta/8,
$$
(2.285)

$$
|x((S + k(\Delta) + 2)\tau) - x_f(0)| \leq \delta.
$$
(2.286)

By (2.285), (2.286), and the choice of δ [see (2.273)],

$$I^f(0, (S + k(\Delta) + 2)\tau, x, u) - \mu(f)(S + k(\Delta) + 2)\tau$$
$$- \pi^f(x(0)) + \pi^f(x((S + k(\Delta) + 2)\tau)) \leq \Delta. \tag{2.287}$$

Since S is any natural number it follows from (2.287) and Proposition 2.15 that for any integer $p \geq 1$,

$$I^f(0, p\tau, x, u) - \mu(f)\tau p - \pi^f(x(0)) + \pi^f(x(p\tau)) \leq \Delta. \tag{2.288}$$

In view of (2.288), the boundedness from below of the function π^f and Theorem 2.2, $(x, u) \in X(A, B, 0, \infty)$ is an (f, A, B)-good pair. Since Δ is any positive number Proposition 2.15 and (2.288) imply that for any integer $p \geq 1$,

$$I^f(0, p\tau, x, u) - \mu(f)\tau p - \pi^f(x(0)) + \pi^f(x(p\tau)) = 0. \tag{2.289}$$

By Proposition 2.22 and (2.289), (x, u) is an (f, A, B)-overtaking optimal pair. Since (x, u) and (ξ, η) are (f, A, B)-overtaking optimal pairs it follows from (2.272), (2.283), and Theorem 2.3 that

$$x(t) = \xi(t), \ u(t) = \eta(t), \ t \in [0, \infty). \tag{2.290}$$

By (2.269), (2.283), and (2.290), for all sufficiently large natural numbers k,

$$|x_k(t) - \xi_k(t)| \leq \epsilon/4, \ t \in [0, L_0\tau].$$

This contradicts (2.266). The contradiction we have reached proves Theorem 2.48.

Chapter 3
Linear Control Systems with Nonconvex Integrands

In this chapter we study the existence and structure of optimal trajectories of linear control systems with autonomous nonconvex integrands. For these control systems we establish the existence of optimal trajectories over an infinite horizon and show that the turnpike phenomenon holds. We also study the structure of approximate optimal trajectories in regions close to the endpoints of the time intervals. It is shown that in these regions optimal trajectories converge to solutions of the corresponding infinite horizon optimal control problem which depend only on the integrand.

3.1 Preliminaries

In this chapter we study the structure of approximate optimal trajectories of linear control systems described by a differential equation

$$x'(t) = Ax(t) + Bu(t) \text{ for almost every (a. e.) } t \in \mathcal{I}, \tag{3.1}$$

where $\mathcal{I}$ is either R^1 or $[T_1, \infty)$ or $[T_1, T_2]$ (here $-\infty < T_1 < T_2 < \infty$), n, m are natural numbers, $x : \mathcal{I} \to R^n$ is an absolutely continuous (a. c.) function and the control function $u : \mathcal{I} \to R^m$ is Lebesgue measurable, and A and B are given matrices of dimensions $n \times n$ and $n \times m$ with integrands $f : R^n \times R^m \to R^1$.

Note that if $\mathcal{I}$ is an unbounded interval, then $x : \mathcal{I} \to R^n$ is an absolutely continuous function if and only if it is an absolutely continuous function on any bounded subinterval of $\mathcal{I}$.

We assume that the linear system (3.1) is controllable and that the integrand f is a continuous function.

We denote by $|\cdot|$ the Euclidean norm and by $\langle \cdot, \cdot \rangle$ the inner product in the k-dimensional Euclidean space R^k. For every $z \in R^1$ denote by $\lfloor z \rfloor$ the largest

© Springer International Publishing Switzerland 2015
A.J. Zaslavski, *Turnpike Theory of Continuous-Time Linear Optimal Control Problems*, Springer Optimization and Its Applications 104,
DOI 10.1007/978-3-319-19141-6_3

integer which does not exceed z: $\lfloor z \rfloor = \max\{i \in \mathbf{Z} : i \leq z\}$. For every $s \in R^1$ set $s_+ = \max\{s, 0\}$. For every nonempty set X and every function $h : X \to R^1 \cup \{\infty\}$ set

$$\inf(h) = \inf\{h(x) : x \in X\}.$$

Let $a_0 > 0$ and $\psi : [0, \infty) \to [0, \infty)$ be an increasing function such that

$$\lim_{t \to \infty} \psi(t) = \infty. \tag{3.2}$$

Suppose that $f : R^n \times R^m \to R^1$ is a continuous function such that the following assumption holds:

(A1)

(i) for each $(x, u) \in R^n \times R^m$,

$$f(x, u) \geq \max\{\psi(|x|), \ \psi(|u|),$$

$$\psi([|Ax + Bu| - a_0|x|]_+)[|Ax + Bu| - a_0|x|]_+\} - a_0; \tag{3.3}$$

(ii) for each $x \in R^n$ the function $f(x, \cdot) : R^m \to R^1$ is convex;
(iii) for each $M, \epsilon > 0$ there exist $\Gamma, \delta > 0$ such that

$$|f(x_1, u_1) - f(x_2, u_2)| \leq \epsilon \max\{f(x_1, u_1), f(x_2, u_2)\}$$

for each $u_1, u_2 \in R^m$ and each $x_1, x_2 \in R^n$ which satisfy

$$|x_i| \leq M, \ |u_i| \geq \Gamma, \ i = 1, 2, \quad \max\{|x_1 - x_2|, |u_1 - u_2|\} \leq \delta;$$

(iv) for each $K > 0$ there exists a constant $a_K > 0$ and an increasing function

$$\psi_K : [0, \infty) \to [0, \infty)$$

such that

$$\psi_K(t) \to \infty \text{ as } t \to \infty$$

and

$$f(x, u) \geq \psi_K(|u|)|u| - a_K$$

for each $u \in R^m$ and each $x \in R^n$ satisfying $|x| \leq K$.

Let $T_1 \in R^1$ and $T_2 > T_1$. A pair of an absolutely continuous function $x : [T_1, T_2] \to R^n$ and a Lebesgue measurable function $u : [T_1, T_2] \to R^m$ is called an (A, B)-trajectory-control pair if (3.1) holds with $\mathcal{I} = [T_1, T_2]$. Denote by $X(A, B, T_1, T_2)$ the set of all (A, B)-trajectory-control pairs $x : [T_1, T_2] \to R^n$, $u : [T_1, T_2] \to R^m$.

Let $T \in R^1$ and $\mathcal{I} = [T, \infty)$ be an infinite closed subinterval of R^1. Denote by $X(A, B, T, \infty)$ the set of all pairs of a.c. functions $x : [T, \infty) \to R^n$ and Lebesgue measurable functions $u : [T, \infty) \to R^m$ satisfying (3.1).

Note that a function h satisfies (A1) if $h \in C^1(R^n \times R^m)$, (A1)(i), (A1)(ii), (A1)(iv) hold, and for each $K > 0$ there exists an increasing function $\tilde{\psi} : [0, \infty) \to [0, \infty)$ such that for each $x \in R^n$ satisfying $|x| \leq K$ and each $u \in R^m$,

$$\max\{|\partial h/\partial x(x, u)|, \; |\partial h/\partial u(x, u)|\} \leq \tilde{\psi}(|x|)(1 + \psi_K(|u|)|u|).$$

The performance of the above control system is measured on any finite interval $[T_1, T_2] \subset [0, \infty)$ and for any $(x, u) \in X(A, B, T_1, T_2)$ by the integral functional

$$I^f(T_1, T_2, x, u) = \int_{T_1}^{T_2} f(x(t), u(t))dt. \tag{3.4}$$

In this chapter we study the existence and structure of optimal trajectories of linear control system (3.1) with the integrand f. For these control systems we establish the existence of optimal trajectories over an infinite horizon and show that the turnpike phenomenon holds. We also study the structure of approximate optimal trajectories in regions close to the endpoints of the time intervals. It is shown that in these regions optimal trajectories converge to solutions of the corresponding infinite horizon optimal control problem which depend only on the integrand.

More precisely, we consider the following optimal control problems

$$I^f(0, T, x, u) \to \min, \tag{P_1}$$

$$(x, u) \in X(A, B, 0, T) \text{ such that } x(0) = y, \; x(T) = z,$$

$$I^f(0, T, x, u) \to \min, \tag{P_2}$$

$$(x, u) \in X(A, B, 0, T) \text{ such that } x(0) = y,$$

$$I^f(0, T, x, u) \to \min, \tag{P_3}$$

$$(x, u) \in X(A, B, 0, T),$$

where $y, z \in R^n$ and $T > 0$. The study of these problems is based on the properties of solutions of the corresponding infinite horizon optimal control problem associated with the control system (3.1) and the integrand f.

In this chapter we establish the turnpike property for the approximate solutions of problems (P_1), (P_2) and (P_3). For problems (P_2) and (P_3) we show that in regions close to the right endpoint T of the time interval their approximate solutions are determined only by the integrand, and are essentially independent of the choice of interval and the endpoint value y. For problems (P_3), approximate solutions are determined only by the integrand also in regions close to the left endpoint 0 of the time interval.

A number

$$\mu(f) := \inf\{\liminf_{T\to\infty} T^{-1} I^f(0, T, x, u) \ : \ (x, u) \in X(A, B, 0, \infty)\} \tag{3.5}$$

is called the minimal long-run average cost growth rate of f. By (A1)(i), $-\infty < \mu(f)$.

Let $T > 0$ and $y, z \in R^n$. Set

$$\sigma(f, y, z, T) = \inf\{I^f(0, T, x, u) \ :$$

$$(x, u) \in X(A, B, 0, T) \text{ and } x(0) = y, \ x(T) = z\}, \tag{3.6}$$

$$\sigma(f, y, T) = \inf\{I^f(0, T, x, u) \ : \ (x, u) \in X(A, B, 0, T) \text{ and } x(0) = y\}, \tag{3.7}$$

$$\hat{\sigma}(f, z, T) = \inf\{I^f(0, T, x, u) \ : \ (x, u) \in X(A, B, 0, T) \text{ and } x(T) = z\}, \tag{3.8}$$

$$\sigma(f, T) = \inf\{I^f(0, T, x, u) \ : \ (x, u) \in X(A, B, 0, T)\}. \tag{3.9}$$

We say that $(\tilde{x}, \tilde{u}) \in X(A, B, 0, \infty)$ is (f, A, B)-overtaking optimal [44, 53] if for each $(x, u) \in X(A, B, 0, \infty)$ satisfying $x(0) = \tilde{x}(0)$,

$$\limsup_{T\to\infty}[I^f(0, T, \tilde{x}, \tilde{u}) - I^f(0, T, x, u)] \leq 0.$$

We say that $(x, u) \in X(A, B, 0, \infty)$ is (f, A, B)-minimal [44, 53] if for each $T > 0$,

$$I^f(0, T, x, u) = \sigma(f, x(0), x(T), T).$$

Let $(x_f, u_f) \in R^n \times R^m$ satisfy

$$Ax_f + Bu_f = 0. \tag{3.10}$$

Clearly, $\mu(f) \leq f(x_f, u_f)$. It is easy to see that the following result holds.

Proposition 3.1. *Assume that $\mu(f) = f(x_f, u_f)$ and let $x(t) = x_f$, $u(t) = u_f$ for all $t \in [0, \infty)$. Then $(x, u) \in X(A, B, 0, \infty)$ is (f, A, B)-minimal.*

The next proposition is proved in Sect. 3.4.

Proposition 3.2. *Assume that $\mu(f) = f(x_f, u_f)$. Then there exists $\Delta_* > 0$ such that for each $T > 0$ and each $(x, u) \in X(A, B, 0, T)$,*

$$I^f(0, T, x, u) \geq Tf(x_f, u_f) - \Delta_*. \tag{3.11}$$

It is easy to see that the following proposition holds.

Proposition 3.3. *Let $\Delta_* > 0$ and let for each $T > 0$ and each $(x,u) \in X(A,B,0,T)$, relation (3.11) holds. Then $\mu(f) = f(x_f, u_f)$.*

We suppose that the following assumption holds.

(A2) $\mu(f) = f(x_f, u_f)$ and if $(x,u) \in R^n \times R^m$ satisfies

$$Ax + Bu = 0, \ \mu(f) = f(x,u),$$

then $x = x_f$.

Proposition 3.2 imply the following result.

Proposition 3.4. *For each $(x,u) \in X(A,B,0,\infty)$ either*

$$I^f(0,T,x,u) - T\mu(f) \to \infty \ as \ T \to \infty$$

or $\sup\{|I^f(0,T,x,u) - T\mu(f)| : T > 0\} < \infty.$

A trajectory-control pair $(x,u) \in X(A,B,0,\infty)$ is called (f,A,B)-good [44, 53] if

$$\sup\{|I^f(0,T,x,u) - T\mu(f)| : T > 0\} < \infty.$$

Proposition 3.4 and Theorem 2.1 of [52] imply the following result.

Proposition 3.5. *For every (f,A,B)-good trajectory-control pair $(x,u) \in X(A,B,0,\infty)$,*

$$\sup\{|x(t)| : t \in [0,\infty)\} < \infty.$$

We suppose that the following assumption holds.

(A3)   For each (f,A,B)-good trajectory-control pair

$$(x,u) \in X(A,B,0,\infty)$$

the equality $\lim_{t\to\infty} x(t) = x_f$ is true.

Let us consider examples of integrands satisfying assumptions (A1)–(A3). First note that if a continuous strictly convex function $h : R^n \times R^m \to R^1$ satisfies assumption (A1) (with $f = h$) and

$$h(x,u)/|u| \to \infty \ as \ |u| \to \infty \ uniformly \ in \ x \in R^n,$$

then in view of Corollary 2.11 of Chap. 2, the function h satisfies assumptions (A2) and (A3) (with $f = h$). Let us consider another example of an integrand which satisfies (A1)–(A3).

Let $c \in R^1$, $a_1 > 0$, $l \in R^n$, $(x_*, u_*) \in R^n \times R^m$ satisfy $Ax_* + Bu_* = 0$ and let $\psi_0 : [0, \infty) \to [0, \infty)$ be an increasing function such that $\lim_{t \to \infty} \psi_0(t) = \infty$. Assume that a continuous function $L : R^n \times R^m \to [0, \infty)$ satisfies for each $(x, u) \in R^n \times R^m$,

$$L(x, u) \geq \max\{\psi_0(|x|), \ \psi_0(|u|)|u|\} - a_1 + |l||Ax + Bu|,$$

$$L(x, u) = 0 \text{ if and only if } x = x_*, \ u = u_*,$$

for each $x \in R^n$, the function $L(x, \cdot) : R^m \to R^1$ is convex and for each $M, \epsilon > 0$ there exist $\Gamma, \delta > 0$ such that

$$|L(x_1, u_1) - L(x_2, u_2)| \leq \epsilon \max\{L(x_1, u_1), \ L(x_2, u_2)\}$$

for each $x_1, x_2 \in R^n$ and each $u_1, u_2 \in R^m$ which satisfy

$$|x_i| \leq M, \ |u_i| \geq \Gamma, \ i = 1, 2, \ |x_1 - x_2|, \ |u_1 - u_2| \leq \delta.$$

For every $(x, u) \in R^n \times R^m$ set

$$h(x, u) = L(x, u) + c + \langle l, Ax + Bu \rangle.$$

It is not difficult to see that for each $(x, u) \in R^n \times R^m$,

$$h(x, u) \geq \max\{\psi_0(|x|), \ \psi_0(|u|)|u|\} - a_1 - |c|$$

and that h satisfies (A1) under the appropriate choice of $a_0 > 0$, ψ. In Sect. 3.12 we prove the following result.

Proposition 3.6. $\mu(h) = h(x_*, u_*) = c$, (A2) holds for $f = h$ and for any (h, A, B)-good trajectory-control pair $(x, u) \in X(A, B, 0, \infty)$,

$$\lim_{t \to \infty} x(t) = x_*.$$

3.2   Turnpike Results

We use the notation, definitions, and assumptions introduced in Sect. 3.1. The following turnpike result is proved in Sect. 3.6.

Theorem 3.7. *Let* ϵ, $M_0, M_1 > 0$. *Then there exist* $L > 0$, $\delta \in (0, \epsilon)$ *such that for each* $T > 2L$ *and each* $(x, u) \in X(A, B, 0, T)$ *which satisfies for each* $S \in [0, T - L]$,

$$I^f(S, S + L, x, u) \leq \sigma(f, x(S), x(S + L), L) + \delta$$

and satisfies at least one of the following conditions:

(a) $|x(0)|$, $|x(T)| \leq M_0$, $I^f(0, T, x, u) \leq \sigma(f, x(0), x(T), T) + M_1$;

(b) $|x(0)| \leq M_0$, $I^f(0, T, x, u) \leq \sigma(f, x(0), T) + M_1$;

(c) $I^f(0, T, x, u) \leq \sigma(f, T) + M_1$

there exist $p_1 \in [0, L]$, $p_2 \in [T - L, T]$ *such that*

$$|x(t) - x_f| \leq \epsilon \text{ for all } t \in [p_1, p_2].$$

Moreover if $|x(0) - x_f| \leq \delta$, *then* $p_1 = 0$ *and if* $|x(T) - x_f| \leq \delta$, *then* $p_2 = T$.

Theorem 3.7 and Theorem 4.1.1 of [52] imply the following result.

Theorem 3.8. *Let* $x_0 \in R^n$. *Then there exists an* (f, A, B)-*overtaking optimal trajectory-control pair* $(x, u) \in X(A, B, 0, \infty)$ *satisfying* $x(0) = x_0$.

In the sequel we show (see Proposition 3.36) that any (f, A, B)-overtaking optimal trajectory-control pair is (f, A, B)-good.

The next result which describes the limit behavior of overtaking optimal trajectories is proved in Sect. 3.7.

Theorem 3.9. *Let* $M, \epsilon > 0$. *Then there exists* $L > 0$ *such that for any* (f, A, B)-*overtaking optimal trajectory-control pair* $(x, u) \in X(A, B, 0, \infty)$ *which satisfies* $|x(0)| \leq M$ *the inequality*

$$|x(t) - x_f| \leq \epsilon$$

holds for all numbers $t \geq L$. *Moreover, there exists* $\delta > 0$ *such that for any* (f, A, B)-*overtaking optimal trajectory-control pair* $(x, u) \in X(A, B, 0, \infty)$ *satisfying* $|x(0) - x_f| \leq \delta$, *the inequality*

$$|x(t) - x_f| \leq \epsilon$$

holds for all numbers $t \geq 0$.

We say that $(x, u) \in X(A, B, 0, \infty)$ is (f, A, B)-minimal [5, 53] if for each $T > 0$,

$$I^f(0, T, x, u) = \sigma(f, x(0), x(T), T). \tag{3.11}$$

The next result which is proved in Sect. 3.8 shows the equivalence of the optimality criterions introduced above.

Theorem 3.10. *Assume that* $(x, u) \in X(A, B, 0, \infty)$. *Then the following conditions are equivalent:*

(i) (x, u) *is* (f, A, B)*-overtaking optimal; (ii)* (x, u) *is* (f, A, B)*-minimal and* (f, A, B)*-good; (iii)* (x, u) *is* (f, A, B)*-minimal and*

$$\lim_{t \to \infty} x(t) = x_f;$$

(iv) (x, u) *is* (f, A, B)*-minimal and* $\liminf_{t \to \infty} |x(t)| < \infty$.

3.3   Structure of Solutions in the Regions Close to the End Points

In this section we state results which describe the structure of solutions of problems (P_2) and (P_3) in the regions close to the end points. Combined with the turnpike results of Sect. 3.2 they provide the full description of the structure of their solutions. We use the notation, definitions, and assumptions introduced in Sects. 3.1 and 3.2.

For each $z \in R^n$ denote by $\Lambda(z)$ the set of all (f, A, B)-overtaking optimal pairs $(x, u) \in X(A, B, 0, \infty)$ such that $x(0) = z$ which is nonempty in view of Theorem 3.8.

Let $z \in R^n$. Set

$$\pi^f(z) = \lim_{T \to \infty} \inf[I^f(0, T, x, u) - T\mu(f)], \tag{3.12}$$

where $(x, u) \in \Lambda(z)$. In view of Proposition 3.4, $\pi^f(z)$ is finite, well defined, and does not depend on the choice of $(x, u) \in \Lambda(z)$. Definition (3.12) and the definition of (f, A, B)-overtaking optimal pairs imply the following result.

Proposition 3.11.

1. *Let* $(x, u) \in X(A, B, 0, \infty)$ *be* (f, A, B)*-good. Then*

$$\pi^f(x(0)) \le \lim_{T \to \infty} \inf[I^f(0, T, x, u) - T\mu(f)]$$

and for each pair of numbers $S > T \ge 0$,

$$\pi^f(x(T)) \le I^f(T, S, x, u) - (S - T)\mu(f) + \pi^f(x(S)). \tag{3.13}$$

2. *Let* $S > T \ge 0$ *and* $(x, u) \in X(A, B, T, S)$. *Then* (3.13) *holds.*

The next result follows from definition (3.12).

Proposition 3.12. *Let* $(x, u) \in X(A, B, 0, \infty)$ *be an* (f, A, B)-*overtaking optimal pair. Then for each pair of numbers* $S > T \geq 0$,

$$\pi^f(x(T)) = I^f(T, S, x, u) - (S - T)\mu(f) + \pi^f(x(S)).$$

Theorems 3.8 and 3.9 and (A2) imply the following result.

Proposition 3.13. $\pi^f(x_f) = 0$.

The following result is proved in Sect. 3.4.

Proposition 3.14. *The function* π^f *is continuous at* x_f.

Proposition 3.15. *Let* $(x, u) \in X(A, B, 0, \infty)$ *be* (f, A, B)-*overtaking optimal. Then*

$$\pi^f(x(0)) = \lim_{T \to \infty} [I^f(0, T, x, u) - T\mu(f)].$$

Proof. It follows from Propositions 3.12, 3.14, and (A3) that

$$\pi^f(x(0)) = \lim_{T \to \infty} (\pi^f(x(0)) - \pi^f(x(T)))$$
$$= \lim_{T \to \infty} [I^f(0, T, x, u) - T\mu(f)].$$

Proposition 3.15 is proved.

The next result is proved in Sect. 3.9.

Proposition 3.16. *For each* $M > 0$ *the set* $\{x \in R^n : \pi^f(x) \leq M\}$ *is bounded.*

In Sect. 3.9 we prove the following proposition.

Proposition 3.17. *The function* π^f *is lower semicontinuous.*

By Propositions 3.13, 3.16, and 3.17, $\inf(\pi^f)$ is finite and there exists $\theta \in R^n$ such that $\pi^f(\theta) = \inf(\pi^f)$.

Proposition 3.18. *Let* $(x, u) \in X(A, B, 0, \infty)$ *be* (f, A, B)-*good pair such that for all* $T > 0$,

$$I^f(0, T, x, u) - T\mu(f) = \pi^f(x(0)) - \pi^f(x(T)). \tag{3.14}$$

Then $(x, u) \in X(A, B, 0, \infty)$ *is* (f, A, B)-*overtaking optimal.*

Proof. Theorem 3.8 implies that there exists an (f, A, B)-overtaking optimal pair $(x_1, u_1) \in X(A, B, 0, \infty)$ such that $x_1(0) = x(0)$. By Proposition 3.12, for each integer $T \geq 1$,

$$I^f(0, T, x_1, u_1) - T\mu(f) = \pi^f(x_1(0)) - \pi^f(x_1(T)).$$

It follows from the equality above, (3.14), (A3) and Propositions 3.13 and 3.14 that for all $T > 0$,

$$I^f(0, T, x, u) - I^f(0, T, x_1, u_1) = \pi^f(x_1(T)) - \pi^f(x(T)) \to 0 \text{ as } T \to \infty.$$

Thus

$$\lim_{T \to \infty} [I^f(0, T, x, u) - I^f(0, T, x_1, u_1)] = 0.$$

Since (x_1, u_1) is an (f, A, B)-overtaking optimal pair the equality above implies that (x, u) is an (f, A, B)-overtaking optimal pair too. Proposition 3.18 is proved.

Consider a linear control system

$$x'(t) = -Ax(t) - Bu(t), \tag{3.15}$$

$$x(0) = x_0$$

which is also controllable. For the triplet $(f, -A, -B)$ we use all the notation and definitions introduced for the triplet (f, A, B). It is clear that assumption (A1) holds for the triplet $(f, -A, -B)$.

Let $T_1 \in R^1$, $T_2 > T_1$. A pair of an absolutely continuous function $x : [T_1, T_2] \to R^n$ and a Lebesgue measurable function $u : [T_1, T_2] \to R^m$ is called an $(-A, -B)$-trajectory-control pair if (3.15) holds for a. e. $t \in [T_1, T_2]$. Denote by $X(-A, -B, T_1, T_2)$ the set of all $(-A-, B)$-trajectory-control pairs $x : [T_1, T_2] \to R^n$, $u : [T_1, T_2] \to R^m$.

Let $T \in R^1$. Denote by $X(-A, -B, T, \infty)$ the set of all pairs of a. c. functions $x : [T, \infty) \to R^n$ and Lebesgue measurable functions $u : [T, \infty) \to R^m$ satisfying (3.15) for a. e. $t \geq T$, which are called $(-A, -B)$-trajectory-control pairs.

Assume that $S_1 \in R^1$, $S_2 > S_1$ and that $(x, u) \in X(A, B, S_1, S_2)$. For all $t \in [S_1, S_2]$ set

$$\bar{x}(t) = x(S_2 - t + S_1), \quad \bar{u}(t) = u(S_2 - t + S_1). \tag{3.16}$$

In view of (3.1) and (3.16) for a. e. $t \in [S_1, S_2]$,

$$\bar{x}'(t) = -x'(S_2 - t + S_1) = -Ax(S_2 - t + S_1) - Bu(S_2 - t + S_1)$$
$$= -A\bar{x}(t) - B\bar{u}(t),$$

$$(\bar{x}, \bar{u}) \in X(-A, -B, S_1, S_2). \tag{3.17}$$

By (3.16),

$$\int_{S_1}^{S_2} f(\bar{x}(t), \bar{u}(t)) dt = \int_{S_1}^{S_2} f(x(S_2 - t + S_1), u(S_2 - t + S_1)) dt$$

$$= \int_{S_1}^{S_2} f(x(t), u(t)) dt. \tag{3.18}$$

For each pair of numbers $T_2 > T_1$ and each $(x, u) \in X(-A, -B, T_1, T_2)$ set

$$I^f(T_1, T_2, x, u) = \int_{T_1}^{T_2} f(x(t), u(t))dt. \tag{3.19}$$

For each $y, z \in R^n$ and each $T > 0$ set

$$\sigma_-(f, y, z, T) = \inf\{I^f(0, T, x, u) :$$
$$(x, u) \in X(-A, -B, 0, T) \text{ and } x(0) = y, \ x(T) = z\}, \tag{3.20}$$

$$\sigma_-(f, y, T) = \inf\{I^f(0, T, x, u) : \ (x, u) \in X(-A, -B, 0, T) \text{ and } x(0) = y\}, \tag{3.21}$$

$$\hat{\sigma}_-(f, z, T) = \inf\{I^f(0, T, x, u) : \ (x, u) \in X(-A, -B, 0, T) \text{ and } x(T) = z\}, \tag{3.22}$$

$$\sigma_-(f, T) = \inf\{I^f(0, T, x, u) : \ (x, u) \in X(-A, -B, 0, T)\}. \tag{3.23}$$

Relations (3.17) and (3.18) imply the following result.

Proposition 3.19. *Let $S_2 > S_1$ be real numbers, $M \geq 0$ and let $(x_i, u_i) \in X(A, B, S_1, S_2)$, $i = 1, 2$. Then*

$$I^f(S_1, S_2, x_1, u_1) \geq I^f(S_1, S_2, x_2, u_2) - M$$

if and only if $I^f(S_1, S_2, \bar{x}_1, \bar{u}_1) \geq I^f(S_1, S_2, \bar{x}_2, \bar{u}_2) - M$.

Proposition 3.19 implies the following result.

Proposition 3.20. *Let $S_2 > S_1$ be real numbers and*

$$(x, u) \in X(A, B, S_1, S_2).$$

Then the following assertions hold:

$$I^f(S_1, S_2, x, u) \leq \sigma(f, S_2 - S_1) + M$$
if and only if $I^f(S_1, S_2, \bar{x}, \bar{u}) \leq \sigma_-(f, S_2 - S_1) + M$;
$$I^f(S_1, S_2, x, u) \leq \sigma(f, x(S_1), x(S_2), S_2 - S_1) + M$$
if and only if $I^f(S_1, S_2, \bar{x}, \bar{u}) \leq \sigma_-(f, \bar{x}(S_1), \bar{x}(S_2), S_2 - S_1) + M$;
$$I^f(S_1, S_2, x, u) \leq \sigma(f, x(S_1), S_2 - S_1) + M$$
if and only if $I^f(S_1, S_2, \bar{x}, \bar{u}) \leq \hat{\sigma}_-(f, \bar{x}(S_2), S_2 - S_1) + M$;
$$I^f(S_1, S_2, x, u) \leq \hat{\sigma}(f, x(S_2), S_2 - S_1) + M$$
if and only if $I^f(S_1, S_2, \bar{x}, \bar{u}) \leq \sigma_-(f, \bar{x}(S_1), S_2 - S_1) + M$.

By (3.10),

$$-Ax_f - Bu_f = 0.$$

Set

$$\mu_-(f) = \inf\{\liminf_{T\to\infty} T^{-1}I^f(0, T, x, u) \in X(-A, -B, 0, \infty)\}. \tag{3.24}$$

It is easy to see that

$$\mu_-(f) \le f(x_f, u_f) = \mu(f). \tag{3.25}$$

In view of (A1), $\mu_-(f) > -\infty$. Thus $\mu_-(f)$ is finite.

Proposition 3.21. $\mu_-(f) = \mu(f) = f(x_f, u_f).$

Proof. By Proposition 3.2, there exists $\Delta_* > 0$ such that for each $T > 0$ and each $(x, u) \in X(A, B, 0, T)$,

$$I^f(0, T, x, u) \ge T\mu(f) - \Delta_*.$$

Together with (3.17) and (3.18) this implies that for each $T > 0$ and each $(x, u) \in X(-A, -B, 0, T)$, $I^f(0, T, x, u) \ge T\mu(f) - \Delta_*$. This implies that $\mu_-(f) \ge \mu(f)$. Combined with (3.25) this completes the proof of Proposition 3.21.

It follows from Proposition 3.21 that (A2) holds for the triplet $(f, -A, -B)$. The next result is proved in Sect. 3.9.

Proposition 3.22. *For any* $(f, -A, -B)$-*good trajectory-control pair* $(x, u) \in X(-A, -B, 0, \infty)$,

$$\lim_{t\to\infty} x(t) = x_f.$$

Therefore $(f, -A, -B)$ satisfies all the assumptions posed for the triplet (f, A, B) and all the results stated above for the triplet (f, A, B) are also true for $(f, -A, -B)$.

For each $z \in R^n$, set

$$\pi_-^f(z) = \liminf_{T\to\infty}[I^f(0, T, x, u) - T\mu(f)],$$

where $(x, u) \in X(-A, -B, 0, \infty)$ an $(f, -A, -B)$-overtaking optimal pair such that $x(0) = z$.

In Sect. 3.11 we prove the following two theorems which describe the structure of solutions of problems (P_2) and (P_3) in the regions closed to the end points.

Theorem 3.23. *Let $L_0 > 0$, $\epsilon \in (0, 1)$, $M > 0$. Then there exist $\delta > 0$ and $L_1 > L_0$ such that for each $T \geq L_1$ and each $(x, u) \in X(A, B, 0, T)$ which satisfies*

$$|x(0)| \leq M, \ I^f(0, T, x, u) \leq \sigma(f, x(0), T) + \delta$$

there exists an $(f, -A, -B)$-overtaking optimal pair

$$(\bar{x}_*, \bar{u}_*) \in X(-A, -B, 0, \infty)$$

such that

$$\pi_-^f(\bar{x}_*(0)) = \inf(\pi_-^f),$$
$$|x(T - t) - \bar{x}_*(t)| \leq \epsilon \ for \ all \ t \in [0, L_0].$$

Theorem 3.24. *Let $L_0 > 0$ and $\epsilon > 0$. Then there exist $\delta > 0$ and $L_1 > L_0$ such that for each $T \geq L_1$ and each $(x, u) \in X(A, B, 0, T)$ which satisfies*

$$I^f(0, T, x, u) \leq \sigma(f, 0, T) + \delta$$

there exist an (f, A, B)-overtaking optimal pair $(x_*, u_*) \in X(A, B, 0, \infty)$ and an $(f, -A, -B)$-overtaking optimal pair $(\bar{x}_*, \bar{u}_*) \in X(-A, -B, 0, \infty)$ such that

$$\pi^f(x_*(0)) = \inf(\pi^f),$$
$$\pi_-^f(\bar{x}_*(0)) = \inf(\pi_-^f)$$

and for all $t \in [0, L_0]$,

$$|x(t) - x_*(t)| \leq \epsilon, \ |x(T - t) - \bar{x}_*(t)| \leq \epsilon.$$

3.4   Auxiliary Results and the Proof of Proposition 3.2

In the sequel we use the following auxiliary results.

Proposition 3.25 (Proposition 4.2 of [52]). *Let $T_2 > T_1$ be real numbers, $\{(x_j, u_j)\}_{j=1}^{\infty} \subset X(A, B, T_1, T_2)$ and let the sequence $\{I^f(T_1, T_2, x_j, u_j)\}_{j=1}^{\infty}$ be bounded. Then there exist a subsequence $\{(x_{j_k}, u_{j_k})\}_{k=1}^{\infty}$ and $(x, u) \in X(A, B, T_1, T_2)$ such that*

$$x_{j_k}(t) \to x(t) \ as \ k \to \infty \ uniformly \ in \ [T_1, T_2],$$
$$u_{j_k} \to u \ as \ k \to \infty \ weakly \ in \ L^1(R^m; (T_1, T_2)),$$
$$I^f(T_1, T_2, x, u) \leq \liminf_{k \to \infty} I^f(T_1, T_2, x_{j_k}, u_{j_k}).$$

Proposition 3.26 (Proposition 4.3 of [52]). *For every* $\tilde{y}, \tilde{z} \in R^n$ *and every* $T > 0$ *there exists a solution* $x(\cdot)$, $y(\cdot)$ *of the system*

$$x' = Ax + BB^t y, \quad y' = x - A^t y$$

with the boundary conditions $x(0) = \tilde{y}$, $x(T) = \tilde{z}$ *(where* B^t *denotes the transpose of* B).

Propositions 3.25 and 3.26 and (A1) imply the following result.

Proposition 3.27. *Let* $T > 0$ *and* $y, z \in R^n$. *Then there exists* $(x, u) \in X(A, B, 0, T)$ *such that*

$$x(0) = y, \quad x(T) = z,$$
$$I^f(0, T, x, u) = \sigma(f, y, z, T).$$

Proposition 3.28 (Proposition 4.5 of [52]). *Let* $M, \tau > 0$. *Then*

$$\sup\{|\sigma(f, y, z, \tau)| : y, z \in R^n, \ |y|, |z| \leq M\} < \infty.$$

Proposition 3.29 (Proposition 4.6 of [52]). *Let* $M, \tau, \epsilon > 0$. *Then there exists a number* $\delta > 0$ *such that for each* $y_1, y_2, z_1, z_2 \in R^n$ *satisfying*

$$|y_i|, |z_i| \leq M, \ i = 1, 2, \ |y_1 - y_2|, \ |z_1 - z_2| \leq \delta$$

the following relation holds:

$$|\sigma(f, y_1, z_1, \tau) - \sigma(f, y_2, z_2, \tau)| \leq \epsilon.$$

Proposition 3.30 (Proposition 2.7 of [52]). *Let* $M_1 > 0$ *and* $0 < \tau_0 < \tau_1$. *Then there exists a positive number* M_2 *such that for each* $T_1 \in R^1$, *each* $T_2 \in [T_1 + \tau_0, T_1 + \tau_1]$ *and each* $(x, u) \in X(A, B, T_1, T_2)$ *satisfying* $I^f(T_1, T_2, x, u) \leq M_1$ *the inequality* $|x(t)| \leq M_2$ *holds for all* $t \in [T_1, T_2]$.

Proposition 3.31. *Let* $M_0 > 0$. *Then there exists* $\Delta > 0$ *such that for each* $T > 0$ *and each* $(x, u) \in X(A, B, 0, T)$ *satisfying*

$$|x(0)|, |x(T)| \leq M_0$$

the following inequality holds:

$$I^f(0, T, x, u) \geq T\mu(f) - \Delta.$$

Proof. We may assume without loss of generality that

$$M_0 > |x_f|. \tag{3.26}$$

In view of Proposition 3.28, there exists $\Delta_0 > 0$ such that

$$\sup\{|\sigma(f, y, z, 1)| : y, z \in R^n, |y|, |z| \leq M_0\} \leq \Delta_0. \tag{3.27}$$

Fix

$$\Delta \geq 2\Delta_0 + 2|\mu(f)|.$$

Let

$$T > 0, \quad (x, u) \in X(A, B, 0, T), \quad |x(0)|, |x(T)| \leq M_0. \tag{3.28}$$

We show that

$$I^f(0, T, x, u) \geq T\mu(f) - \Delta.$$

By Propositions 3.27 and 3.28 there exists $(x_1, u_1) \in X(A, B, 0, T + 2)$ such that

$$x_1(0) = x_f, \quad I^f(0, 1, x_1, u_1) = \sigma(f, x_f, x(0), 1),$$
$$x_1(t) = x(t - 1), \quad u_1(t) = u(t - 1), \quad t \in [1, T + 1],$$

$$x_1(T + 2) = x_f, \quad I^f(T + 1, T + 2, x_1, u_1) = \sigma(f, x(T), x_f, 1). \tag{3.29}$$

It follows from Proposition 3.1 and (3.26)–(3.29) that

$$(T + 2)\mu(f) \leq I^f(0, T + 2, x_1, u_1)$$
$$= I^f(0, T, x, u) + \sigma(f, x_f, x(0), 1) + \sigma(f, x(T), x_f, 1)$$
$$\leq I^f(0, T, x, u) + 2\Delta_0.$$

By the relation above and the choice of Δ,

$$I^f(0, T, x, u) \geq T\mu(f) + 2\mu(f) - 2\Delta_0 \geq T\mu(f) - \Delta.$$

Proposition 3.31 is proved.

Proof of Proposition 3.2. By (A1) there exists $M_0 > 0$ such that

$$\psi(M_0) > a_0 + 2 + |\mu(f)|. \tag{3.30}$$

By Proposition 3.31, there exists $\Delta_* > 0$ such that for each $T > 0$ and each $(x, u) \in X(A, B, 0, T)$ satisfying

$$|x(0)|, |x(T)| \leq M_0$$

we have

$$I^f(0, T, x, u) \geq T\mu(f) - \Delta_*. \tag{3.31}$$

Assume that $T > 0$ and $(x, u) \in X(A, B, 0, T)$. If $|x(t)| \geq M_0$ for all $t \in [0, T]$, then it follows from (A1) and (3.30) that

$$I^f(0, T, x, u) - T\mu(f) \geq T(\psi(M_0) - a_0 - \mu(f)) \geq 2T > -\Delta_*.$$

Therefore we may assume without loss of generality that there exist $S_1, S_2 \in [0, T]$ such that

$$S_1 < S_2, \ |x(S_i)| \leq M_0, \ i = 1, 2,$$

$$|x(t)| \geq M_0 \text{ for all } t \in [0, S_1] \cup [S_2, T].$$

It follows from the relation above, (A1), (3.30) and the choice of Δ_* (see (3.31)) that

$$\begin{aligned}
I^f(0, T, x, u) &\geq (\psi(M_0) - a_0)S_1 + (\psi(M_0) - a_0)(T - S_2) + I^f(S_1, S_2, x, u) \\
&\geq (\psi(M_0) - a_0)(S_1 + T - S_2) + (S_2 - S_1)\mu(f) - \Delta_* \\
&= T\mu(f) - (S_1 + T - S_2)\mu(f) + (\psi(M_0) - a_0)(S_1 + T - S_2) - \Delta_* \\
&\geq T\mu(f) - \Delta_*.
\end{aligned}$$

Proposition 3.2 is proved.

3.5   Auxiliary Results for Theorems 3.7, 3.9 and 3.10

Proposition 3.32. *Let $\epsilon > 0$. Then there exists $\delta > 0$ such that for each $T \geq 1$ and each $y, z \in R^n$ satisfying $|y - x_f|, \ |z - x_f| \leq \delta$,*

$$\sigma(f, y, z, T) \leq T\mu(f) + \epsilon.$$

Proof. By Proposition 3.29, there exists $\delta \in (0, 4^{-1})$ such that for each $y_1, y_2, z_1, z_2 \in R^n$ satisfying

$$|y_i|, |z_i| \leq |x_f| + 1, \ i = 1, 2, \ |y_1 - y_2|, \ |z_1 - z_2| \leq \delta$$

we have

$$|\sigma(f, y_1, z_1, 4^{-1}) - \sigma(f, y_2, z_2, 4^{-1})| \leq \epsilon/6. \tag{3.32}$$

Assume that

$$T \geq 1, \ y, z \in R^n, \ |y - x_f|, \ |z - x_f| \leq \delta. \tag{3.33}$$

In view of Proposition 3.27, there exists $(x, u) \in X(A, B, 0, T)$ such that

$$x(0) = y, \ x(t) = x_f, \ u(t) = u_f, \ t \in [4^{-1}, T - 4^{-1}], \ x(T) = z,$$
$$I^f(0, 4^{-1}, x, u) = \sigma(f, y, x_f, 4^{-1}),$$

$$I^f(T - 4^{-1}, T, x, u) = \sigma(f, x_f, z, 4^{-1}). \tag{3.34}$$

It follows from (3.34) that

$$\sigma(f, y, z, T) \leq I^f(0, T, x, u)$$
$$= \sigma(f, y, x_f, 4^{-1}) + \sigma(f, x_f, z, 4^{-1}) + (T - 1/2)\mu(f). \tag{3.35}$$

By (3.33) and the choice of δ (see (3.32)),

$$\sigma(f, y, x_f, 4^{-1}), \ \sigma(f, x_f, z, 4^{-1})$$
$$\leq \epsilon/6 + \sigma(f, x_f, x_f, 4^{-1}) = \epsilon/6 + 4^{-1}\mu(f).$$

Together with (3.35) this implies that

$$\sigma(f, y, z, T) \leq T\mu(f) + \epsilon/3.$$

Proposition 3.32 is proved.

Proposition 3.33. *Let $M_0 > 0$. Then there exists $M > 0$ such that for each $T \geq 1$ and each $y, z \in R^n$ satisfying $|y|, |z| \leq M_0$,*

$$\sigma(f, y, z, T) \leq T\mu(f) + M.$$

Proof. We may assume without loss of generality that

$$M_0 \geq |x_f| + 1.$$

By Proposition 3.28, there exists

$$M_1 > \sup\{|\sigma(f, y, z, 4^{-1})| : \ y, z \in R^n, \ |y|, |z| \leq M_0\}.$$

Choose a number

$$M > 2M_1 + 2|\mu(f)|.$$

Assume that

$$T \geq 1, \ y, z \in R^n, \ |y|, |z| \leq M_0. \tag{3.36}$$

In view of Proposition 3.27, there exists $(x, u) \in X(A, B, 0, T)$ such that

$$x(0) = y, \ x(t) = x_f, \ u(t) = u_f, \ t \in [4^{-1}, T - 4^{-1}], \ x(T) = z,$$
$$I^f(0, 4^{-1}, x, u) = \sigma(f, y, x_f, 4^{-1}),$$
$$I^f(T - 4^{-1}, T, x, u) = \sigma(f, x_f, z, 4^{-1}). \tag{3.37}$$

It follows from (3.36), the choice of M_1, and the relations $M_0 > |x_f| + 1$ that

$$|\sigma(f, y, x_f, 4^{-1})|, \ |\sigma(f, x_f, z, 4^{-1})| \leq M_1.$$

By the inequalities above, (3.37) and the choice of M,

$$\sigma(f, y, z, T) \leq I^f(0, T, x, u) = (T - 2^{-1})\mu(f) + 2M_1$$
$$= T\mu(f) + |\mu(f)| + 2M_1 < T\mu(f) + M.$$

Proposition 3.33 is proved.

Proposition 3.34. *Let $M, \epsilon > 0$. Then there exists a natural number L such that for each $(x, u) \in X(A, B, 0, L)$ satisfying*

$$I^f(0, L, x, u) \leq L\mu(f) + M$$

the inequality

$$\min\{|x(t) - x_f| : \ t \in [0, L]\} \leq \epsilon$$

holds.

Proof. Assume that the proposition does not hold. Then there exist a strictly increasing sequence of natural numbers $\{L_k\}_{k=1}^{\infty}$ such that $L_k \geq k$ for all natural numbers k and a sequence $(x_k, u_k) \in X(A, B, 0, L_k)$, $k = 1, 2, \ldots$ such that for each natural number k we have

$$I^f(0, L_k, x_k, u_k) \leq L_k\mu(f) + M, \tag{3.38}$$
$$\min\{|x_k(t) - x_f| : \ t \in [0, L_k]\} > \epsilon. \tag{3.39}$$

Proposition 3.2 implies that there exists a positive constant Δ_* such that for every positive number T and every trajectory-control pair $(x, u) \in X(A, B, 0, T)$,

$$I^f(0, T, x, u) \geq T\mu(f) - \Delta_*. \tag{3.40}$$

Let p be a natural number. By (3.38) and (3.40), for each natural number $k > p$ we have

$$I^f(0, p, x_k, u_k) = I^f(0, L_k, x_k, u_k) - I^f(p, L_k, x_k, u_k)$$
$$\leq L_k \mu(f) + M - (L_k - p)\mu(f)$$
$$+ \Delta_* \leq p\mu(f) + M + \Delta_*. \tag{3.41}$$

In view of (3.41) and Proposition 3.25, extracting a subsequence and re-indexing if necessary, we may assume without loss of generality that there exists a trajectory-control pair $(x, u) \in X(A, B, 0, \infty)$ such that for every natural number p,

$$x_k(t) \to x(t) \text{ as } k \to \infty \text{ uniformly on } [0, p], \tag{3.42}$$

$$I^f(0, p, x, u) \leq p\mu(f) + M + \Delta_*. \tag{3.43}$$

It follows from (3.43) and Proposition 3.4 that $(x, u) \in X(A, B, 0, \infty)$ is an (f, A, B)-good trajectory-control pair and there exists a positive number τ_0 such that

$$|x(t) - x_f| \leq \epsilon/4 \text{ for all } t \geq \tau_0. \tag{3.44}$$

Relation (3.42) implies that there exists an integer $k_0 > \tau_0 + 8$ such that for each integer $k \geq k_0$,

$$|x_k(t) - x(t)| \leq \epsilon/4 \text{ for all } t \in [\tau_0, \tau_0 + 4]. \tag{3.45}$$

Relations (3.44) and (3.45) imply that for every integer $k \geq k_0$ and every number $t \in [\tau_0, \tau_0 + 4]$,

$$|x_f - x_k(t)| \leq |x_f - x(t)| + |x(t) - x_k(t)| \leq \epsilon/2.$$

This contradicts (3.39). The contradiction we have reached proves Proposition 3.34.

Proposition 3.35. *Let $M, \epsilon > 0$. Then there exists a natural number L such that for each $T \geq L$, each $(x, u) \in X(A, B, 0, T)$ satisfying*

$$I^f(0, T, x, u) \leq T\mu(f) + M \tag{3.46}$$

and each number S satisfying

$$[S, S + L] \subset [0, T] \tag{3.47}$$

the following inequality holds:

$$\min\{|x(t) - x_f| : t \in [S, S + L]\} \leq \epsilon.$$

Proof. Proposition 3.2 implies that there exists a positive constant Δ_* such that for every positive number T and every trajectory-control pair $(x, u) \in X(A, B, 0, T)$ we have

$$I^f(0, T, x, u) \geq T\mu(f) - \Delta_*.$$

It follows from Proposition 3.34 that there exists an integer $L \geq 1$ such that for every trajectory-control pair $(x, u) \in X(A, B, 0, L)$ satisfying

$$I^f(0, L, x, u) \leq L\mu(f) + M + 2\Delta_*$$

the following inequality holds:

$$\min\{|x(t) - x_f| : t \in [0, L]\} \leq \epsilon. \tag{3.48}$$

Assume that $T \geq L$, $(x, u) \in X(A, B, 0, T)$ satisfies (3.46) and a real number S satisfies (3.47). In view of the choice of Δ_* we have

$$I^f(0, S, x, u) \geq S\mu(f) - \Delta_*, \ I^f(S + L, T, x, u) \geq (T - S - L)\mu(f) - \Delta_*.$$

Combined with (3.46) this implies that

$$I^f(S, S + L, x, u) = I^f(0, T, x, u) - I^f(0, S, x, u) - I^f(S + L, T, x, u)$$
$$\leq T\mu(f) + M - S\mu(f) + \Delta_* - (T - (S + L))\mu(f)$$
$$+ \Delta_* \leq L\mu(f) + 2\Delta_*.$$

By the inequality above and the choice of L (see (3.48)), there exists $t \in [S, S + L]$ such that $|x(t) - x_f| \leq \epsilon$. Proposition 3.35 is proved.

Proposition 3.36. *Any (f, A, B)-overtaking optimal pair*

$$(x, u) \in X(A, B, 0, \infty)$$

is (f, A, B)-good.

Proof. Let $(x, u) \in X(A, B, 0, \infty)$ be (f, A, B)-overtaking optimal. By Proposition 3.27, there exists $(x_1, u_1) \in X(A, B, 0, \infty)$ such that

$$x_1(0) = x(0), \ x_1(t) = x_f, \ u_1(t) = u_f, \ t \in [1, \infty),$$
$$I^f(0, 1, x_1, u_1) = \sigma(f, x_1(0), x_f, 1).$$

Since the pair (x, u) is (f, A, B)-overtaking optimal it follows from the relations above that

$$0 \geq \limsup_{T \to \infty}[I^f(0, T, x, u) - I^f(0, T, x_1, u_1)]$$

$$= \limsup_{T \to \infty}[I^f(0, T, x, u) - \sigma(f, x(0), x_f, 1) - (T-1)\mu(f)].$$

Together with Proposition 3.4 this implies that the pair (x, u) is (f, A, B)-good. Proposition 3.36 is proved.

Proposition 3.37. *Let $\epsilon \in (0, 1)$. Then there exists $\delta > 0$ such that for each $T \geq 1$ and each $(x, u) \in X(A, B, 0, T)$ satisfying*

$$|x(0) - x_f|, \ |x(T) - x_f| \leq \delta,$$

$$I^f(0, T, x, u) \leq \sigma(f, x(0), x(T), T) + \delta$$

the inequality $|x(t) - x_f| \leq \epsilon$ holds for all $t \in [0, T]$.

Proof. By Propositions 3.29 and 3.32, for each integer $k \geq 1$, there is

$$\delta_k \in (0, 4^{-k}\epsilon) \tag{3.49}$$

such that the following properties hold:

(i) for each pair of points $y, z \in R^n$ satisfying $|y - x_f|, |z - x_f| \leq \delta_k$,

$$|\sigma(f, y, z, 1) - \mu(f)| \leq 4^{-k};$$

(ii) for each number $T \geq 1$ and each pair of points $y, z \in R^n$ satisfying $|y - x_f|, |z - x_f| \leq \delta_k$,

$$\sigma(f, y, z, T) \leq T\mu(f) + 4^{-k}\epsilon.$$

We may assume without loss of generality that the sequence $\{\delta_k\}_{k=1}^{\infty}$ is decreasing. Assume that the proposition does not hold. Then for each integer $k \geq 1$ there exist $T_k \geq 1$ and a trajectory-control pair $(x_k, u_k) \in X(A, B, 0, T_k)$ such that

$$|x_k(0) - x_f| \leq \delta_k, \ |x_k(T_k) - x_f| \leq \delta_k, \tag{3.50}$$

$$I^f(0, T_k, x_k, u_k) \leq \sigma(f, x_k(0), x_k(T_k), T_k) + \delta_k, \tag{3.51}$$

$$\sup\{\{|x_k(t) - x_f| : t \in [0, T_k]\} > \epsilon. \tag{3.52}$$

It follows from property (ii) and (3.49)–(3.51) that for every natural number k we have

$$I^f(0, T_k, x_k, u_k) \leq T_k\mu(f) + 2 \cdot 4^{-k}. \tag{3.53}$$

Proposition 3.27 implies that there exists a trajectory-control pair $(x, u) \in X(A, B, 0, \infty)$ such that

$$x(t) = x_1(t), \; u(t) = u_1(t), \; t \in [0, T_1], \; x(T_1 + 1) = x_2(0), \tag{3.54}$$

$$I^f(T_1, T_1 + 1, x, u) = \sigma(f, x_1(T_1), x_2(0), 1) \tag{3.55}$$

and for every natural number k we have

$$x\left(\sum_{i=1}^{k}(T_i + 1) + t\right) = x_{k+1}(t),$$

$$u\left(\sum_{i=1}^{k}(T_i + 1) + t\right) = u_{k+1}(t),$$

$$I^f\left(\sum_{i=1}^{k+1}(T_i + 1) - 1, \sum_{i=1}^{k+1}(T_i + 1), x, u\right) = \sigma(f, x_{k+1}(T_k), x_{k+2}(0), 1). \tag{3.56}$$

In view of (3.53), (3.56) and property (i), for each integer $k \geq 2$,

$$I^f\left(0, \sum_{i=1}^{k}(T_i + 1), x, u\right) = \sum_{i=1}^{k}(I^f(0, T_i, x_i, u_i) + \sigma(f, x_i(T_i), x_{i+1}(0), 1))$$

$$\leq \sum_{i=1}^{k}[T_i \mu(f) + 2 \cdot 4^{-i} + \mu(f) + 4^{-i}]$$

$$\leq \mu(f) \sum_{i=1}^{k}(T_i + 1) + 6.$$

Since the relation above holds for any integer $k \geq 2$ it follows from Proposition 3.4 that the trajectory-control pair (x, u) is (f, A, B)-good and

$$\lim_{t \to \infty} x(t) = x_f.$$

Thus there exists an integer $i_0 \geq 1$ such that for each integer $k \geq i_0$ and all $t \in [0, T_k]$,

$$|x_k(t) - x_f| \leq \epsilon/2.$$

This contradicts (3.52). The contradiction we have reached proves Proposition 3.37.

3.6   Proof of Theorem 3.7

By Proposition 3.33, there exists $M_2 > 0$ such that for each $\tau \geq 1$ and each $y, z \in R^n$ satisfying $|y|, |z| \leq M_0 + M_1$,

$$\sigma(f, y, z, \tau) \leq \tau\mu(f) + M_2. \tag{3.57}$$

By Proposition 3.37, there exists $\delta \in (0, \epsilon)$ such that the following property holds:

(i) for each $T \geq 1$ and each $(x, u) \in X(A, B, 0, T)$ satisfying

$$|x(0) - x_f|, \ |x(T) - x_f| \leq \delta,$$
$$I^f(0, T, x, u) \leq \sigma(f, x(0), x(T), T) + \delta$$

the inequality $|x(t) - x_f| \leq \epsilon$ holds for all $t \in [0, T]$.
   By Proposition 3.35, there exists $L_0 > 0$ such that the following property holds:

(ii) for each $T \geq L_0$, each $(x, u) \in X(A, B, 0, T)$ satisfying

$$I^f(0, T, x, u) \leq T\mu(f) + M_1 + M_2$$

and each number S satisfying $[S, S + L_0] \subset [0, T]$ we have

$$\min\{|x(t) - x_f| : t \in [S, S + L_0]\} \leq \delta.$$

Fix

$$L > 4L_0 + 4.$$

Assume that $T > 2L$, $(x, u) \in X(A, B, 0, T)$, for each $S \in [0, T - L]$,

$$I^f(S, S + L, x, u) \leq \sigma(f, x(S), x(S + L), L) + \delta \tag{3.58}$$

and that at least one of the conditions (a), (b), and (c) of Theorem 3.7 holds. It follows from conditions (a)–(c) and the choice of M_2 (see (3.57)) that

$$I^f(0, T, x, u) \leq T\mu(f) + M_2 + M_1. \tag{3.59}$$

It follows from (3.59) and property (ii) that there exists a sequence $\{S_i\}_{i=1}^q$ such that

$$0 \leq S_1 \leq L_0,$$

$$1 \leq S_{i+1} - S_i \leq 2 + L_0 \text{ for all integers } i \text{ satisfying } 1 \leq i < q,$$

$$T_q \geq S_q \geq T_q - L_0 - 1, \tag{3.60}$$

$$|x(S_i) - x_f| \leq \delta, \ i = 1, \ldots, q. \tag{3.61}$$

Clearly, if $|x(0) - x_f| \leq \delta$, then we may assume that $S_1 = 0$ and if $|x(T) - x_f| \leq \delta$, then we may assume that $S_q = T$.

Assume that $t \in [S_1, S_q]$. Then there is an integer $j \in \{1, \ldots, q-1\}$ such that

$$t \in [S_j, S_{j+1}].$$

In view of (3.60), there exists a number S such that

$$[S_j, S_{j+1}] \subset [S, S+L] \subset [0, T].$$

Combined with (3.58) this implies that

$$I^f(S_j, S_{j+1}, x, u) \leq \sigma(f, x(S_j), x(S_{j+1}), S_{j+1} - S_j) + \delta.$$

Together with (3.61) and property (i) this implies that $|x(t) - x_f| \leq \epsilon$. Theorem 3.7 is proved.

3.7   Proof of Theorem 3.9

We may assume without loss of generality that

$$M > |x_f| + 1, \ \epsilon < 2^{-1}.$$

By Theorem 3.7, there exist $L > 0$, $\delta > 0$ such that the following property holds:

(i) for each $T > 2L$ and each $(x, u) \in X(A, B, 0, T)$ which satisfies

$$|x(0)|, \ |x(T)| \leq M, \ I^f(0, T, x, u) = \sigma(f, x(0), x(T), T)$$

we have

$$|x(t) - x_f| \leq \epsilon \text{ for all } t \in [L, T - L]$$

and if $|x(0) - x_f| \leq \delta$, then

$$|x(t) - x_f| \leq \epsilon \text{ for all } t \in [0, T - L]$$

Assume that $(x, u) \in X(A, B, 0, \infty)$ is an (f, A, B)-overtaking optimal pair and

$$|x(0)| \leq M. \tag{3.62}$$

By (A3) and Proposition 3.36,

$$\lim_{t \to \infty} x(t) = x_f.$$

Thus there exists $T_0 > 0$ such that for all $t \geq T_0$,

$$|x(t)| < M. \tag{3.63}$$

Let

$$T > T_0 + 2L. \tag{3.64}$$

By property (i), (3.62) and (3.64),

$$|x(t) - x_f| \leq \epsilon \text{ for all } t \in [L, T - L]$$

and if $|x(0) - x_f| \leq \delta$, then

$$|x(t) - x_f| \leq \epsilon \text{ for all } t \in [0, T - L].$$

Since T is any number satisfying (3.64) we conclude that Theorem 3.9 is proved.

3.8 Proof of Theorem 3.10

In view of Proposition 3.36, (i) implies (ii). By (A3), (ii) implies (iii). Clearly, (iv) follows from (iii). Assume that (iv) holds. We show that (i) is true. It follows from (iv), (f, A, B)-minimality of (x, u) and Theorem 3.7 that

$$\lim_{t \to \infty} x(t) = x_f. \tag{3.65}$$

Assume that the pair (x, u) is not (f, A, B)-overtaking optimal. By Theorem 3.8, there exists an (f, A, B)-overtaking optimal pair $(\tilde{x}, \tilde{u}) \in X(A, B, 0, \infty)$ satisfying

$$\tilde{x}(0) = x(0). \tag{3.66}$$

Clearly,

$$\limsup_{T \to \infty} [I^f(0, T, \tilde{x}, \tilde{u}) - I^f(0, T, x, u)] \leq 0.$$

Since the pair (x, u) is not (f, A, B)-overtaking optimal we have

$$\limsup_{T \to \infty} [I^f(0, T, x, u) - I^f(0, T, \tilde{x}, \tilde{u})] > 0.$$

Thus there exist $\epsilon > 0$ and a strictly increasing sequence of positive numbers $T_k \to \infty$ as $k \to \infty$ such that for all integers $k \geq 1$,

$$I^f(0, T_k, x, u) - I^f(0, T_k, \tilde{x}, \tilde{u}) \geq 2\epsilon. \tag{3.67}$$

Proposition 3.29 and (A2) imply that there exists $\delta > 0$ such that

$$|\sigma(f, z_1, z_2, 1) - \mu(f)| \leq \epsilon/4 \tag{3.68}$$

for all $z_1, z_2 \in R^n$ satisfying $|z_i - x_f| \leq \delta$, $i = 1, 2$. It follows from Proposition 3.36 and (A3) that the pair $(\tilde{x}, \tilde{u})$ is (f, A, B)-good and

$$\lim_{t \to \infty} \tilde{x}(t) = x_f. \tag{3.69}$$

By property (iv), the (f, A, B)-minimality of (x, u), Proposition 3.33, and (3.65),

$$\limsup_{T \to \infty} [I^f(0, T, x, u) - T\mu(f)] < \infty.$$

Thus the pair (x, u) is (f, A, B)-good. In view of (3.65) and (3.69) there exists $S_0 > 0$ such that for all $t \geq S_0$,

$$|x(t) - x_f|, \ |\tilde{x}(t) - x_f| \leq \delta. \tag{3.70}$$

Choose a natural number k such that $T_k > S_0$. By Proposition 3.27, there exists $(x_1, u_1) \in X(A, B, 0, T_k + 1)$ such that

$$x_1(t) = \tilde{x}(t), \ u_1(t) = \tilde{u}(t), t \in [0, T_k],$$
$$x_1(T_k + 1) = x(T_k + 1),$$

$$I^f(T_k, T_k + 1, x_1, u_1) = \sigma(f, \tilde{x}(T_k), x(T_k + 1), 1). \tag{3.71}$$

It follows from (3.66) and (3.71) that

$$x_1(0) = x(0), \ x_1(T_k + 1) = x(T_k + 1). \tag{3.72}$$

By the relation $T_k > S_0$, (3.68), and (3.70),

$$|\sigma(f, \tilde{x}(T_k), x(T_k + 1), 1) - \sigma(f, x(T_k), x(T_k + 1), 1)| \leq \epsilon/2. \tag{3.73}$$

In view of (3.67), (3.71), (3.73), and the (f, A, B)-minimality of (x, u),

$$I^f(0, T_k + 1, x_1, u_1) = I^f(0, T_k, \tilde{x}, \tilde{u}) + \sigma(f, \tilde{x}(T_k), x(T_k + 1), 1)$$
$$\leq I^f(0, T_k, x, u) - \epsilon + \sigma(f, x(T_k), x(T_k + 1), 1) + \epsilon/2$$
$$= I^f(0, T_{k+1}, x, u) - \epsilon/2.$$

Combined with (3.72) this contradicts the (f, A, B)-minimality of (x, u). The contradiction we have reached proves that the pair (x, u) is (f, A, B)-overtaking optimal. Theorem 3.10 is proved.

3.9 Proofs of Propositions 3.14, 3.16, 3.17, and 3.22

Proof of Proposition 3.14. Let $\epsilon > 0$. By Proposition 3.29, there exists $\delta > 0$ such that

$$|\sigma(f, z_1, z_2, 1) - \mu(f)| \leq \epsilon/4 \tag{3.74}$$

for all $z_1, z_2 \in R^n$ satisfying $|z_i - x_f| \leq \delta$, $i = 1, 2$.

By Theorem 3.9, exists $\gamma \in (0, \delta)$ such that for any (f, A, B)-overtaking optimal trajectory-control pair $(x, u) \in X(A, B, 0, \infty)$ satisfying $|x(0) - x_f| \leq \gamma$, we have

$$|x(t) - x_f| \leq \delta \text{ for all numbers } t \geq 0. \tag{3.75}$$

Let

$$z_1, z_2 \in R^n, \ |z_i - x_f| \leq \gamma, \ i = 1, 2, \ (x, u) \in \Lambda(z_1). \tag{3.76}$$

By (3.76) and the choice of γ (see (3.75)) relation (3.75) is true. It follows from (3.12) that

$$\pi^f(z_1) = \liminf_{T \to \infty}[I^f(0, T, x, u) - T\mu(f)]$$
$$= I^f(0, 1, x, u) - \mu(f)$$
$$+ \liminf_{T \to \infty}[I^f(1, T, x, u) - (T - 1)\mu(f)]. \tag{3.77}$$

By Proposition 3.27 there exists $(x_1, u_1) \in X(A, B, 0, \infty)$ such that

$$x_1(0) = z_2, \ x_1(t) = x(t), \ u_1(t) = u(t) \text{ for all } t \geq 1,$$

$$I^f(0, 1, x_1, u_1) = \sigma(f, z_2, x(1), 1). \tag{3.78}$$

It follows from (3.75), (3.76), and the choice of δ (see (3.74)) that

$$|\sigma(f, z_2, x(1), 1) - \sigma(f, z_1, x(1), 1)| \leq \epsilon/2. \tag{3.79}$$

By (3.76)–(3.79) and Proposition 3.11,

$$\pi^f(z_2) \leq \liminf_{T \to \infty}[I^f(0, T, x_1, u_1) - T\mu(f)]$$

$$= I^f(0, 1, x_1, u_1) - \mu(f)$$

$$+ \liminf_{T \to \infty}[I^f(1, T, x, u) - (T - 1)\mu(f)]$$

$$\leq \pi^f(z_1) + \sigma(f, z_2, x(1), 1) - \sigma(f, z_1, x(1), 1)$$

$$\leq \pi^f(z_1) + \epsilon/2.$$

This implies that $|\pi^f(z_1) - \pi^f(z_2)| \leq \epsilon/2$ for all $z_1, z_2 \in R^n$ satisfying (3.76), Proposition 3.14 is proved.   □

Proof of Proposition 3.16. Let $M > 0$. We show that the set $\{x \in R^n : \pi^f(x) \leq M\}$ is bounded. Assume the contrary. Then there exists a sequence $\{z_k\}_{k=1}^{\infty}$ such that

$$\lim_{k \to \infty} |z_k| = \infty, \tag{3.80}$$

$$\pi^f(z_k) \leq M, \; k = 1, 2, \dots. \tag{3.81}$$

By Proposition 3.2, there is $\Delta_* > 0$ such that for each $T > 0$ and each $(x, u) \in X(A, B, 0, T)$,

$$I^f(0, T, x, u) \geq T\mu(f) - \Delta_*. \tag{3.82}$$

In view of Theorem 3.8 and (3.12), for each integer $k \geq 1$ there exists (f, A, B)-overtaking optimal trajectory-control pair $(x_k, u_k) \in X(A, B, 0, \infty)$ such that

$$x_k(0) = z_k,$$

$$\pi^f(z_k) = \liminf_{T \to \infty}[I^f(0, T, x_k, u_k) - T\mu(f)]. \tag{3.83}$$

Let $k \geq 1$ be an integer and $T > 0$. It follows from (3.81) and (3.83) that there is $S > T$ such that

$$I^f(0, S, x_k, u_k) - S\mu(f) \leq \pi^f(z_k) + 1 \leq M + 1. \tag{3.84}$$

Relations (3.82) and (3.84) imply that

$$I^f(0, T, x_k, u_k) - T\mu(f)$$

$$= I^f(0, S, x_k, u_k) - S\mu(f) - I^f(T, S, x_k, u_k) - (S - T)\mu(f) \leq M + 1 + \Delta_*.$$

Thus for each integer $k \geq 1$ and each $T > 0$,

$$I^f(0, T, x_k, u_k) - T\mu(f) \leq M + 1 + \Delta_*.$$

By the inequality above and Proposition 3.25, extracting a subsequence and re-indexing if necessary we may assume without loss of generality that there exists $(x, u) \in X(A, B, 0, \infty)$ such that for each integer $p \geq 1$,

$$x_k(t) \to x(t) \text{ as } k \to \infty \text{ uniformly on } [0, p],$$

$$I^f(0, p, x, u) \leq \liminf_{k \to \infty} I^f(0, p, x_k, u_k).$$

Combined with (3.83) this implies that $\lim_{k \to \infty} z_k = x(0)$. This contradicts (3.80). The contradiction we have reached proves Proposition 3.16. □

Proof of Proposition 3.17. Assume that $\{z_k\}_{k=1}^{\infty} \subset R^n$, $z \in R^n$ and that

$$\lim_{k \to \infty} z_k = z. \tag{3.85}$$

We show that

$$\pi^f(z) \leq \liminf_{k \to \infty} \pi^f(z_k).$$

By Proposition 3.2, the sequence $\{\pi^f(z_k)\}_{k=1}^{\infty}$ is bounded from below. We may assume without loss of generality there exists $\lim_{k \to \infty} \pi^f(z_k) < \infty$. Clearly,

$$\Delta := \lim_{k \to \infty} \pi^f(z_k)$$

is finite. By Theorem 3.8 and (3.12), for each integer $k \geq 1$ there exists (f, A, B)-overtaking optimal trajectory-control pair $(x_k, u_k) \in X(A, B, 0, \infty)$ such that

$$x_k(0) = z_k,$$

$$\pi^f(z_k) = \liminf_{T \to \infty}[I^f(0, T, x_k, u_k) - T\mu(f)]. \tag{3.86}$$

Proposition 3.12 implies that for each integer $k \geq 1$ and each $T > 0$,

$$I^f(0, T, x_k, u_k) - T\mu(f) = \pi^f(x_k(0)) - \pi^f(x_k(T)). \tag{3.87}$$

Proposition 3.2 implies that there is $\Delta_* > 0$ such that for each $T > 0$ and each $(x, u) \in X(A, B, 0, T)$,

$$I^f(0, T, x, u) \geq T\mu(f) - \Delta_*. \tag{3.88}$$

Let $k \geq 1$ be an integer and $T > 0$. It follows from (3.86), (3.88), and the choice of Δ that there is $S > T$ such that

$$I^f(0, S, x_k, u_k) - S\mu(f) \leq \Delta + 1.$$

The relations above and (3.88) imply that

$$I^f(0, T, x_k, u_k) - T\mu(f)$$
$$= I^f(0, S, x_k, u_k) - S\mu(f) - I^f(T, S, x_k, u_k) - (S - T)\mu(f) \leq \Delta + 1 + \Delta_*.$$

Thus for each integer $k \geq 1$ and each $T > 0$,

$$I^f(0, T, x_k, u_k) - T\mu(f) \leq \Delta + 1 + \Delta_*.$$

By the inequality above and Proposition 3.25, extracting a subsequence and re-indexing if necessary we may assume without loss of generality that there exists $(x, u) \in X(A, B, 0, \infty)$ such that for each integer $p \geq 1$,

$$x_k(t) \to x(t) \text{ as } k \to \infty \text{ uniformly on } [0, p],$$

$$I^f(0, p, x, u) \leq \liminf_{k \to \infty} I^f(0, p, x_k, u_k). \tag{3.89}$$

Let $\epsilon > 0$. In view of Proposition 3.14, there exists a positive number δ such that for each $\xi \in R^n$ satisfying $|\xi - x_f| \leq \delta$,

$$|\pi^f(\xi)| \leq \epsilon/2. \tag{3.90}$$

By (3.85) and Theorem 3.9, there exists $L_0 > 0$ such that for each integer $k \geq 1$ and $t \geq L_0$,

$$|x_k(t) - x_f| \leq \delta. \tag{3.91}$$

Assume that an integer $k \geq 1$ and $T \geq L_0$. In view of (3.90), (3.91) and Proposition 3.12,

$$|\pi^f(x_k(T))| \leq \epsilon/2$$

and

$$I^f(0, T, x_k, u_k) = T\mu(f) + \pi^f(x_k(0)) - \pi^f(x_k(T))$$
$$\leq T\mu(f) + \pi^f(x_k(0)) + \epsilon/2.$$

By the relation above, (3.86) and (3.89), for each integer $T > L_0$,

$$I^f(0, T, x, u) \leq \liminf_{k \to \infty} I^f(0, T, x_k, u_k)$$
$$\leq T\mu(f) + \lim_{k \to \infty} \pi^f(z_k) + \epsilon/2$$

and

$$I^f(0, T, x_k, u_k) - T\mu(f) \leq \lim_{k \to \infty} \pi^f(z_k) + \epsilon/2.$$

In view of Proposition 3.11, (3.85), (3.86) and (3.89),

$$\pi^f(z) \leq \lim_{k \to \infty} \pi^f(z_k) + \epsilon/2.$$

Since ϵ is any positive number this completes the proof of Proposition 3.17.   □

Proof of Proposition 3.22. Assume that $(x, u) \in X(-A, -B, 0, \infty)$ is an $(f, -A, -B)$-good pair. Proposition 3.5 implies that there exists a number

$$M > \sup\{x(t)| : t \in [0, \infty)\}. \tag{3.92}$$

Let $\epsilon > 0$. By Theorem 3.7, there exist $L, \delta > 0$ such that the following property holds:

(i) for each number $T > 2L$ and each $(y, v) \in X(A, B, 0, \infty)$ which satisfies

$$|y(0)|, \ |y(T)| \leq M,$$
$$I^f(0, T, y, v) \leq \sigma(f, y(0), y(T), T) + \delta$$

the inequality $|y(t) - x_f| \leq \epsilon$ is true for all numbers $t \in [L, T - L]$.

In view of Proposition 3.4, there exists $T_0 > 0$ such that for each pair of numbers $S_2 > S_1 \geq T_0$,

$$I^f(S_1, S_2, x, u) \leq \sigma_-(f, x(S_1), x(S_2), S_2 - S_1) + \delta. \tag{3.93}$$

Let $T > T_0 + 2L$. Set

$$y(t) = x(T - t + T_0), \ v(t) = u(T - t + T_0), \ t \in [T_0, T]. \tag{3.94}$$

In view of (3.17) and (3.18), $(y, v) \in X(A, B, T_0, T)$. By (3.92) and (3.94),

$$|y(T_0)|, \ |y(T)| \leq M. \tag{3.95}$$

It follows from (3.93), (3.94), and Proposition 3.20 that

$$I^f(T_0, T, y, v) \leq \sigma(f, y(T_0), y(T), T - T_0) + \delta.$$

Together with (3.95) and property (i) this implies that

$$|y(t) - x_f| \leq \epsilon, \ t \in [T_0 + L, T - L]$$

and

$$|x(t) - x_f| \leq \epsilon, \; t \in [T_0 + L, T - L].$$

Since T is any natural number satisfying $T > T_0 + 2L$ we conclude that

$$|x(t) - x_f| \leq \epsilon \text{ for all } t \geq T_0.$$

Since ϵ is any positive number we conclude that $\lim_{t \to \infty} x(t) = x_f$. Proposition 3.22 is proved. □

3.10   The Basic Lemma for Theorem 3.23

Lemma 3.38. *Let $S_0 > 0$, $\epsilon \in (0, 1)$. Then there exists $\delta \in (0, \epsilon)$ such that for each $(x, u) \in X(A, B, 0, S_0)$ which satisfies*

$$\pi^f(x(0)) \leq \inf(\pi^f) + \delta,$$

$$I^f(0, S_0, x, u) - S_0\mu(f) - \pi^f(x(0)) + \pi^f(x(S_0)) \leq \delta$$

there exists an (f, A, B)-overtaking optimal pair $(x_*, u_*) \in X(A, B, 0, \infty)$ such that

$$\pi^f(x_*(0)) = \inf(\pi^f),$$

$$|x(t) - x_*(t)| \leq \epsilon \text{ for all } t \in [0, S_0].$$

Proof. Assume that the lemma does not hold. Then there exist a sequence $\{\delta_k\}_{k=1}^{\infty} \subset (0, 1]$ and a sequence $\{(x_k, u_k)\}_{k=1}^{\infty} \subset X(A, B, 0, S_0)$ such that

$$\lim_{k \to \infty} \delta_k = 0 \tag{3.96}$$

and that for all integer $k \geq 1$,

$$\pi^f(x_k(0)) \leq \inf(\pi^f) + \delta_k, \tag{3.97}$$

$$I^f(0, S_0, x_k, u_k) - S_0\mu(f) - \pi^f(x_k(0)) + \pi^f(x_k(S_0)) \leq \delta_k \tag{3.98}$$

and that the following property holds:

(i) for each (f, A, B)-overtaking optimal pair $(y, v) \in X(A, B, 0, \infty)$ satisfying

$$\pi^f(y(0)) = \inf(\pi^f)$$

we have

$$\sup\{|x_k(t) - y(t)| : t \in [0, S_0]\} > \epsilon.$$

In view of (3.97) and (3.98) and the boundedness from below of the function π^f, the sequence $\{I^f(0, S_0, x_k, u_k)\}_{k=1}^{\infty}$ is bounded. By Proposition 3.25, extracting a subsequence and re-indexing if necessary, we may assume without loss of generality that there exists $(x, u) \in X(A, B, 0, S_0)$ such that

$$x_k(t) \to x(t) \text{ as } k \to \infty \text{ uniformly on } [0, S_0], \tag{3.99}$$

$$I^f(0, S_0, x, u) \leq \liminf_{k\to\infty} I^f(0, S_0, x_k, u_k). \tag{3.100}$$

It follows from (3.96), (3.97), (3.99), and the lower semicontinuity of π^f that

$$\pi^f(x(0)) \leq \liminf_{k\to\infty} \pi^f(x_k(0)) = \inf(\pi^f), \quad \pi^f(x(0)) = \inf(\pi^f). \tag{3.101}$$

By (3.99) and the lower semicontinuity of π^f,

$$\pi^f(x(S_0)) \leq \liminf_{k\to\infty} \pi^f(x_k(S_0)). \tag{3.102}$$

It follows from (3.96)–(3.98), (3.100)–(3.102) that

$$I^f(0, S_0, x, u) - S_0\mu(f) - \pi^f(x(0)) + \pi^f(x(S_0))$$
$$\leq \liminf_{k\to\infty}[I^f(0, S_0, x_k, u_k) - S_0\mu(f)] - \lim_{k\to\infty} \pi^f(x_k(0)) + \lim_{k\to\infty} \pi^f(x_k(S_0))$$
$$\leq \liminf_{k\to\infty}[I^f(0, S_0, x_k, u_k) - S_0\mu(f) - \pi^f(x_k(0)) + \pi^f(x_k(S_0))] \leq 0.$$

In view of the inequality above and Proposition 3.11,

$$I^f(0, S_0, x, u) - S_0\mu(f) - \pi^f(x(0)) + \pi^f(x(S_0)) = 0. \tag{3.103}$$

Theorem 3.8 implies that there exists an (f, A, B)-overtaking optimal pair $(\tilde{x}, \tilde{u}) \in X(A, B, 0, \infty)$ such that

$$\tilde{x}(0) = x(S_0). \tag{3.104}$$

For all $t > S_0$ set

$$x(t) = \tilde{x}(t - S_0), \quad u(t) = \tilde{u}(t - S_0). \tag{3.105}$$

It is not difficult to see that the pair $(x, u) \in X(A, B, 0, \infty)$ is an (f, A, B)-good pair. By (3.13), (3.105), and Propositions 3.11 and 3.12,

$$I^f(0, S, x, u) - S\mu(f) - \pi^f(x(0)) + \pi^f(x(S)) = 0 \text{ for all } S > 0.$$

Combined with Proposition 3.18 and (3.10) this implies that

$$(x, u) \in X(A, B, 0, \infty)$$

is an (f, A, B)-overtaking optimal pair satisfying

$$\pi^f(x(0)) = \inf(\pi^f).$$

By (3.99), for all sufficiently large natural numbers k,

$$|x_k(t) - x(t)| \le \epsilon/2 \text{ for all } t \in [0, S_0].$$

This contradicts the property (i). The contradiction we have reached proves Lemma 3.38.

Note that Lemma 3.38 can also be applied for the triplet $(f, -A, -B)$.

3.11   Proofs of Theorems 3.23 and 3.24

Proof of Theorem 3.23. By Lemma 3.38 applied to the triplet $(f, -A - B)$ there exist

$$\delta_1 \in (0, \epsilon/4)$$

such that the following property holds:

(P1)   for each $(x, u) \in X(-A, -B, 0, L_0)$ which satisfies

$$\pi^f_-(x(0)) \le \inf(\pi^f_-) + \delta_1,$$

$$I^f(0, L_0, x, u) - L_0\mu(f) - \pi^f_-(x(0)) + \pi^f_-(x(L_0)) \le \delta_1$$

there exists an $(f, -A, -B)$-overtaking optimal pair

$$(x_*, u_*) \in X(-A, -B, 0, \infty)$$

such that

$$\pi^f_-(x_*(0)) = \inf(\pi^f_-),$$

$$|x(t) - x_*(t)| \le \epsilon \text{ for all } t \in [0, L_0].$$

In view of Propositions 3.13, 3.14, and 3.29, there exists $\delta_2 \in (0, \delta_1)$ such that for each $z \in R^n$ satisfying $|z - x_f| \le 2\delta_2$,

$$|\pi^f_-(z)| = |\pi^f_-(z) - \pi^f_-(x_f)| \le \delta_1/8; \tag{3.106}$$

for each $y, z \in R^n$ satisfying $|y - x_f| \leq 2\delta_2$, $|z - x_f| \leq 2\delta_2$,

$$|\sigma(f, y, z, 1) - \mu(f)| \leq \delta_1/8. \tag{3.107}$$

By Theorem 3.7, there exist $l_0 > 0$, $\delta_3 \in (0, \delta_2/8)$ such that the following property holds:

(P2)   for each $T > 2l_0$ and each

$$(x, u) \in X(A, B, 0, T)$$

such that

$$|x(0)| \leq M, \ I^f(0, T, x, u) \leq \sigma(f, x(0), T) + \delta_3$$

we have

$$|x(t) - x_f| \leq \delta_2 \text{ for all } t \in [l_0, T - l_0]. \tag{3.108}$$

By Theorem 2.8, there exists an $(f, -A, -B)$-overtaking optimal pair $(\bar{x}_*, \bar{u}_*) \in X(-A, -B, 0, \infty)$ such that

$$\pi_-^f(\bar{x}_*(0)) = \inf(\pi_-^f). \tag{3.109}$$

Proposition 3.36 and (A3) imply that there exists $l_1 > 0$ such that

$$|\bar{x}^*(t) - x_f| \leq \delta_2 \text{ for all } t \geq l_1. \tag{3.110}$$

Choose $\delta > 0$ and $L_1 > 0$ such that

$$\delta \leq \delta_3/4, \tag{3.111}$$

$$L_1 > 2L_0 + 2l_0 + 2l_1 + 8. \tag{3.112}$$

Assume that

$$T \geq L_1, \ (x, u) \in X(A, B, 0, T) \tag{3.113}$$

and that

$$|x(0)| \leq M, \ I^f(0, T, x, u) \leq \sigma(f, x(0), T) + \delta. \tag{3.114}$$

In view of property (P2) and (3.112)–(3.114), relation (3.108) holds. It follows from (3.112) and (3.113) that

$$[T - l_0 - l_1 - L_0 - 4, T - l_0 - l_1 - L_0] \subset [l_0, T - l_0 - l_1 - L_0]. \tag{3.115}$$

Relations (3.108) and (3.115) imply that

$$|x(t) - x_f| \le \delta_2 \text{ for all } t \in [T - l_0 - l_1 - L_0 - 4, T - l_0 - l_1 - L_0]. \qquad (3.116)$$

By Proposition 3.27, there exists a trajectory-control pair

$$(x_1, u_1) \in X(A, B, 0, T)$$

such that

$$x_1(t) = x(t), \ u_1(t) = u(t), \ t \in [0, T - l_0 - l_1 - L_0 - 4],$$
$$x_1(t) = \bar{x}_*(T - t), \ u_1(t) = \bar{u}_*(T - t), \ t \in [T - l_0 - l_1 - L_0 - 3, T],$$
$$I^f(T - l_0 - l_1 - L_0 - 4, T - l_0 - l_1 - L_0 - 3, x_1, u_1)$$

$$= \sigma(f, x(T - l_0 - l_1 - L_0 - 4), \bar{x}_*(l_0 + l_1 + L_0 + 3), 1). \qquad (3.117)$$

It follows from (3.114) and (3.117) that

$$-\delta \le I^f(0, T, x_1, u_1) - I^f(0, T, x, u)$$
$$= I^f(T - l_0 - l_1 - L_0 - 4, T - l_0 - l_1 - L_0 - 3, x_1, u_1)$$
$$+ I^f(T - l_0 - l_1 - L_0 - 3, T, x_1, u_1)$$
$$- I^f(T - l_0 - l_1 - L_0 - 4, T - l_0 - l_1 - L_0 - 3, x, u)$$
$$- I^f(T - l_0 - l_1 - L_0 - 3, T, x, u). \qquad (3.118)$$

In view of (3.110), (3.116), (3.117), and the choice of δ_2 (see (3.107)),

$$I^f(T - l_0 - l_1 - L_0 - 4, T - l_0 - l_1 - L_0 - 3, x_1, u_1) \le \mu(f) + \delta_1/8. \qquad (3.119)$$

By (3.116) and the choice of δ_2 (see (3.107)),

$$I^f(T - l_0 - l_1 - L_0 - 4, T - l_0 - l_1 - L_0 - 3, x, u) \ge \mu(f) - \delta_1/8. \qquad (3.120)$$

It follows from (3.118)–(3.120) that

$$I^f(T - l_0 - l_1 - L_0 - 3, T, x_1, u_1) - I^f(T - l_0 - l_1 - L_0 - 3, T, x, u)$$
$$\ge -\delta - \delta_1/4. \qquad (3.121)$$

Since $(\bar{x}_*, \bar{u}_*)$ is an $(f, -A, -B)$-overtaking optimal pair it follows from (3.117) and Proposition 3.12 that

$$I^f(T - l_0 - l_1 - L_0 - 3, T, x_1, u_1) = I^f(0, l_0 + l_1 + L_0 + 3, \bar{x}_*, \bar{u}_*)$$
$$= \mu(f)(l_0 + l_1 + L_0 + 3) + \pi_-^f(\bar{x}_*(0))$$
$$- \pi_-^f(\bar{x}_*(l_0 + l_1 + L_0 + 3)). \qquad (3.122)$$

In view of (3.110) and the choice of δ_2 (see (3.106)) we have

$$|\pi_-^f(\bar{x}_*(l_0 + l_1 + L_0 + 3))| \le \delta_1/8.$$

Combined with (3.121) and (3.122) this implies that

$$I^f(T - l_0 - l_1 - L_0 - 3, T, x, u) \le \mu(f)(l_0 + l_1 + L_0 + 3) + \pi_-^f(\bar{x}_*(0)) + \delta + 3\delta_1/8.$$
$$(3.123)$$

Set

$$\tilde{x}(t) = x(T - t), \ \tilde{u}(t) = u(T - t), \ t \in [0, T]. \qquad (3.124)$$

Evidently, $(\tilde{x}, \tilde{u}) \in X(-A, -B, 0, T)$ and by (3.123) and (3.124) we have

$$I^f(0, l_0 + l_1 + L_0 + 3, \tilde{x}, \tilde{u}) = I^f(T - l_0 - l_1 - L_0 - 3, T, x, u)$$
$$\le \pi_-^f(\bar{x}_*(0)) + \mu(f)(l_0 + l_1 + L_0 + 3) + \delta + 3\delta_1/8.$$
$$(3.125)$$

In view of (3.124) and (3.116),

$$|\tilde{x}(l_0 + l_1 + L_0 + 3) - x_f| \le \delta_2.$$

By the relation above and the choice of δ_2 (see (3.106)),

$$|\pi_-^f(\tilde{x}(l_0 + l_1 + L_0 + 3))| \le \delta_1/8. \qquad (3.126)$$

It follows from (3.125), (3.126), and Proposition 3.11 that

$$\pi_-^f(\tilde{x}(0)) - \pi_-^f(\bar{x}_*(0))$$
$$+ I^f(0, L_0, \tilde{x}, \tilde{u}) - L_0\mu(f) - \pi_-^f(\tilde{x}(0)) + \pi_-^f(\tilde{x}(L_0))$$
$$\le \pi_-^f(\tilde{x}(0)) - \pi_-^f(\bar{x}_*(0)) + I^f(0, l_0 + l_1 + L_0 + 3, \tilde{x}, \tilde{u})$$
$$- \mu(f)(l_0 + l_1 + L_0 + 3) - \pi_-^f(\tilde{x}(0)) + \pi_-^f(\tilde{x}(l_0 + l_1 + L_0 + 3))$$

$$\leq \pi^f_-(\tilde{x}(0)) - \pi^f_-(\tilde{x}_*(0)) + \pi^f_-(\tilde{x}_*(0)) + \mu(f)(l_0 + l_1 + L_0 + 3) + \delta + 3\delta_1/8$$
$$-\mu(f)(l_0 + l_1 + L_0 + 3) - \pi^f_-(\tilde{x}(0)) + \delta_1/8$$
$$\leq \delta + \delta_1/2 \leq \delta_1.$$

By the relation above, Proposition 3.11 and the relation $\pi^f_-(\tilde{x}_*(0)) = \inf(\pi^f_-)$,

$$\pi^f_-(\tilde{x}(0)) \leq \inf(\pi^f_-) + \delta_1,$$
$$I^f(0, L_0, \tilde{x}, \tilde{u}) - L_0\mu(f) - \pi^f_-(\tilde{x}(0)) + \pi^f_-(\tilde{x}(L_0)) \leq \delta_1.$$

It follows from the two inequalities above and property (P1) that there exists an $(f, -A, -B)$-overtaking optimal pair $(x_*, u_*) \in X(-A, -B, 0, \infty)$ such that

$$\pi^f_-(x_*(0)) = \inf(\pi^f_-),$$
$$|\tilde{x}(t) - x_*(t)| \leq \epsilon \text{ for all } t \in [0, L_0].$$

Together with (3.124) this implies that

$$|x(T - t) - x_*(t)| \leq \epsilon \text{ holds for all } t \in [0, L_0].$$

Theorem 3.23 is proved.   □

Theorems 3.7 and 3.23 imply the following result.

Theorem 3.39. *Let $L_0 > 0$, $\epsilon \in (0, 1)$. Then there exist $\delta > 0$ and $L_1 > L_0$ such that for each $T \geq L_1$ and each $(x, u) \in X(A, B, 0, T)$ which satisfies*

$$I^f(0, T, x, u) \leq \sigma(f, T) + \delta$$

there exists an $(f, -A, -B)$-overtaking optimal pair

$$(\bar{x}_*, \bar{u}_*) \in X(-A, -B, 0, \infty)$$

such that

$$\pi^f_-(\bar{x}_*(0)) = \inf(\pi^f_-),$$
$$|x(T - t) - \bar{x}_*(t)| \leq \epsilon \text{ for all } t \in [0, L_0].$$

Theorem 3.39 and Proposition 3.20 imply Theorem 3.24.

3.12 Proof of Proposition 3.6

Since (x_*, u_*) is the unique minimizer of the function L we have

$$\mu(h) \leq c. \tag{3.127}$$

We show that $\mu(h) = c$. Assume the contrary. Then $\mu(h) < c$ and there exist $\Delta > 0$ and $(x, u) \in X(A, B, 0, \infty)$ such that

$$\liminf_{T \to \infty} T^{-1} I^h(0, T, x, u) < c - \Delta. \tag{3.128}$$

It is not difficult to see that for each $T > 0$,

$$
\begin{aligned}
I^h(0, T, x, u) - T\mu(h) &= \int_0^T [h(x(t), u(t)) - \mu(h)]dt \\
&= \int_0^T L(x(t), u(t))dt + cT - \mu(h)T + \int_0^T \langle l, Ax(t) + Bu(t) \rangle dt \\
&= \int_0^T L(x(t), u(t))dt + (c - \mu(h))T + \int_0^T \langle l, x'(t) \rangle dt \\
&= \int_0^T L(x(t), u(t))dt + (c - \mu(h))T + \langle l, x(T) - x(0) \rangle. \tag{3.129}
\end{aligned}
$$

By (3.129), there exists a sequence of numbers $\{T_k\}_{k=1}^{\infty}$ such that

$$T_k \geq 10, \ T_{k+1} - T_k \geq 10, \ k = 1, 2, \ldots,$$

$$T_k^{-1} I^h(0, T_k, x, u) < c - \Delta, k = 1, 2, \ldots. \tag{3.130}$$

In view of (3.2) there exists $M_0 > 1$ such that

$$\psi(M_0) > a_0 + 4 + |c|. \tag{3.131}$$

It follows from (3.130), (3.131), and (A1) that for each natural number k there exists $S_k \in [0, T_k]$ such that

$$|x(S_k)| \leq M_0,$$

$$|x(t)| > M_0 \text{ for all } t \text{ satisfying } S_k < t \leq T_k. \tag{3.132}$$

Let $k \geq 1$ be an integer. By (3.129), (3.131), (3.132), and (A1),

$$I^h(0, T_k, x, u) = I^h(0, S_k, x, u) + I^h(S_k, T_k, x, u)$$

$$= \int_0^{S_k} L(x(t), u(t))dt + cS_k + \langle l, x(S_k) - x(0) \rangle$$

$$+ I^h(S_k, T_k, x, u)$$

$$\geq cS_k - |l|(M_0 + |x(0)|) + (T_k - S_k)(\psi(M_0) - a_0)$$

$$\geq cS_k - (T_k - S_k)|c| - |l|(M_0 + |x(0)|)$$

$$\geq cT_k - |l|(M_0 + |x(0)|).$$

This implies that

$$\liminf_{k \to \infty} T_k^{-1} I^h(0, T_k, x, u) \geq c.$$

This contradicts (3.128). The contradiction we have reached proves that

$$\mu(h) = c = h(x_*, u_*).$$

Now it is easy to see that (A2) holds.

Let us show that (A3) holds. Let $(x, u) \in X(A, B, 0, \infty)$ be (h, A, B)-good. We show that

$$\lim_{t \to \infty} x(t) = x_*.$$

Assume the contrary. Then there exist $\epsilon > 0$ and a sequence of numbers $\{t_k\}_{k=1}^{\infty}$ such that for all integers $k \geq 1$,

$$t_{k+1} - t_k \geq 2, \ |x(t_k) - x_*| \geq \epsilon. \tag{3.133}$$

For each integer $k \geq 1$ set

$$x_k(t) = x(t_k + t), \ u_k(t) = u(t_k + t), \ t \in [0, 1]. \tag{3.134}$$

Clearly, $\{(x_k, u_k)\}_{k=1}^{\infty} \subset X(A, B, 0, 1)$. (A1) and (3.134) imply that the sequence $\{I^h(0, 1, x_k, u_k)\}_{k=1}^{\infty}$ is bounded. By the lower semicontinuity of the integral functionals (see Proposition 4.2 of [52]), extracting a subsequence and re-indexing if necessary, we may assume without loss of generality that there exists $(\hat{x}, \hat{u}) \in X(A, B, 0, 1)$ such that

$$x_k(t) \to \hat{x}(t) \text{ as } k \to \infty \text{ uniformly in } [0, 1], \tag{3.135}$$

$$I^h(0, 1, \hat{x}, \hat{u}) \leq \liminf_{k \to \infty} I^h(0, 1, x_k, u_k), \tag{3.136}$$

In view of the equality $\mu(h) = c$, for each integer $k \geq 1$,

$$I^h(0, 1, x_k, u_k) - \mu(h)$$

$$= \int_0^1 L(x_k(t), u_k(t))dt + \int_0^1 \langle l, Ax_k(t) + Bu_k(t)\rangle dt$$

$$= \int_0^1 L(x_k(t), u_k(t))dt + \langle l, x_k(1) - x_k(0)\rangle, \qquad (3.137)$$

$$I^h(0, 1, \hat{x}, \hat{u}) - \mu(h)$$

$$= \int_0^1 L(\hat{x}(t), \hat{u}(t))dt + \int_0^1 \langle l, A\hat{x}(t) + B\hat{u}(t)\rangle dt$$

$$= \int_0^1 L(\hat{x}(t), \hat{u}(t))dt + \langle l, \hat{x}(1) - \hat{x}(0)\rangle. \qquad (3.138)$$

By (3.135)–(3.138),

$$\int_0^1 L(\hat{x}(t), \hat{u}(t))dt \leq \liminf_{k \to \infty} \int_0^1 L(x_k(t), u_k(t))dt. \qquad (3.139)$$

Proposition 3.5 implies that the function x is bounded. Together with (3.129) this implies that

$$\sup\left\{ \int_0^T L(x(t), u(t))dt : T > 0 \right\} < \infty.$$

Combined with (3.134) and (3.139) this implies that

$$\int_0^1 L(\hat{x}(t), \hat{u}(t))dt = 0,$$

$$x(t) = x_* \text{ for all } t \in [0, 1]$$

and

$$\lim_{k \to \infty} x(t_k) = \lim_{k \to \infty} x_k(0) = \hat{x}(0) = x_*.$$

This contradicts (3.133). The contradiction we have reached proves that (A3) holds and completes the proof of Proposition 3.6.   □

Chapter 4
Stability Properties

In this chapter we continue to study the structure of optimal trajectories of linear control systems with autonomous nonconvex integrands on large intervals. We show that the turnpike property and the convergence of solutions in regions close to the endpoints of the time intervals, which was established in Chap. 3, are stable under small perturbations of objective functions (integrands).

4.1 Preliminaries and Main Results

We use the notation, definitions, and assumptions introduced in Sects. 3.1–3.3. Recall that $a_0 > 0$ and $\psi : [0, \infty) \to [0, \infty)$ is an increasing function such that

$$\lim_{t \to \infty} \psi(t) = \infty.$$

We continue to study the structure of optimal trajectories of the controllable linear control system

$$x' = Ax + Bu,$$

where A and B are given matrices of dimensions $n \times n$ and $n \times m$, with the continuous integrand $f : R^n \times R^m \to R^1$ which satisfy assumptions (A1)–(A3) and (3.10).

Denote by $\mathfrak{M}$ the set of all borelian functions $g : R^{n+m+1} \to R^1$ which satisfy

$$g(t, x, u) \geq \max\{\psi(|x|), \ \psi(|u|),$$

$$\psi([|Ax + Bu| - a_0|x|]_+)[|Ax + Bu| - a_0|x|]_+\} - a_0 \tag{4.1}$$

for each $(t, x, u) \in R^{n+m+1}$.

© Springer International Publishing Switzerland 2015

A.J. Zaslavski, *Turnpike Theory of Continuous-Time Linear Optimal Control Problems*, Springer Optimization and Its Applications 104,

DOI 10.1007/978-3-319-19141-6_4

We equip the set $\mathfrak{M}$ with the uniformity which is determined by the following base:

$$E(N, \epsilon, \lambda) = \{(f, g) \in \mathfrak{M} \times \mathfrak{M} : |f(t, x, u) - g(t, x, u)| \leq \epsilon$$

$$\text{for each } (t, x, u) \in R^{n+m+1} \text{ satisfying } |x|, |u| \leq N\}$$

$$\cap \{(f, g) \in \mathfrak{M} \times \mathfrak{M} : (|f(t, x, u)| + 1)(|g(t, x, u)| + 1)^{-1} \in [\lambda^{-1}, \lambda]$$

$$\text{for each } (t, x, u) \in R^{n+m+1} \text{ satisfying } |x| \leq N\},$$

where $N > 0$, $\epsilon > 0$ and $\lambda > 1$.

Clearly, the uniform space $\mathfrak{M}$ is Hausdorff and has a countable base. Therefore $\mathfrak{M}$ is metrizable. It is not difficult to show that the uniform space $\mathfrak{M}$ is complete.

Denote by $\mathfrak{M}_b$ the set of all functions $g \in \mathfrak{M}$ which are bounded on bounded subsets of R^{n+m+1}. Clearly, $\mathfrak{M}_b$ is a closed subset of $\mathfrak{M}$. We consider the topological subspace $\mathfrak{M}_b \subset \mathfrak{M}$ equipped with the relative topology.

For each pair of numbers $T_1 \in R^1$, $T_2 > T_1$, each $(x, u) \in X(A, B, T_1, T_2)$ and each borelian bounded from below function $g : [T_1, T_2] \times R^n \times R^m$ set

$$I^g(T_1, T_2, x, u) = \int_{T_1}^{T_2} g(t, x(t), u(t))dt.$$

We consider the following optimal control problems

$$I^g(T_1, T_2, x, u) \to \min,$$

$$(x, u) \in X(A, B, T_1, T_2) \text{ such that } x(T_1) = y, \ x(T_2) = z,$$

$$I^g(T_1, T_2, x, u) \to \min,$$

$$(x, u) \in X(A, B, T_1, T_2) \text{ such that } x(T_1) = y,$$

$$I^g(T_1, T_2, x, u) \to \min,$$

$$(x, u) \in X(A, B, T_1, T_2),$$

where $y, z \in R^n$, $\infty > T_2 > T_1 > -\infty$ and $g \in \mathfrak{M}$.

Let $y, z \in R^n$, $T_1 \in R^1$, $T_2 > T_1$ and $g : [T_1, T_2] \times R^n \times R^m$ be a borelian bounded from below function. Set

$$\sigma(g, y, z, T_1, T_2) = \inf\{I^g(T_1, T_2, x, u) :$$

$$(x, u) \in X(A, B, T_1, T_2) \text{ and } x(T_1) = y, \ x(T_2) = z\}, \tag{4.2}$$

$$\sigma(g, y, T_1, T_2) = \inf\{I^g(T_1, T_2, x, u) :$$

$$(x, u) \in X(A, B, T_1, T_2) \text{ and } x(T_1) = y\}, \tag{4.3}$$

$$\hat{\sigma}(g, z, T_1, T_2) = \inf\{I^g(T_1, T_2, x, u) :$$

$$(x, u) \in X(A, B, T_1, T_2) \text{ and } x(T_2) = z\}, \tag{4.4}$$

$$\sigma(g, T_1, T_2) = \inf\{I^g(T_1, T_2, x, u) : (x, u) \in X(A, B, T_1, T_2)\}. \tag{4.5}$$

Recall that $f : R^n \times R^m \to R^1$ is a continuous function which satisfies (3.10) and assumptions (A1)–(A3). For each $(t, x, u) \in R^{n+m+1}$ set

$$F(t, x, u) = f(x, u). \tag{4.6}$$

In this chapter we prove the following three stability results. They show that the turnpike phenomenon, for approximate solutions on large intervals, is stable under small perturbations of the objective function (integrand) f.

Theorem 4.1. *Let $\epsilon, M > 0$. Then there exist $L_0 \geq 1$ and $\delta_0 > 0$ such that for each $L_1 \geq L_0$ there exists a neighborhood $\mathcal{U}$ of F in $\mathfrak{M}_b$ such that the following assertion holds.*

Assume that $T > 2L_1$, $g \in \mathcal{U}$, $(x, u) \in X(A, B, 0, T)$ and that a finite sequence of numbers $\{S_i\}_{i=0}^q$ satisfy

$$S_0 = 0, \ S_{i+1} - S_i \in [L_0, L_1], \ i = 0, \dots, q - 1, \ S_q \in (T - L_1, T],$$

$$I^g(S_i, S_{i+1}, x, u) \leq (S_{i+1} - S_i)\mu(f) + M$$

for each integer $i \in [0, q - 1]$,

$$I^g(S_i, S_{i+2}, x, u) \leq \sigma(g, x(S_i), x(S_{i+2}), S_i, S_{i+2}) + \delta_0$$

for each nonnegative integer $i \leq q - 2$ and

$$I^g(S_{q-2}, T, x, u) \leq \sigma(g, x(S_{q-2}), x(T), S_{q-2}, T) + \delta_0.$$

The there exist $p_1, p_2 \in [0, T]$ such that $p_1 \leq p_2$, $p_1 \leq 2L_0$, $p_2 > T - 2L_1$ and that

$$|x(t) - x_f| \leq \epsilon \text{ for all } t \in [p_1, p_2].$$

Moreover if $|x(0) - x_f| \leq \delta$, then $p_1 = 0$ and if $|x(T) - x_f| \leq \delta$, then $p_2 = T$.

Theorem 4.2. *Let $\epsilon \in (0, 1)$, $M_0, M_1 > 0$. Then there exist $L > 0$, $\delta \in (0, \epsilon)$ and a neighborhood $\mathcal{U}$ of F in $\mathfrak{M}_b$ such that for each $T > 2L$, each $g \in \mathcal{U}$ and each $(x, u) \in X(A, B, 0, T)$ which satisfies for each $S \in [0, T - L]$,*

$$I^g(S, S + L, x, u) \leq \sigma(g, x(S), x(S + L), S, S + L) + \delta$$

and satisfies at least one of the following conditions:

(a) $|x(0)|$, $|x(T)| \leq M_0$, $I^g(0, T, x, u) \leq \sigma(g, x(0), x(T), 0, T) + M_1$;

(b) $|x(0)| \leq M_0$, $I^g(0, T, x, u) \leq \sigma(g, x(0), 0, T) + M_1$;

(c) $I^g(0, T, x, u) \leq \sigma(g, 0, T) + M_1$

there exist $p_1 \in [0, L]$, $p_2 \in [T - L, T]$ such that

$$|x(t) - x_f| \leq \epsilon \text{ for all } t \in [p_1, p_2].$$

Moreover if $|x(0) - x_f| \leq \delta$, then $p_1 = 0$ and if $|x(T) - x_f| \leq \delta$, then $p_2 = T$.

Denote by Card(A) the cardinality of the set A.

Theorem 4.3. *Let $\epsilon \in (0, 1)$, $M_0, M_1 > 0$. Then there exist $l > 0$, an integer $Q \geq 1$ and a neighborhood $\mathcal{U}$ of F in $\mathfrak{M}_b$ such that for each $T > lQ$, each $g \in \mathcal{U}$ and each $(x, u) \in X(A, B, 0, T)$ which satisfies and satisfies at least one of the following conditions:*

(a) $|x(0)|$, $|x(T)| \leq M_0$, $I^g(0, T, x, u) \leq \sigma(g, x(0), x(T), 0, T) + M_1$;

(b) $|x(0)| \leq M_0$, $I^g(0, T, x, u) \leq \sigma(g, x(0), 0, T) + M_1$;

(c) $I^g(0, T, x, u) \leq \sigma(g, 0, T) + M_1$

there exist strictly increasing sequences of numbers $\{a_i\}_{i=1}^q$, $\{b_i\}_{i=1}^q \subset [0, T]$ such that $q \leq Q$, for all $i = 1, \ldots, q$,

$$0 \leq b_i - a_i \leq l,$$

$b_i \leq a_{i+1}$ for all integers i satisfying $1 \leq i < q$ and that

$$|x(t) - x_f| \leq \epsilon \text{ for all } t \in [0, T] \setminus \cup_{i=1}^q [a_i, b_i].$$

In this chapter we also prove the following three stability results. They show that the convergence of approximate solutions on large intervals, in the regions close to the end points, is stable under small perturbations of the objective function (integrand) f.

Theorem 4.4. *Let $L_0 > 0$, $\epsilon \in (0, 1)$, $M > 0$. Then there exist $\delta > 0$, a neighborhood $\mathcal{U}$ of F in $\mathfrak{M}_b$ and $L_1 > L_0$ such that for each $T \geq L_1$, each $g \in \mathcal{U}$, and each $(x, u) \in X(A, B, 0, T)$ which satisfies*

$$|x(0)| \leq M, \quad I^g(0, T, x, u) \leq \sigma(g, x(0), 0, T) + \delta$$

there exists an $(f, -A, -B)$-overtaking optimal pair

$$(x_*, u_*) \in X(-A, -B, 0, \infty)$$

such that

$$\pi^f_-(x_*(0)) = \inf(\pi^f_-),$$

$$|x(T - t) - x_*(t)| \leq \epsilon \text{ for all } t \in [0, L_0].$$

Theorem 4.5. *Let $L_0 > 0$, $\epsilon \in (0, 1)$. Then there exist $\delta > 0$, a neighborhood $\mathcal{U}$ of F in $\mathfrak{M}_b$ and $L_1 > L_0$ such that for each $T \geq L_1$, each $g \in \mathcal{U}$, and each $(x, u) \in X(A, B, 0, T)$ which satisfies*

$$I^g(0, T, x, u) \leq \sigma(g, 0, T) + \delta$$

there exist an (f, A, B)-overtaking optimal pair $(x_*, u_*) \in X(A, B, 0, \infty)$ and an $(f, -A, -B)$-overtaking optimal pair $(\bar{x}_*, \bar{u}_*) \in X(-A, -B, 0, \infty)$ such that

$$\pi^f(x_*(0)) = \inf(\pi^f),$$

$$\pi^f_-(\bar{x}_*(0)) = \inf(\pi^f_-)$$

and for all $t \in [0, L_0]$,

$$|x(t) - x_*(t)| \leq \epsilon, \ |x(T - t) - \bar{x}_*(t)| \leq \epsilon.$$

Theorem 4.6. *Let $y, z \in R^n$, $L_0 > 0$, $\epsilon \in (0, 1)$. Then there exist $\delta > 0$, a neighborhood $\mathcal{U}$ of F in $\mathfrak{M}_b$ and $L_1 > L_0$ such that for each $T \geq L_1$, each $g \in \mathcal{U}$, and each $(x, u) \in X(A, B, 0, T)$ which satisfies*

$$x(0) = y, \ x(T) = z,$$

$$I^g(0, T, x, u) \leq \sigma(g, x(0), x(T), 0, T) + \delta$$

there exist an (f, A, B)-overtaking optimal pair $(x_1, u_1) \in X(A, B, 0, \infty)$ and an $(f, -A, -B)$-overtaking optimal pair $(x_2, u_2) \in X(-A, -B, 0, \infty)$ such that

$$x_1(0) = y, \ x_2(0) = z$$

and for all $t \in [0, L_0]$,

$$|x(t) - x_1(t)| \leq \epsilon, \ |x(T - t) - x_2(t)| \leq \epsilon.$$

This chapter is organized as follows. Section 4.2 contains two auxiliary results and the proof of Theorem 4.1. A basic lemma for Theorem 4.2 is proved in Sect. 4.3

while Theorem 4.2 itself is proved in Sect. 4.4. Section 4.5 contains the proof of Theorem 4.3. Auxiliary results for Theorem 4.4 are given in Sect. 4.6. Theorems 4.4 and 4.5 are proved in Sect. 4.7. Section 4.8 contains the proof of Theorem 4.6.

4.2 Two Auxiliary Results and Proof of Theorem 4.1

In the sequel we use the following auxiliary results.

Proposition 4.7 (Proposition 2.7 of [52]). *Let $M_1 > 0$ and $0 < \tau_0 < \tau_1$. Then there exists $M_2 > 0$ such that for each $g \in \mathfrak{M}$, each $T_1 \in R^1$, each $T_2 \in [T_1 + \tau_0, T_1 + \tau_1]$ and each $(x, u) \in X(A, B, T_1, T_2)$ which satisfies*

$$I^g(T_1, T_2, x, u) \le M_1$$

the inequality $|x(t)| \le M_2$ holds for all $t \in [T_1, T_2]$.

Proposition 4.8 (Proposition 2.9 of [52]). *Let $g \in \mathfrak{M}$, $0 < c_1 < c_2$ and $D, \epsilon > 0$. Then there exists a neighborhood $\mathcal{U}$ of g in $\mathfrak{M}$ such that for each $h \in \mathcal{U}$, each $T_1 \in R^1$, each $T_2 \in [T_1 + c_1, T_1 + c_2]$, and each trajectory-control pair $(x, u) \in X(A, B, T_1, T_2)$ which satisfies*

$$\min\{I^g(T_1, T_2, x, u), \ I^h(T_1, T_2, x, u)\} \le D$$

the inequality

$$|I^g(T_1, T_2, x, u) - I^h(T_1, T_2, x, u)| \le \epsilon$$

holds.

Proof of Theorem 4.1. For every $z \in R^1$ set

$$\lfloor z \rfloor = \max\{j : j \text{ is an integer and } j \le z\}.$$

By Proposition 3.37, there exists $\delta_0 \in (0, 1/8)$ such that the following property holds:

(P1)   for each $T \ge 1$, each $(x, u) \in X(A, B, 0, T)$ satisfying

$$|x(0) - x_f)|, \ |x(T) - x_f| \le 4\delta_0,$$

$$I^f(0, T, x, u) \le \sigma(f, x(0), x(T), T) + 4\delta_0$$

we have

$$|x(t) - x_f| \le \epsilon \text{ for all } t \in [0, T].$$

By Proposition 3.35, there exists $L_0 \geq 5$ such that the following property holds:

(P2)   for each $T \geq (L_0 - 4)$, each $(x, u) \in X(A, B, 0, T)$ satisfying

$$I^f(0, T, x, u) \leq T\mu(f) + M + 4$$

and each number S satisfying $[S, S + L_0 - 4] \subset [0, T]$ we have

$$\max\{|x(t) - x_f| : t \in [S, S + L_0 - 4]\} \leq \delta_0.$$

Let

$$L_1 \geq L_0. \tag{4.7}$$

In view of Proposition 3.33, there exists $M_0 > 0$ such that for each $\tau \geq 1$ and each pair of points $y, z \in R^n$ satisfying $|y|, |z| \leq |x_f| + 4$ we have

$$\sigma(f, y, z, \tau) \leq \tau\mu(f) + M_0. \tag{4.8}$$

By Proposition 4.8, there exists a neighborhood $\mathcal{U}$ of F in $\mathfrak{M}$ such that the following property holds:

(P3)   for each $g \in \mathcal{U}$, each $T_1 \in R^1$, each $T_2 \in [T_1 + 1, T_1 + 4L_1]$ and each trajectory-control pair $(x, u) \in X(A, B, T_1, T_2)$ satisfying

$$\min\{I^f(T_1, T_2, x, u), I^g(T_1, T_2, x, u)\} \leq 4L_1|\mu(f)| + M + M_0 + 1$$

we have

$$|I^f(T_1, T_2, x, u) - I^g(T_1, T_2, x, u)| \leq \delta_0.$$

Assume that

$$T > 2L_1, \quad g \in \mathcal{U}, \quad (x, u) \in X(A, B, 0, T), \tag{4.9}$$

$$\{S_i\}_{i=0}^q \subset [0, T], \quad S_0 = 0, \tag{4.10}$$

$$S_{i+1} - S_i \in [L_0, L_1], \quad i = 0, \ldots, q - 1, \quad S_q \in (T - L_1, T], \tag{4.11}$$

$$I^g(S_i, S_{i+1}, x, u) \leq (S_{i+1} - S_i)\mu(f) + M \tag{4.12}$$

for each integer $i \in [0, q - 1]$,

$$I^g(S_i, S_{i+2}, x, u) \leq \sigma(g, x(S_i), x(S_{i+2}), S_i, S_{i+2}) + \delta_0 \tag{4.13}$$

for each nonnegative integer $i \leq q - 2$ and

$$I^g(S_{q-2}, T, x, u) \leq \sigma(g, x(S_{q-2}), x(T), S_{q-2}, T) + \delta_0. \tag{4.14}$$

Let $i \in [0, q-1]$ be an integer. It follows from (4.11), (4.12) and the choice of $\mathcal{U}$ (see property (P3)) that

$$I^f(S_i, S_{i+1}, x, u) \leq I^g(S_i, S_{i+1}, x, u) + \delta_0 \leq (S_{i+1} - S_i)\mu(f) + M + 1.$$

By the inequality above, (4.11) and property (P2), there exists a number τ_i such that

$$\tau_i \in [S_i + 3, S_i + L_0], \; |x(\tau_i) - x_f| \leq \delta_0. \tag{4.15}$$

Let a nonnegative integer $i \leq q - 2$. Relations (4.11) and (4.15) imply that

$$\tau_i, \; \tau_{i+1} \in [S_i + 3, S_{i+2}], \; 3 \leq \tau_{i+1} - \tau_i \leq 2L_1. \tag{4.16}$$

In view of (4.13) and (4.16),

$$I^g(\tau_i, \tau_{i+1}, x, u) \leq \sigma(g, x(\tau_i), x(\tau_{i+1}), \tau_i, \tau_{i+1}) + \delta_0. \tag{4.17}$$

Thus we have shown that there exists a strictly increasing sequence of numbers $\{\tau_i\}_{i=0}^{k}$ where k is a natural number such that

$$\tau_0 \leq L_0, \; \tau_k > T - 2L_1, \tag{4.18}$$

$$|x(\tau_i) - x_f| \leq \delta_0, \; i = 0, \ldots, k, \tag{4.19}$$

$$3 \leq \tau_{i+1} - \tau_i \leq 2L_1, \; i = 0, \ldots, k-1 \tag{4.20}$$

and (4.17) holds for all $i = 0, \ldots, k-1$.

Clearly, if $|x(0) - x_f| \leq \delta_0$, then we may assume that $\tau_0 = 0$ and if $|x(T) - x_f| \leq \delta_0$, then we may assume that $\tau_k = T$.

Let $i \in \{0, \ldots, k-1\}$. By (4.19), (4.20), and the choice of M_0 (see (4.8)),

$$\sigma(f, x(\tau_i), x(\tau_{i+1}), \tau_{i+1} - \tau_i) \leq \mu(f)(\tau_{i+1} - \tau_i) + M_0. \tag{4.21}$$

By the relation above, property (P3), (4.9), and (4.20),

$$|\sigma(f, x(\tau_i), x(\tau_{i+1}), \tau_{i+1} - \tau_i) - \sigma(g, x(\tau_i), x(\tau_{i+1}), \tau_i, \tau_{i+1})| \leq \delta_0. \tag{4.22}$$

It follows from (4.17), (4.21), and (4.22) that

$$I^g(\tau_i, \tau_{i+1}, x, u) \leq \sigma(f, x(\tau_i), x(\tau_{i+1}), \tau_{i+1} - \tau_i) + 2\delta_0$$
$$\leq \mu(f)(\tau_{i+1} - \tau_i) + M_0 + 1.$$

By the relation above, (P3), (4.9), and (4.10),

$$I^f(\tau_i, \tau_{i+1}, x, u) \leq I^g(\tau_i, \tau_{i+1}, x, u) + \delta_0$$
$$\leq \sigma(f, x(\tau_i), x(\tau_{i+1}), \tau_{i+1} - \tau_i) + 3\delta_0.$$

It follows from the relation above, property (P1), (4.10), and (4.19) that for all $i = 0, \ldots, k - 1$,

$$|x(t) - x_f| \leq \epsilon, \ t \in [\tau_i, \tau_{i+1}].$$

This completes the proof of Theorem 4.1.   □

4.3   Basic Lemma for Theorem 4.2

Lemma 4.9. *Let $\epsilon \in (0, 1)$, $M_0, M_1 > 0$. Then there exist $L > 0$ and a neighborhood $\mathcal{U}$ of F in $\mathfrak{M}_b$ such that the following assertion holds.*
   Assume that $T > L$, $g \in \mathcal{U}$,

$$0 \leq S_1 \leq S_2 - L, \ [S_1, S_2] \subset [0, T] \tag{4.23}$$

and $(x, u) \in X(A, B, 0, T)$ satisfies at least one of the following conditions:

(a) $|x(0)|, \ |x(T)| \leq M_0$, $I^g(0, T, x, u) \leq \sigma(g, x(0), x(T), 0, T) + M_1$;
(b) $|x(0)| \leq M_0$, $I^g(0, T, x, u) \leq \sigma(g, x(0), 0, T) + M_1$;
(c) $I^g(0, T, x, u) \leq \sigma(g, 0, T) + M_1$.

Then

$$\min\{|x(t) - x_f| : t \in [S_1, S_2]\} \leq \epsilon. \tag{4.24}$$

Proof. By Proposition 3.35 there exists an integer $L_0 > 0$ such that the following property holds:

(P4)   for each $T \geq L_0$, each trajectory-control pair $(x, u) \in X(A, B, 0, T)$ satisfying

$$I^f(0, T, x, u) \leq T\mu(f) + 16(1 + a_0)$$

and each number S satisfying $[S, S + L_0] \subset [0, T]$ we have

$$\min\{|x(t) - x_f| : \ t \in [S, S + L_0]\} \leq \epsilon/2.$$

We may assume without loss of generality that

$$M_0 > |x_f| + 4 \tag{4.25}$$

and that

$$M_1 > \sup\{|\sigma(f, z_1, z_2, 1)| : z_1, z_2 \in R^n, |z_1|, |z_2| \le M_0\} \qquad (4.26)$$

(see Proposition 3.28). Proposition 3.33 implies that there exists $M_2 > M_1 + M_0$ such that for each $S \ge 1$ and each pair of points $y, z \in R^n$ satisfying $|y|, |z| \le M_0$,

$$\sigma(f, y, z, S) \le S\mu(f) + M_2. \qquad (4.27)$$

Fix an integer $l \ge 1$ such that

$$l > 28 + M_1 + |\mu(f)|(2L_0 + M_2 + 18)$$
$$+ 4(2L_0 + 18)(1 + a_0) + 4M_2 + a_0(L_0 + 8) \qquad (4.28)$$

and set

$$L = (L_0 + 1)l. \qquad (4.29)$$

By Proposition 4.8, there exists a neighborhood $\mathcal{U}$ of F in $\mathfrak{M}_b$ such that the following property holds:

(P5)   for each $g \in \mathcal{U}$, each $T_1 \ge 0$, each $T_2 \in [T_1 + 1, T_1 + 4L]$ and each trajectory-control pair $(x, u) \in X(A, B, T_1, T_2)$ which satisfies

$$\min\{I^f(T_1, T_2, x, u), I^g(T_1, T_2, x, u)\}$$
$$\le (4L + 2)|\mu(f)| + 4M_2 + 4 + 16(1 + a_0)$$

we have

$$|I^f(T_1, T_2, x, u) - I^g(T_1, T_2, x, u)| \le (8L)^{-1}.$$

Assume that

$$T > L, \ g \in \mathcal{U}, \qquad (4.30)$$

numbers S_1, S_2 satisfy (4.23) and a trajectory-control pair

$$(x, u) \in X(A, B, 0, T)$$

satisfies at least one of the conditions (a), (b), (c). We claim that (4.24) holds. Assume the contrary. Then

$$|x(t) - x_f| > \epsilon \text{ for all } t \in [S_1, S_2]. \qquad (4.31)$$

We may assume without loss of generality that at least one of the following conditions hold:

$$S_1 \leq 1, \ S_2 \geq T - 1; \tag{4.32}$$

$$S_1 > 1, \ \text{there is } \hat{S}_1 \in [S_1 - 1, S_1] \text{ such that } |x(\hat{S}_1) - x_f| = \epsilon, \ S_2 \geq T - 1; \tag{4.33}$$

$$S_1 \leq 1, \ S_2 < T - 1, \ \text{there is } \hat{S}_2 \in [S_2, S_2 + 1] \text{ such that } |x(\hat{S}_2) - x_f| = \epsilon, \tag{4.34}$$

$$S_1 > 1, \ S_2 < T - 1, \ \text{there are } \hat{S}_1 \in [S_1 - 1, S_1], \hat{S}_2 \in [S_2, S_2 + 1]$$

$$\text{such that } |x(\hat{S}_i) - x_f| = \epsilon, \ i = 1, 2. \tag{4.35}$$

In view of (4.1), (4.23), and (4.29),

$$
\begin{aligned}
I^g(S_1, S_2, x, u) &= I^g(S_1, S_1 + L_0 \lfloor (S_2 - S_1) L_0^{-1} \rfloor, x, u) \\
&\quad + I^g(S_1 + L_0 \lfloor (S_2 - S_1) L_0^{-1} \rfloor, S_2, x, u) \\
&\geq I^g(S_1, S_1 + L_0 \lfloor (S_2 - S_1) L_0^{-1} \rfloor, x, u) - L_0 a_0 \\
&= \sum_{i=0}^{\lfloor (S_2 - S_1) L_0^{-1} \rfloor - 1} I^g(S_1 + iL_0, S_1 + (i+1)L_0, x, u) - a_0 L_0.
\end{aligned} \tag{4.36}
$$

Let

$$j \in \{0, \ldots, \lfloor (S_2 - S_1) L_0^{-1} \rfloor - 1\}. \tag{4.37}$$

It follows from (4.37), (4.31), and property (P4) that

$$I^f(S_1 + jL_0, S_1 + (j+1)L_0, x, u) > L_0 \mu(f) + 16(1 + a_0). \tag{4.38}$$

We claim that

$$I^g(S_1 + jL_0, S_1 + (j+1)L_0, x, u) \geq L_0 \mu(f) + 16(1 + a_0) - 1. \tag{4.39}$$

Assume the contrary. Then

$$I^g(S_1 + jL_0, S_1 + (j+1)L_0, x, u) < L_0 \mu(f) + 16(1 + a_0) - 1.$$

Combined with (P5), (4.30), and (4.39) this implies that

$$I^f(S_1 + jL_0, S_1 + (j+1)L_0, x, u)$$
$$\leq 1 + I^g(S_1 + jL_0, S_1 + (j+1)L_0, x, u)$$
$$< L_0\mu(f) + 16(1 + a_0).$$

This contradicts (4.38). The contradiction we have reached proves (4.39). Thus (4.39) holds for all $j \in \{0, \ldots, \lfloor(S_2 - S_1)L_0^{-1}\rfloor - 1\}$. Put

$$z_0 = x(0) \text{ if } |x(0)| \leq M_0, \ z_0 = 0 \text{ if } |x(0)| > M_0, ;$$
$$z_1 = x(T) \text{ if } |x(T)| \leq M_0, \ z_1 = 0 \text{ if } |x(T)| > M_0. \tag{4.40}$$

It is not difficult to see that there exists $(x_1, u_1) \in X(A, B, 0, T)$ such that:
  if (4.32) holds, then

$$x_1(0) = z_0, \ x_1(t) = x_f, \ u_1(t) = u_f, \ t \in [1, T-1], \ x_1(T) = z_1,$$
$$I^f(0, 1, x_1, u_1) \leq \sigma(f, z_0, x_f, 1) + 1, \ I^f(T-1, T, x_1, u_1) \leq \sigma(f, x_f, z_1, 1) + 1;$$

if (4.33) holds, then

$$x_1(t) = x(t), \ u_1(t) = u(t), \ t \in [0, \hat{S}_1],$$
$$x_1(t) = x_f, \ u_1(t) = u_f, \ t \in [\hat{S}_1 + 1, T-1],$$
$$I^f(\hat{S}_1, \hat{S}_1 + 1, x_1, u_1) \leq \sigma(f, x(\hat{S}_1), x_f, 1) + 1,$$
$$x_1(T) = z_1, \ I^f(T-1, T, x_1, u_1) \leq \sigma(f, x_f, z_1, 1) + 1;$$

if (4.34) holds, then

$$x_1(0) = z_0, \ x_1(t) = x_f, \ u_1(t) = u_f, \ t \in [1, \hat{S}_2 - 1],$$
$$x_1(t) = x(t), \ u_1(t) = u(t), \ t \in [\hat{S}_2, T],$$
$$I^f(0, 1, x_1, u_1) \leq \sigma(f, z_0, x_f, 1) + 1,$$
$$I^f(\hat{S}_2 - 1, \hat{S}_2, x_1, u_1) \leq \sigma(f, x_f, x(\hat{S}_2), 1) + 1;$$

if (4.45) holds, then

$$x_1(t) = x(t), \ u_1(t) = u(t), \ t \in [0, \hat{S}_1] \cup [\hat{S}_2, T],$$
$$x_1(t) = x_f, \ u_1(t) = u_f, \ t \in [\hat{S}_1 + 1, \hat{S}_2 - 1],$$
$$I^f(\hat{S}_1, \hat{S}_1 + 1, x_1, u_1) \leq \sigma(f, x(\hat{S}_1), x_f, 1) + 1,$$
$$I^f(\hat{S}_2 - 1, \hat{S}_2, x_1, u_1) \leq \sigma(f, x_f, x(\hat{S}_2), 1) + 1.$$

It follows from conditions (a)–(c) and the choice of (x_1, u_1) that

$$I^g(0, T, x, u) \leq I^g(0, T, x_1, u_1) + M. \tag{4.41}$$

We consider the cases (4.32)–(4.35) separately and obtain a lower bound for $I^g(0, T, x, u) - I^g(0, T, x_1, u_1)$.

Assume that (4.32) holds. In view of (4.1), (4.32), and (4.39),

$$I^g(0, T, x, u) \geq I^g(S_1, S_2, x, u) - 2a_0$$

$$\geq -2a_0 - a_0 L_0 + \sum_{i=0}^{\lfloor (S_2 - S_1)L_0^{-1} \rfloor - 1} I^g(S_1 + iL_0, S_1 + (i+1)L_0, x, u)$$

$$\geq -a_0(2 + L_0) + \lfloor (S_2 - S_1)L_0^{-1} \rfloor (L_0 \mu(f) + 16(1 + a_0) - 1)$$

$$\geq (S_2 - S_1)\mu(f) - L_0|\mu(f)| - a_0(2 + L_0)$$

$$+ ((S_2 - S_1)L_0^{-1} - 1)(16(a_0 + 1) - 1). \tag{4.42}$$

By (4.25), (4.32), (4.40), and the choice of M_2 (see (4.27)),

$$I^f(0, 1, x_1, u_1) \leq \mu(f) + M_2, \ I^f(T - 1, T, x_1, u_1) \leq \mu(f) + M_2 + 1.$$

It follows from these inequalities, (4.30), and property (P5) that

$$I^g(0, 1, x_1, u_1), \ I^g(T - 1, T, x_1, u_1) \leq \mu(f) + M_2 + 5/4. \tag{4.43}$$

In view of (4.1),

$$I^g(1, T - 1, x_1, u_1) \leq \int_1^{\lfloor T \rfloor} g(t, x_f, u_f)dt + a_0. \tag{4.44}$$

By (4.30) and property (P5), for each $i \in \{1, \ldots, \lfloor T - 1 \rfloor\}$,

$$\int_i^{i+1} g(t, x_f, u_f)dt \leq \mu(f) + (8L)^{-1}.$$

Together with (4.1) and (4.44) this implies that

$$I^g(1, T - 1, x_1, u_1) \leq a_0 + \lfloor T - 1 \rfloor (\mu(f) + (8L)^{-1}). \tag{4.45}$$

By (4.29), (4.30), (4.32), (4.41)–(4.43), and (4.45),

$$M_1 \geq I^g(0, T, x, u) - I^g(0, T, x_1, u_1)$$

$$\geq T\mu(f) - 2|\mu(f)| - L_0|\mu(f)| - a_0(2 + L_0)$$

$$+(16(a_0 + 1) - 1)(TL_0^{-1} - 4) - 2\mu(f) - 2M_2 - 5/2 - a_0$$
$$-\lfloor T - 1 \rfloor \mu(f) - \lfloor T - 1 \rfloor (8L)^{-1}$$
$$\geq T(L_0^{-1}(16(a_0 + 1) - 1) - (8L)^{-1}) - 64(a_0 + 1)$$
$$-6|\mu(f)| - L_0|\mu(f)| - a_0(2 + L_0) - 2M_2 - 3 - a_0$$
$$\geq l - 1 - 64(a_0 + 1) - |\mu(f)|(L_0 + 6) - a_0(3 + L_0) - 2M_2 - 3.$$

This contradicts (4.28). Thus if (4.32) holds we have reached a contradiction.
    Assume that (4.33) holds. In view of (4.33) and the choice of (x_1, u_1),

$$I^g(0, T, x, u) - I^g(0, T, x_1, u_1) = I^g(\hat{S}_1, T, x, u) - I^g(\hat{S}_1, T, x_1, u_1). \qquad (4.46)$$

By (4.1), (4.33), and (4.39),

$$I^g(\hat{S}_1, T, x, u) \geq I^g(S_1, T, x, u) - a_0$$
$$\geq \sum_{j=0}^{\lfloor (S_2 - S_1)L_0^{-1} \rfloor - 1} I^g(S_1 + jL_0, S_1 + (j+1)L_0, x, u) - a_0(L_0 + 2)$$
$$\geq -a_0(L_0 + 2) + \lfloor (S_2 - S_1)L_0^{-1} \rfloor (L_0\mu(f) + 16(1 + a_0) - 1)$$
$$\geq -a_0(L_0 + 2) - (S_2 - S_1)\mu(f) - |\mu(f)|(1 + L_0)$$
$$+ (\lfloor (S_2 - S_1)L_0^{-1} \rfloor - 1)(16a_0 + 15). \qquad (4.47)$$

By (4.25), (4.33), (4.40), and the choice of M_2 (see (4.27)) and (x_1, u_1),

$$I^f(\hat{S}_1, \hat{S}_1 + 1, x_1, u_1) \leq \mu(f) + M_2 + 1, \ I^f(T - 1, T, x_1, u_1) \leq \mu(f) + M_2 + 1.$$

In view of these inequalities, (4.30) and property (P5),

$$I^g(\hat{S}_1, \hat{S}_1 + 1, x_1, u_1), \ I^g(T - 1, T, x_1, u_1) \leq \mu(f) + M_2 + 3/2. \qquad (4.48)$$

It follows from (4.1) that

$$I^g(\hat{S}_1 + 1, T - 1, x_1, u_1) \leq I^g(\hat{S}_1 + 1, \lfloor T - \hat{S}_1 - 1 \rfloor + \hat{S}_1 + 1, x_1, u_1) + a_0$$
$$= a_0 + \sum_{i=0}^{\lfloor T - \hat{S}_1 - 2 \rfloor} I^g(\hat{S}_1 + i + 1, \hat{S}_1 + i + 2, x_1, u_1).$$

$$(4.49)$$

By (4.30), (4.33), the choice of (x_1, u_1) and property (P5), for each $i \in \{0, \ldots, \lfloor T - \hat{S}_1 - 3 \rfloor\}$,

$$I^g(\hat{S}_1 + i + 1, \hat{S}_1 + i + 2, x_1, u_1)$$
$$\leq I^f(\hat{S}_1 + i + 1, \hat{S}_1 + i + 2, x_1, u_1) + (8L)^{-1} = \mu(f) + (8L)^{-1}, \quad (4.50)$$

$$I^g(T - 2, T - 1, x_1, u_1) \leq \mu(f) + (8L)^{-1}. \quad (4.51)$$

It follows from (4.1), (4.33), (4.48)–(4.51), and the choice of (x_1, u_1) that

$$I^g(\hat{S}_1 + 1, T - 1, x_1, u_1)$$
$$\leq a_0 + \lfloor T - \hat{S}_1 - 2 \rfloor (\mu(f) + (8L)^{-1})$$
$$+ I^g(\hat{S}_1 + 1 + \lfloor T - \hat{S}_1 - 2 \rfloor, S_1 + 2 + \lfloor T - \hat{S}_1 - 2 \rfloor, x_1, u_1)$$
$$\leq a_0 + \lfloor T - \hat{S}_1 - 2 \rfloor (\mu(f) + (8L)^{-1}) + I^g(T - 2, T, x_1, u_1) + 2a_0$$
$$\leq a_0 + \lfloor T - \hat{S}_1 - 2 \rfloor (\mu(f) + (8L)^{-1}) + M_2 + \mu(f) + 3/2$$
$$+ I^g(T - 2, T - 1, x_1, u_1) + 2a_0$$
$$\leq 3a_0 + M_2 + \mu(f) + 3/2 + (\mu(f) + (8L)^{-1}) \lfloor T - \hat{S}_1 - 1 \rfloor.$$

By the relation above, (4.23), (4.29), (4.41), and (4.46)–(4.48),

$$M_1 \geq I^g(0, T, x, u) - I^g(0, T, x_1, u_1)$$
$$= I^g(\hat{S}_1, T, x, u) - I^g(\hat{S}_1, T, x_1, u_1)$$
$$\geq -a_0(2 + L_0) + (S_2 - S_1)\mu(f) - |\mu(f)|(1 + L_0)$$
$$+ ((S_2 - S_1)L_0^{-1} - 2)(16a_0 + 15) - 2M_2 - 2|\mu(f)| - 3 - 3a_0$$
$$- (T - \hat{S}_1 - 2)(\mu(f) + (8L)^{-1}) - 2|\mu(f)| - 3 - M_2$$
$$\geq -a_0(L_0 + 5) - |\mu(f)|(11 + L_0) - 3M_2 - 6$$
$$+ (S_2 - S_1)(L_0^{-1}(16a_0 + 15) - (8L)^{-1}) - 2(16a_0 + 15) - 1$$
$$\geq 1 - a_0(L_0 + 5) - |\mu(f)|(11 + L_0) - 3M_2 - 8 - 2(16a_0 + 15).$$

This contradicts (4.28). Thus if (4.33) holds we have reached a contradiction. Assume that (4.34) holds. In view of (4.34) and the choice of (x_1, u_1),

$$I^g(0, T, x, u) - I^g(0, T, x_1, u_1)$$
$$= I^g(0, \hat{S}_2, x, u) - I^g(0, \hat{S}_2, x_1, u_1). \quad (4.52)$$

By (4.1), (4.34), and (4.39),

$$I^g(0, \hat{S}_2, x, u) \geq I^g(S_1, S_2, x, u) - 2a_0$$

$$= \sum_{j=0}^{\lfloor (S_2-S_1)L_0^{-1} \rfloor - 1} I^g(S_1 + jL_0, S_1 + (j+1)L_0, x, u) - a_0(L_0 + 2)$$

$$\geq -a_0(L_0 + 2) + (\lfloor (S_2 - S_1)L_0^{-1} \rfloor - 1)(L_0\mu(f) + 16a_0 + 15)$$

$$\geq -a_0(L_0 + 2) - 2L_0|\mu(f)| - 32a_0 - 30$$

$$+(S_2 - S_1)\mu(f) + (S_2 - S_1)L_0^{-1}(16a_0 + 15). \tag{4.53}$$

By (4.34), (4.25), and the choice of M_2 (see (4.27)) and (x_1, u_1),

$$I^f(0, 1, x_1, u_1) \leq \mu(f) + M_2 + 1, \quad I^f(\hat{S}_2 - 1, \hat{S}_2, x_1, u_1) \leq \mu(f) + M_2 + 1.$$

In view of these inequalities, (4.30) and property (P5),

$$I^g(0, 1, x_1, u_1), \ I^g(\hat{S}_2 - 1, \hat{S}_2, x_1, u_1) \leq \mu(f) + M_2 + 5/4. \tag{4.54}$$

It follows from (4.1), (4.34), and the choice of (x_1, u_1) that

$$I^g(1, \hat{S}_2 - 1, x_1, u_1) = \int_1^{\hat{S}_2-1} g(t, x_f, u_f)dt$$

$$\leq \int_1^{\lfloor \hat{S}_2 \rfloor} g(t, x_f, u_f)dt + a_0. \tag{4.55}$$

By (4.30) and property (P5), for each $j \in \{1, \ldots, \lfloor \hat{S}_2 - 1 \rfloor\}$,

$$\int_j^{j+1} g(t, x_f, u_f)dt \leq \mu(f) + (8L)^{-1}.$$

Together with (4.55) this implies that

$$I^g(1, \hat{S}_2 - 1, x_1, u_1) \leq a_0 + \lfloor \hat{S}_2 - 1 \rfloor(\mu(f) + (8L)^{-1}).$$

By the relation above, (4.23), (4.29), (4.34), (4.41), (4.52)–(4.54), and the choice of (x_1, u_1),

$$M_1 \geq I^g(0, T, x, u) - I^g(0, T, x_1, u_1)$$

$$= I^g(0, \hat{S}_2, x, u) - I^g(0, \hat{S}_2, x_1, u_1)$$

$$\geq -a_0(2 + L_0) + 32a_0 - 30 - 2L_0|\mu(f)| + (S_2 - S_1)\mu(f)$$

$$+(S_2 - S_1)L_0^{-1}(16a_0 + 15) - 2M_2 - 2|\mu(f)| - 3 - a_0$$
$$-\lfloor \hat{S}_2 - 1 \rfloor (\mu(f) + (8L)^{-1})$$
$$\geq -2|\mu(f)| + (S_2 - S_1)(L_0^{-1}(16a_0 + 15) - (8L)^{-1}) - 5$$
$$-a_0(L_0 + 3) - 32a_0 - 30 - (2 + 2L_0)|\mu(f)| - 2M_2$$
$$\geq l - 6 - |\mu(f)|(4 + 2L_0) - a_0(L_0 + 3) - 32a_0 - 30 - 2M_2.$$

This contradicts (4.28). Thus if (4.34) holds we have reached a contradiction. Assume that (4.35) holds. By (4.35) and the choice of (x_1, u_1),

$$I^g(0, T, x, u) - I^g(0, T, x_1, u_1)$$
$$= I^g(\hat{S}_1, \hat{S}_2, x, u) - I^g(\hat{S}_1, \hat{S}_2, x_1, u_1). \tag{4.56}$$

It follows from (4.1), (4.35), (4.39), and the choice of (x_1, u_1) that

$$I^g(\hat{S}_1, \hat{S}_2, x, u) \geq I^g(S_1, S_2, x, u) - 2a_0$$

$$\geq \sum_{i=0}^{\lfloor (S_2 - S_1)L_0^{-1} \rfloor - 1} I^g(S_1 + iL_0, S_1 + (i + 1)L_0, x, u) - a_0(L_0 + 2)$$

$$\geq -a_0(L_0 + 2) + \lfloor (S_2 - S_1)L_0^{-1} \rfloor (L_0\mu(f) + 16a_0 + 15). \tag{4.57}$$

By (4.35), (4.25), and the choice of M_2 (see (4.27)) and (x_1, u_1),

$$I^f(\hat{S}_1, \hat{S}_1 + 1, x_1, u_1) \leq \mu(f) + M_2 + 1, \quad I^f(\hat{S}_2 - 1, \hat{S}_2, x_1, u_1) \leq \mu(f) + M_2 + 1.$$

In view of these inequalities, (4.30) and property (P5),

$$I^g(\hat{S}_1, \hat{S}_1 + 1, x_1, u_1), \quad I^g(\hat{S}_2 - 1, \hat{S}_2, x_1, u_1) \leq \mu(f) + M_2 + 5/4. \tag{4.58}$$

It follows from (4.1) that

$$I^g(\hat{S}_1 + 1, \hat{S}_2 - 1, x_1, u_1) = \int_{\hat{S}_1+1}^{\hat{S}_2-1} g(t, x_f, u_f)dt$$

$$\leq \int_{S_1}^{S_2} g(t, x_f, u_f)dt + 2a_0$$

$$\leq \int_{S_1}^{S_1+\lfloor S_2 - S_1 \rfloor + 1} g(t, x_f, u_f)dt + 3a_0. \tag{4.59}$$

By (4.30) and property (P5), for each $i \in \{0, \dots, \lfloor S_2 - S_1 \rfloor\}$,

$$\int_i^{i+1} g(t, x_f, u_f)dt \le \mu(f) + (8L)^{-1}.$$

Together with (4.58) and (4.59) this implies that

$$I^g(\hat{S}_1, \hat{S}_2, x_1, u_1) \le 2\mu(f) + 2M_2 + 3 + 3a_0 + \lfloor S_2 - S_1 + 1 \rfloor (\mu(f) + (8L)^{-1}).$$

By the relation above, (4.23), (4.29), (4.41), (4.56), and (4.57),

$$
\begin{aligned}
M_1 &\ge I^g(\hat{S}_1, \hat{S}_2, x, u) - I^g(\hat{S}_1, \hat{S}_2, x_1, u_1) \\
&\ge (S_2 - S_1)\mu(f) - L_0|\mu(f)| + (S_2 - S_1)L_0^{-1}(16a_0 + 15) \\
&\quad -16a_0 - 15 - a_0(2 + L_0) - \lfloor S_2 - S_1 \rfloor (\mu(f) + (8L)^{-1}) \\
&\quad -4|\mu(f)| - 2M_2 - 4 - 3a_0 \\
&\ge l - (5 + L_0)|\mu(f)| - 19a_0 - 19 - a_0(L_0 + 2) - 2M_2.
\end{aligned}
$$

This contradicts (4.28). Thus if (4.35) holds we have reached a contradiction. Thus in all the cases we have reached a contradiction which proves (4.24) and Lemma 4.9 itself.

4.4   Proof of Theorem 4.2

By Proposition 3.27, there exists $\delta_0 \in (0, \epsilon)$ such that the following property holds:

(P6)   for each $\tau \ge 1$ and each trajectory-control pair $(x, u) \in X(A, B, 0, \tau)$ satisfying

$$|x(0) - x_f|, \; |x(\tau) - x_f| \le 4\delta_0,$$
$$I^f(0, \tau, x, u) \le \sigma(f, x(0), x(\tau), \tau) + 4\delta_0$$

we have

$$|x(t) - x_f)| \le \epsilon, \; t \in [0, \tau].$$

By Lemma 4.9, there exist $L_0 > 0$ and a neighborhood $\mathcal{U}_0$ of F in $\mathfrak{M}_b$ such that the following property holds:

(P7)   for each $T > L_0$, each $g \in \mathcal{U}_0$, each pair of numbers S_1, S_2 satisfying $0 \le S_1 \le S_2 - L_0$, $S_2 \le T$ and each trajectory-control pair $(x, u) \in X(A, B, 0, T)$ for which at least one of the conditions (a), (b), (c) of Theorem 4.2 holds, we have

$$\min\{|x(t) - x_f| : t \in [S_1, S_2]\} \le \delta_0.$$

Fix

$$L \geq 4(L_0 + 1), \ \delta \in (0, 4^{-1}\delta_0). \tag{4.60}$$

In view of assumption (A1) and Proposition 3.33, there exists a number $M_2 > 0$ such that the following property holds:

(P8)   for each $S \in [1, L]$ and each pair of points $y, z \in R^n$ satisfying $|y|, |z| < 4 + |x_f|$ we have $|\sigma(f, y, z, S)| < M_2$.

By Proposition 4.8, there exists a neighborhood $\mathcal{U}$ of F in $\mathfrak{M}_b$ such that $\mathcal{U} \subset \mathcal{U}_0$ and that the following property holds:

(P9)   for each $g \in \mathcal{U}$, each $T_1 \in R^1$, each $T_2 \in [T_1 + 1, T_1 + 4L]$ and each trajectory-control pair $(x, u) \in X(A, B, T_1, T_2)$ which satisfies

$$\min\{I^f(T_1, T_2, x, u), I^g(T_1, T_2, x, u)\} \leq M_2 + 4$$

we have

$$|I^f(T_1, T_2, x, u) - I^g(T_1, T_2, x, u)| \leq \delta.$$

Assume that

$$T > 2L, \ g \in \mathcal{U}, \ (x, u) \in X(A, B, 0, T), \tag{4.61}$$

for each $S \in [0, T - L]$,

$$I^g(S, S + L, x, u) \leq \sigma(g, x(S), x(S + L), S, S + L) + \delta \tag{4.62}$$

and at least one of the following conditions below holds:

$$\text{(a) } |x(0)|, \ |x(T)| \leq M_0, \ I^g(0, T, x, u) \leq \sigma(g, x(0), x(T), 0, T) + M_1;$$
$$\text{(b) } |x(0)| \leq M_0, \ I^g(0, T, x, u) \leq \sigma(g, x(0), 0, T) + M_1;$$
$$\text{(c) } I^g(0, T, x, u) \leq \sigma(g, 0, T) + M_1.$$

By conditions (a)–(c) and property (P7), there exists a finite strictly increasing sequence of integers $S_i \in [0, T], \ i = 1, \ldots, q$ such that

$$0 \leq S_1 \leq L_0, \ S_q \geq T - (1 + L_0),$$
$$1 \leq S_{i+1} - S_i \leq L_0 + 2, \ i = 1, \ldots, q - 1,$$
$$|x(S_i) - x_f| \leq \delta_0, \ i = 1, \ldots, q. \tag{4.63}$$

We may assume without loss of generality that

$$\text{if } |x(0) - x_f| \leq \delta, \text{ then } S_1 = 0$$
$$\text{and if } |x(T) - x_f| \leq \delta, \text{ then } S_q = T.$$

Assume that a number

$$\tau \in [S_1, S_q]. \tag{4.64}$$

Then there exists a natural number $k \in \{1, \ldots, q - 1\}$ such that

$$S_k \leq \tau < S_{k+1}. \tag{4.65}$$

In view of (4.60), (4.61), and (4.63), there is $S \in [0, T - L]$ such that

$$[S_k, S_{k+1}] \subset [S, S + L].$$

Together with (4.62) this implies that

$$I^g(S_k, S_{k+1}, x, u) \leq \sigma(g, x(S_k), x(S_{k+1}), S_k, S_{k+1}) + \delta. \tag{4.66}$$

In view of (4.63) and property (P8),

$$\sigma(f, x(S_k), x(S_{k+1}), S_{k+1} - S_k) < M_2.$$

By the relation above, (4.61), (4.63), (4.66), and property (P9),

$$\sigma(g, x(S_k), x(S_{k+1}), S_k, S_{k+1})$$
$$\leq \sigma(f, x(S_k), x(S_{k+1}), S_{k+1} - S_k) + \delta \tag{4.67}$$

and

$$I^g(S_k, S_{k+1}, x, u) \leq M_2 + 2.$$

It follows from the relation above, (4.61), (4.63), (4.66), (4.67), and property (P9) that

$$I^f(S_k, S_{k+1}, x, u) \leq I^g(S_k, S_{k+1}, x, u) + \delta$$
$$\leq \sigma(f, x(S_k), x(S_{k+1}), S_{k+1} - S_k) + 3\delta.$$

Combined with (4.63), (4.65), (4.64), and property (P6) this implies that $|x(\tau) - x_f| \leq \epsilon$ for all number $\tau \in [S_1, S_q]$. Theorem 4.2 is proved.   □

4.5 Proof of Theorem 4.3

By Proposition 3.37, there exists $\delta_0 \in (0, \epsilon)$ such that the following property holds:

(P10) for each $\tau \geq 1$, each $(x, u) \in X(A, B, 0, \tau)$ satisfying

$$|x(0) - x_f|, \; |x(\tau) - x_f| \leq 4\delta_0,$$
$$I^f(0, \tau, x, u) \leq \sigma(f, x(0), x(\tau), \tau) + 4\delta_0$$

we have

$$|x(t) - x_f| \leq \epsilon, \; t \in [0, \tau].$$

By Lemma 4.9, there exist a number $L_0 > 0$ and a neighborhood $\mathcal{U}_0$ of F in $\mathfrak{M}_b$ such that the following property holds:

(P11) for each $T > L_0$, each $g \in \mathcal{U}_0$, each pair of numbers S_1, S_2 satisfying $0 \leq S_1 \leq S_2 - L_0$, $S_2 \leq T$ and each trajectory-control pair $(x, u) \in X(A, B, 0, T)$ which satisfies at least one of the conditions (a), (b), and (c) of Theorem 4.3, we have

$$\min\{|x(t) - x_f| : \; t \in [S_1, S_2]\} \leq \delta_0.$$

Fix numbers

$$l \geq 4(L_0 + 1), \; \delta \in (0, 4^{-1}\delta_0) \tag{4.68}$$

and an integer

$$Q > 2\delta_0^{-1}M_1 + 6. \tag{4.69}$$

By (A1) and Proposition 3.33, there is $M_2 > 0$ such that the following property holds:

(P12) for each $S \in [1, l]$ and each $y, z \in R^n$ satisfying $|y|, |z| \leq 4 + |x_f|$ we have

$$|\sigma(f, y, z, S)| < M_2.$$

By Proposition 4.8, there exists a neighborhood $\mathcal{U}$ of F in $\mathfrak{M}_b$ such that $\mathcal{U} \subset \mathcal{U}_0$ and that the following property holds:

(P13) for each $g \in \mathcal{U}$, each $T_1 \in R^1$, each $T_2 \in [T_1 + 1, T_1 + 4l]$ and each trajectory-control pair $(x, u) \in X(A, B, T_1, T_2)$ which satisfies

$$\min\{I^f(T_1, T_2, x, u), I^g(T_1, T_2, x, u)\} \leq M_2 + 4$$

we have

$$|I^f(T_1, T_2, x, u) - I^g(T_1, T_2, x, u)| \le \delta.$$

Assume that

$$T > lQ, \ g \in \mathcal{U}, \ (x, u) \in X(A, B, 0, T) \tag{4.70}$$

and that at least one of the conditions (a), (b), (c) of Theorem 4.3 holds.

By conditions (a)–(c), (4.70), and property (P11), there exists a finite strictly increasing sequence of numbers $S_i \in [0, T]$, $i = 1, \ldots, q$ such that

$$0 \le S_1 \le L_0, \ S_q \ge T - (1 + L_0), \ 1 \le S_{i+1} - S_i \le L_0 + 1, \ i = 1, \ldots, q - 1, \tag{4.71}$$

$$|x(S_i) - x_f| \le \delta_0, \ i = 1, \ldots, q. \tag{4.72}$$

Define by induction a finite strictly increasing sequence of natural numbers $i_1, \ldots, i_k \in \{1, \ldots, q\}$. Set

$$i_1 = 1. \tag{4.73}$$

Assume that $p \ge 1$ is an integer and that we defined integers $i_1 < \ldots < i_p$ belonging to $\{1, \ldots, q\}$ such that for each natural number $m < p$ the following properties hold:

(i)

$$I^g(S_{i_m}, S_{i_{m+1}}, x, u)$$
$$> \sigma(g, x(S_{i_m}), x(S_{i_{m+1}}), S_{i_m}, S_{i_{m+1}}) + \delta_0; \tag{4.74}$$

(ii) if $i_{m+1} > i_m + 1$, then

$$I^g(S_{i_m}, S_{i_{m+1}-1}, x, u) \le \sigma(g, x(S_{i_m}), x(S_{i_{m+1}-1}), S_{i_m}, S_{i_{m+1}-1}) + \delta_0. \tag{4.75}$$

(Note that by (4.73) our assumption holds for $p = 1$.) Let us define i_{p+1}. If $i_p = q$, then our construction is completed, $k = p$, $i_k = q$ and for each natural number $m < p = k$, properties (i) and (ii) hold.

Assume that $i_p < q$. There are two cases:

$$I^g(S_{i_p}, S_q, x, u) \le \sigma(g, x(S_{i_p}), x(S_q), S_{i_p}, S_q) + \delta_0; \tag{4.76}$$

$$I^g(S_{i_p}, S_q, x, u) > \sigma(g, x(S_{i_p}), x(S_q), S_{i_p}, S_q) + \delta_0. \tag{4.77}$$

Assume that (4.76) holds. Then we set $k = p + 1$, $i_k = q$ the construction is completed, for each natural number $m < k - 1$, (4.74) is true and for each natural number $m < k$, property (ii) holds.

Assume that (4.77) holds. Then we set

$$i_{p+1} = \min\{j > S_{i_p} : j \text{ is an integer and}$$

$$I^g(S_{i_p}, S_j, x, u) > \sigma(g, x(S_{i_p}), x(S_j), S_{i_p}, S_j) + \delta_0\}. \tag{4.78}$$

It is easy to see that the assumption made for p also holds for $p + 1$. As a result we obtain a finite strictly increasing sequence of integers $i_1, \ldots, i_k \in \{1, \ldots, q\}$ such that $i_k = q$, for all integers m satisfying $1 \leq m < k-1$, (4.74) holds and each natural number $m < k$, property (ii) holds. It follows from conditions (a)–(c) and (4.74) that

$$M_1 \geq I^g(0, T, x, u) - \sigma(g, x(0), x(T), 0, T)$$

$$\geq \sum\{I^g(S_{i_j}, S_{i_{j+1}}, x, u) - \sigma(g, x(S_{i_j}), x(S_{i_{j+1}}), S_{i_j}, S_{i_{j+1}}) :$$

$$j \text{ is an integer satisfying } 1 \leq j < k - 1\} \geq \delta_0(k - 2),$$

$$k \leq \delta_0^{-1} M_1 + 2. \tag{4.79}$$

Define

$$A = \{j \in \{1, \ldots, k\} : j < k \text{ and } S_{i_{j+1}} - S_{i_j} \geq 4(L_0 + 1)\}. \tag{4.80}$$

Let

$$j \in A. \tag{4.81}$$

Relations (4.81), (4.80), (4.71), (4.75), and property (ii) imply that

$$I^g(S_{i_j}, S_{i_{j+1}-1}, x, u)$$

$$\leq \sigma(g, x(S_{i_j}), x(S_{i_{j+1}-1}), S_{i_j}, S_{i_{j+1}-1}) + \delta_0, \tag{4.82}$$

$$i_{j+1} > i_j + 3. \tag{4.83}$$

Let

$$p \in \{i_j, \ldots, i_{j+1} - 2\}. \tag{4.84}$$

In view of (4.84),

$$\{S_p, S_{p+1}\} \subset \{S_{i_j}, \ldots, S_{i_{j+1}-1}\}$$

and by (4.82),

$$I^g(S_p, S_{p+1}, x, u) \leq \sigma(g, x(S_p), x(S_{p+1}), S_p, S_{p+1}) + \delta_0. \tag{4.85}$$

It follows from the choice of M_2 (see (P12)), (4.71), (4.68), and (4.72) that

$$\sigma(f, x(S_p), x(S_{p+1}), S_p, S_{p+1}) \leq M_2.$$

Combined with (4.68), (4.70), (4.71), and property (P13) this implies that

$$\sigma(g, x(S_p), x(S_{p+1}), S_p, S_{p+1})$$
$$\leq \sigma(f, x(S_p), x(S_{p+1}), S_{p+1} - S_p) + \delta_0 \leq M_2 + \delta_0.$$

It follows from (4.70), (4.71), (4.84), (4.85), the inequality above, and property (P13) that

$$I^f(S_p, S_{p+1}, x, u) \leq \sigma(f, x(S_p), x(S_{p+1}), S_{p+1} - S_p) + 3\delta_0.$$

By the inequality above, (4.71), (4.72), and property (P10), for all integers $p \in \{i_j, \ldots, i_{j+1} - 2\}$,

$$|x(t) - x_f| \leq \epsilon, \; t \in [S_p, S_{p+1}].$$

Since j is an any integer belonging to A we conclude that

$$\{t \in [0, T] : \; |x(t) - x_f| > \epsilon\}$$
$$\subset [0, S_1] \cup [S_q, T]$$
$$\cup \{[S_{i_j}, S_{i_{j+1}}] : j \in \{1, \ldots, k-1\}, \; S_{i_{j+1}} - S_{i_j} < 4(L_0 + 1)\}$$
$$\cup \{[S_{i_{j+1}-2}, S_{i_{j+1}}] : j \in A\}.$$

It is not difficult to see that the right-hand side of the inclusion is a finite union of closed intervals, by (4.68) and (4.71) their lengths does not exceed $4(L_0 + 1) \leq l$ and by (4.69) and (4.79) their number does not exceed

$$2 + 2k \leq 2\delta_0^{-1}M_1 + 6 < Q.$$

Theorem 4.3 is proved.   □

4.6 Auxiliary Results for Theorem 4.4

Assume that $S_1 \in R^1$, $S_2 > S_1$ and $g \in \mathfrak{M}$. For each $(x, u) \in R^n \times R^m$ and each $t \in [S_1, S_2]$ set

$$\mathcal{L}_{S_1, S_2}(g)(t, x, u) = g(S_2 - t + S_1, x, u), \tag{4.86}$$

for each $(t, x, u) \in (-\infty, S_1) \times R^n \times R^m$ set

$$\mathcal{L}_{S_1, S_2}(g)(t, x, u) = \mathcal{L}_{S_1, S_2}(g)(S_1, x, u)$$

and for each $(t, x, u) \in (S_2, \infty) \times R^n \times R^m$ set

$$\mathcal{L}_{S_1, S_2}(g)(t, x, u) = \mathcal{L}_{S_1, S_2}(g)(S_2, x, u).$$

It is clear that $\mathcal{L}_{S_1, S_2}(g) \in \mathfrak{M}$, if $g \in \mathfrak{M}_b$, then $\mathcal{L}_{S_1, S_2}(g) \in \mathfrak{M}_b$ and that $\mathcal{L}_{S_1, S_2}$ is a self-mapping of $\mathfrak{M}$ and of $\mathfrak{M}_b$.

It is not difficult to see that the following result holds.

Proposition 4.10. *Let V be a neighborhood of F in $\mathfrak{M}$. Then there exists a neighborhood $\mathcal{U}$ of F in $\mathfrak{M}$ such that $\mathcal{L}_{S_1, S_2}(g) \in V$ for each $g \in \mathcal{U}$, each $S_1 \in R^1$ and each $S_2 > S_1$.*

Let $S_1 \in R^1$, $S_2 > S_1$, $g \in \mathfrak{M}$ and

$$(x, u) \in X(A, B, S_1, S_2) \ (X(-A, -B, S_1, S_2) \text{ respectively}).$$

Recall that (see (3.16))

$$\bar{x}(t) = x(S_2 - t + S_1), \ \bar{u}(t) = u(S_2 - t + S_1), \ t \in [S_1, S_2]. \tag{4.87}$$

In view of (4.86) and (4.87),

$$\int_{S_1}^{S_2} \mathcal{L}_{S_1, S_2}(g)(t, \bar{x}(t), \bar{u}(t))dt = \int_{S_1}^{S_2} g(S_2 - t + S_1, x(S_2 - t + S_1), u(S_2 - t + S_1))dt$$

$$= \int_{S_1}^{S_2} g(t, x(t), u(t))dt. \tag{4.88}$$

Let $T_2 > T_1$ be a pair of real numbers, $y, z \in R^n$ and $g \in \mathfrak{M}_b$. For each $(x, u) \in X(-A, -B, T_1, T_2)$ set

$$I^g(T_1, T_2, x, u) = \int_{T_1}^{T_2} g(t, x(t), u(t))dt$$

and set

$$\sigma_-(g, y, z, T_1, T_2) = \inf\{I^g(T_1, T_2, x, u) :$$

$$(x, u) \in X(-A, -B, T_1, T_2) \text{ and } x(T_1) = y, \ x(T_2) = z\}, \tag{4.89}$$

$$\sigma_-(g, y, T_1, T_2) = \inf\{I^g(T_1, T_2, x, u) :$$

$$(x, u) \in X(-A, -B, T_1, T_2) \text{ and } x(T_1) = y\}, \tag{4.90}$$

$$\hat{\sigma}_-(g, z, T_1, T_2) = \inf\{I^g(T_1, T_2, x, u) :$$

$$(x, u) \in X(-A, -B, T_1, T_2) \text{ and } x(T_2) = z\}, \tag{4.91}$$

$$\sigma_-(g, T_1, T_2) = \inf\{I^g(T_1, T_2, x, u) : \ (x, u) \in X(-A, -B, T_1, T_2)\}. \tag{4.92}$$

Relations (4.88) implies the following result.

Proposition 4.11. *Let $S_2 > S_1$ be real numbers, $M \geq 0$, $g \in \mathfrak{M}_b$ and let $(x_i, u_i) \in X(A, B, S_1, S_2)$, $i = 1, 2$. Then*

$$I^g(S_1, S_2, x_1, u_1) \geq I^g(S_1, S_2, x_2, u_2) - M$$

if and only if $I^{\bar{g}}(S_1, S_2, \bar{x}_1, \bar{u}_1) \geq I^{\bar{g}}(S_1, S_2, \bar{x}_2, \bar{u}_2) - M$, where $\bar{g} = \mathcal{L}_{S_1, S_2}(g)$.

Proposition 4.11 implies the following result.

Proposition 4.12. *Let $S_2 > S_1$ be real numbers, $M \geq 0$, $g \in \mathfrak{M}_b$, $\bar{g} = \mathcal{L}_{S_1, S_2}(g)$ and*

$$(x, u) \in X(A, B, S_1, S_2).$$

Then the following assertions hold:

$$I^g(S_1, S_2, x, u) \leq \sigma(g, S_1, S_2) + M$$

if and only if $I^{\bar{g}}(S_1, S_2, \bar{x}, \bar{u}) \leq \sigma_-(\bar{g}, S_1, S_2) + M$;

$$I^g(S_1, S_2, x, u) \leq \sigma(g, x(S_1), x(S_2), S_1, S_2) + M$$

if and only if $I^{\bar{g}}(S_1, S_2, \bar{x}, \bar{u}) \leq \sigma_-(\bar{g}, \bar{x}(S_1), \bar{x}(S_2), S_1, S_2) + M$;

$$I^g(S_1, S_2, x, u) \leq \hat{\sigma}(g, x(S_2), S_1, S_2) + M$$

if and only if $I^{\bar{g}}(S_1, S_2, \bar{x}, \bar{u}) \leq \sigma_-(\bar{g}, \bar{x}(S_1), S_1, S_2) + M$;

$$I^g(S_1, S_2, x, u) \leq \sigma(g, x(S_1), S_1, S_2) + M$$

if and only if $I^{\bar{g}}(S_1, S_2, \bar{x}, \bar{u}) \leq \hat{\sigma}_-(\bar{g}, \bar{x}(S_2), S_1, S_2) + M$.

4.7 Proofs of Theorems 4.4 and 4.5

Proof of Theorem 4.4. By Lemma 3.38 applied to the triplet $(f, -A - B)$ there exist

$$\delta_1 \in (0, \epsilon/4)$$

such that the following property holds:

(P14)   for each $(x, u) \in X(-A, -B, 0, L_0)$ which satisfies

$$\pi_-^f(x(0)) \le \inf(\pi_-^f) + \delta_1,$$

$$I^f(0, L_0, x, u) - L_0\mu(f) - \pi_-^f(x(0)) + \pi_-^f(x(L_0)) \le \delta_1$$

there exists an $(f, -A, -B)$-overtaking optimal pair

$$(x_*, u_*) \in X(-A, -B, 0, \infty)$$

such that

$$\pi_-^f(x_*(0)) = \inf(\pi_-^f),$$

$$|x(t) - x_*(t)| \le \epsilon \text{ holds for all } t \in [0, L_0].$$

In view of Propositions 3.13, 3.14, and 3.24, there exists $\delta_2 \in (0, \delta_1)$ such that: for each $z \in R^n$ satisfying $|z - x_f| \le 2\delta_2$,

$$|\pi_-^f(z)| = |\pi_-^f(z) - \pi_-^f(x_f)| \le \delta_1/8; \tag{4.93}$$

for each $y, z \in R^n$ satisfying

$$|y - x_f| \le 2\delta_2, \ |z - x_f| \le 2\delta_2$$

we have

$$|\sigma(f, y, z, 1) - \mu(f)| \le \delta_1/8. \tag{4.94}$$

By Theorem 4.2, there exist $l_0 > 0$, $\delta_3 \in (0, \delta_2/8)$ and a neighborhood $\mathcal{U}_1$ of F in $\mathfrak{M}_b$ such that the following property holds:

(P15)   for each $T > 2l_0$, each $g \in \mathcal{U}_1$ and each

$$(x, u) \in X(A, B, 0, T)$$

such that

$$|x(0)| \leq M,$$

$$I^g(0, T, x, u) \leq \sigma(g, x(0), 0, T) + \delta_3$$

we have

$$|x(t) - x_f| \leq \delta_2 \text{ for all } t \in [l_0, T - l_0]. \tag{4.95}$$

By Theorem 3.8, there exists an $(f, -A, -B)$-overtaking optimal pair

$$(\bar{x}_*, \bar{u}_*) \in X(-A, -B, 0, \infty)$$

such that

$$\pi_-^f(\bar{x}_*(0)) = \inf(\pi_-^f). \tag{4.96}$$

Assumption (A3) implies that there exists $l_1 > 0$ such that

$$|\bar{x}_*(t) - x_f| \leq \delta_2 \text{ for all } t \geq l_1. \tag{4.97}$$

By Proposition 4.8, there exists a neighborhood $\mathcal{U} \subset \mathcal{U}_1$ of F in $\mathfrak{M}_b$ such that the following property holds:

(P16)   for each $g \in \mathcal{U}$, each $T_1 \in R^1$, each $T_2 \in [T_1 + 1, T_1 + 2L_0 + 2l_0 + 2l_1 + 4]$ and each trajectory-control pair $(x, u) \in X(A, B, T_1, T_2)$ which satisfies

$$\min\{I^f(T_1, T_2, x, u), I^g(T_1, T_2, x, u)\}$$

$$\leq (|\mu(f)| + 2)(2L_0 + 2l_0 + 2l_1 + 4) + 2 + |\pi_-^f(\bar{x}_*(0))|$$

the inequality $|I^f(T_1, T_2, x, u) - I^g(T_1, T_2, x, u)| \leq \delta_3/8$ holds.

Choose $\delta > 0$ and $L_1 > 0$ such that

$$\delta \leq \delta_3/4, \ L_1 \geq 2L_0 + 2l_0 + 2l_1 + 4. \tag{4.98}$$

Assume that

$$T \geq L_1, \ g \in \mathcal{U}, \ (x, u) \in X(A, B, 0, T), \tag{4.99}$$

$$|x(0)| \leq M, \ I^g(0, T, x, u) \leq \sigma(g, x(0), 0, T) + \delta. \tag{4.100}$$

It follows from property (P15) and (4.98)–(4.100) that relation (4.95) is true. By (4.98) and (4.99),

$$[T - l_0 - l_1 - L_0 - 4, T - l_0 - l_1 - L_0] \subset [l_0, T - l_0 - l_1 - L_0]. \tag{4.101}$$

Relations (4.95) and (4.101) imply that

$$|x(t) - x_f| \leq \delta_2 \text{ for all } t \in [T - l_0 - l_1 - L_0 - 4, T - l_0 - l_1 - L_0]. \tag{4.102}$$

By Proposition 3.27, there exists a trajectory-control pair

$$(x_1, u_1) \in X(A, B, 0, T)$$

such that

$$x_1(t) = x(t), \ u_1(t) = u(t), \ t \in [0, T - l_0 - l_1 - L_0 - 4],$$
$$x_1(t) = \bar{x}_*(T - t), \ u_1(t) = \bar{u}_*(T - t), \ t \in [T - l_0 - l_1 - L_0 - 3, T],$$
$$I^f(T - l_0 - l_1 - L_0 - 4, T - l_0 - l_1 - L_0 - 3, x_1, u_1)$$
$$= \sigma(f, x(T - l_0 - l_1 - L_0 - 4), \bar{x}_*(l_0 + l_1 + L_0 + 3)). \tag{4.103}$$

It follows from (4.103) and (4.100) that

$$-\delta \leq I^g(0, T, x_1, u_1) - I^g(0, T, x, u)$$
$$= I^g(T - l_0 - l_1 - L_0 - 4, T - l_0 - l_1 - L_0 - 3, x_1, u_1)$$
$$+ I^g(T - l_0 - l_1 - L_0 - 3, T, x_1, u_1)$$
$$- I^g(T - l_0 - l_1 - L_0 - 4, T - l_0 - l_1 - L_0 - 3, x, u)$$
$$- I^g(T - l_0 - l_1 - L_0 - 3, T, x, u). \tag{4.104}$$

It follows from (4.94), (4.97), (4.102), and (4.103) that

$$I^f(T - l_0 - l_1 - L_0 - 4, T - l_0 - l_1 - L_0 - 3, x_1, u_1) \leq \mu(f) + \delta_1/8.$$

Together with (4.99) and property (P16) this implies that

$$I^g(T - l_0 - l_1 - L_0 - 4, T - l_0 - l_1 - L_0 - 3, x_1, u_1) \leq \mu(f) + \delta_1/8 + \delta_3/8. \tag{4.105}$$

By (4.102) and the choice of δ_2 (see (4.94)),

$$I^f(T - l_0 - l_1 - L_0 - 4, T - l_0 - l_1 - L_0 - 3, x, u) \geq \mu(f) - \delta_1/8. \tag{4.106}$$

If

$$I^g(T - l_0 - l_1 - L_0 - 4, T - l_0 - l_1 - L_0 - 3, x, u)$$
$$< \mu(f) - \delta_1/2,$$

then by property (P16) and (4.99),

$$I^f(T - l_0 - l_1 - L_0 - 4, T - l_0 - l_1 - L_0 - 3, x, u)$$
$$< \mu(f) - \delta_1/2 + \delta_3/8 < \mu(f) - 3\delta_1/8$$

and this contradicts (4.106). Hence

$$I^g(T - l_0 - l_1 - L_0 - 4, T - l_0 - l_1 - L_0 - 3, x, u) \geq \mu(f) - \delta_1/2.$$

Together with (4.104) and (4.105) this implies that

$$I^g(T - l_0 - l_1 - L_0 - 3, T, x_1, u_1) - I^g(T - l_0 - l_1 - L_0 - 3, T, x, u)$$
$$\geq -\delta - \delta_1/8 - \delta_3/8 - \delta_1/2. \tag{4.107}$$

Since $(\bar{x}_*, \bar{u}_*)$ is an $(\bar{f}, -A, -B)$-overtaking optimal pair it follows from (3.18), (4.103), and Proposition 3.12 that

$$I^f(T - l_0 - l_1 - L_0 - 3, T, x_1, u_1) = I^f(0, l_0 + l_1 + L_0 + 3, \bar{x}_*, \bar{u}_*)$$
$$= \mu(f)(l_0 + l_1 + L_0 + 3) + \pi_-^f(\bar{x}_*(0))$$
$$- \pi_-^f(\bar{x}_*(l_0 + l_1 + L_0 + 3)). \tag{4.108}$$

In view of (4.97) and the choice of δ_2 (see (4.93),

$$|\pi_-^f(\bar{x}_*(l_0 + l_1 + L_0 + 3))| \leq \delta_1/8.$$

Combined with (4.108) this implies that

$$I^f(T - l_0 - l_1 - L_0 - 3, T, x_1, u_1) \leq \pi_-^f(\bar{x}_*(0)) + \mu(f)(l_0 + l_1 + L_0 + 3) + \delta_1/8. \tag{4.109}$$

By (P16), (4.99), and (4.109),

$$I^g(T - l_0 - l_1 - L_0 - 3, T, x_1, u_1)$$
$$\leq \pi_-^f(\bar{x}_*(0)) + \mu(f)(l_0 + l_1 + L_0 + 3) + \delta_1/8 + \delta_3/8. \tag{4.110}$$

It follows from (4.107) and (4.110) that

$$I^g(T - l_0 - l_1 - L_0 - 3, T, x, u)$$
$$\leq \pi_-^f(\bar{x}_*(0)) + \mu(f)(l_0 + l_1 + L_0 + 3) + \delta + 3\delta_1/4 + \delta_3/4. \tag{4.111}$$

By property (P16), (4.99), and (4.111),

$$I^f(T - l_0 - l_1 - L_0 - 3, T, x, u)$$
$$\leq \pi_-^f(\tilde{x}_*(0)) + \mu(f)(l_0 + l_1 + L_0 + 3) + \delta + 3\delta_1/4 + 3\delta_3/8. \quad (4.112)$$

Define

$$\tilde{x}(t) = x(T - t), \ \tilde{u}(t) = u(T - t), \ t \in [0, T]. \quad (4.113)$$

It is clear that $(\tilde{x}, \tilde{u}) \in X(-A, -B, 0, T)$ and by (4.112), (4.113),

$$I^f(0, l_0 + l_1 + L_0 + 3, \tilde{x}, \tilde{u}) = I^f(T - l_0 - l_1 - L_0 - 3, T, x, u)$$
$$\leq \pi_-^f(\tilde{x}_*(0)) + \mu(f)(l_0 + l_1 + L_0 + 3)$$
$$+ \delta + 3\delta_1/4 + 3\delta_3/8. \quad (4.114)$$

In view of (4.102) and (4.113),

$$|\tilde{x}(l_0 + l_1 + L_0 + 3) - x_f| \leq \delta_2.$$

By the relation above and the choice of δ_2 (see (4.93)),

$$|\pi_-^f(\tilde{x}(l_0 + l_1 + L_0 + 3))| \leq \delta_1/8. \quad (4.115)$$

By Propositions 3.11 and 3.12, (4.114), (4.115), and $(f, -A, -B)$-overtaking optimality of $(\tilde{x}_*, \tilde{u}_*)$,

$$\pi_-^f(\tilde{x}(0)) - \pi_-^f(\tilde{x}_*(0))$$
$$+ I^f(0, L_0, \tilde{x}, \tilde{u}) - L_0\mu(f) - \pi_-^f(\tilde{x}(0)) + \pi_-^f(\tilde{x}(L_0))$$
$$\leq \pi_-^f(\tilde{x}(0)) - \pi_-^f(\tilde{x}_*(0)) + I^f(0, l_0 + l_1 + L_0 + 3, \tilde{x}, \tilde{u})$$
$$- \mu(f)(l_0 + l_1 + L_0 + 3) - \pi_-^f(\tilde{x}(0)) + \pi_-^f(\tilde{x}(l_0 + l_1 + L_0 + 3))$$
$$\leq \pi_-^f(\tilde{x}(0)) - \pi_-^f(\tilde{x}_*(0))$$
$$+ \mu(f)(l_0 + l_1 + L_0 + 3) + \delta + 3\delta_3/8 + 3\delta_1/4$$
$$+ \pi_-^f(\tilde{x}_*(0)) - \mu(f)(l_0 + l_1 + L_0 + 3) - \pi_-^f(\tilde{x}(0)) + \delta_1/8$$
$$\leq \delta + 3\delta_1/4 + \delta_3/2 \leq \delta_1.$$

It follows from the relation above, (4.96), and Proposition 3.11 that

$$\pi_-^f(\tilde{x}(0)) \leq \inf(\pi_-^f) + \delta_1,$$
$$I^f(0, L_0, \tilde{x}, \tilde{u}) - L_0\mu(f) + \pi_-^f(\tilde{x}(0)) + \pi_-^f(\tilde{x}(L_0)) \leq \delta_1.$$

By the inequalities above and property (P14), there exists an $(f, -A, -B)$-overtaking optimal pair

$$(x_*, u_*) \in X(-A, -B, 0, \infty)$$

such that

$$\pi_-^f(x_*(0)) = \inf(\pi_-^f),$$

$$|\tilde{x}(t) - x_*(t)| \le \epsilon \text{ holds for all } t \in [0, L_0].$$

Combined with (4.113) this implies that

$$|x(T - t) - \tilde{x}_*(t)| \le \epsilon \text{ holds for all } t \in [0, L_0].$$

Theorem 4.4 is proved.   □

Proof of Theorem 4.5. Theorems 4.2 and 4.4 imply the following result.

Theorem 4.13. *Let $L_0 > 0$, $\epsilon \in (0, 1)$. Then there exist $\delta > 0$, a neighborhood $\mathcal{U}$ of F in $\mathfrak{M}_b$ and $L_1 > L_0$ such that for each $T \ge L_1$, each $g \in \mathcal{U}$, and each $(x, u) \in X(A, B, 0, T)$ which satisfies*

$$I^g(0, T, x, u) \le \sigma(g, 0, T) + \delta$$

there exists an $(f, -A, -B)$-overtaking optimal pair

$$(x_*, u_*) \in X(-A, -B, 0, \infty)$$

such that

$$\pi_-^f(x_*(0)) = \inf(\pi_-^f)$$

and for all $t \in [0, L_0]$

$$|x(T - t) - x_*(t)| \le \epsilon.$$

Theorem 4.5 follows from Propositions 4.10, 4.12, and Theorem 4.13.

4.8   Proof of Theorem 4.6

Theorem 4.6 follows from Propositions 4.10 and 4.12 and the next result.

Theorem 4.14. *Let $y \in R^n$, $L_0 > 0$, $\epsilon \in (0, 1)$, $M > 0$. Then there exist $\delta > 0$, a neighborhood $\mathcal{U}$ of F in $\mathfrak{M}_b$ and $L_1 > L_0$ such that for each $T \ge L_1$, each $g \in \mathcal{U}$*

and each $(x, u) \in X(A, B, 0, T)$ *which satisfies*

$$x(0) = y, \ |x(T)| \leq M, \ I^g(0, T, x, u) \leq \sigma(g, x(0), x(T), 0, T) + \delta$$

there exists an (f, A, B)-*overtaking optimal pair*

$$(\tilde{x}, \tilde{u}) \in X(A, B, 0, \infty)$$

such that $\tilde{x}(0) = y$ *and for all* $t \in [0, L_0]$,

$$|x(t) - \tilde{x}(t)| \leq \epsilon.$$

Proof. Denote by d the metric of the space $\mathfrak{M}$. Assume that Theorem 4.14 does not hold. Then there exist a sequence $\{\delta_k\}_{k=1}^{\infty} \subset (0, 1)$ such that

$$\delta_k < 4^{-k}, \ k = 1, 2, \ldots, \tag{4.116}$$

a sequence

$$T_k > L_0 + 2k, \ k = 1, 2, \ldots \tag{4.117}$$

a sequence $\{g_k\}_{k=1}^{\infty} \subset \mathfrak{M}_b$ such that

$$d(g_k, f) \leq k^{-1}, \ k = 1, 2, \ldots \tag{4.118}$$

and a sequence $(x_k, u_k) \in X(A, B, 0, T_k), \ k = 1, 2, \ldots$ such that for each natural number k,

$$x_k(0) = y, \ |x_k(T_k)| \leq M, \tag{4.119}$$

$$I^{g_k}(0, T_k, x_k, u_k) \leq \sigma(g_k, x_k(0), x_k(T_k), 0, T_k) + \delta_k \tag{4.120}$$

and that the following property holds:

(i) for each (f, A, B)-overtaking optimal pair

$$(\xi, v) \in X(A, B, 0, \infty)$$

satisfying $\xi(0) = y$ we have

$$\max\{|x_k(t) - \xi(t)| : \ t \in [0, L_0]\} > \epsilon. \tag{4.121}$$

In view of (4.116)–(4.120) and Theorem 4.2, the following property holds:

(ii) for each $\gamma > 0$ there exists $l_\gamma > 0$ and an integer $k_\gamma \geq 1$ and each integer $k \geq k_\gamma + 2l_\gamma$,

$$|x_k(t) - x_f| \leq \gamma, \ t \in [l_\gamma, T_k - l_\gamma].$$

Let $S \geq l_1$. By Proposition 3.28 and (4.119), the sequence

$$\{\sigma(f, y, x_k(S), 0, S)\}_{k=k_1}^{\infty}$$

is bounded. Proposition 4.8 and (4.118) imply that the sequence

$$\{\sigma(g_k, y, x_k(S), 0, S)\}_{k=k_1}^{\infty}$$

is bounded and

$$\lim_{k \to \infty} |\sigma(g_k, y, x_k(S), 0, S) - \sigma(f, y, x_k(S), S)| = 0. \tag{4.122}$$

Together with (4.116) and (4.120) this implies that the sequence

$$\{I^{g_k}(0, S, x_k, u_k)\}_{k=k_1}^{\infty}$$

is bounded. Combined with Proposition 4.8 and (4.118) this implies that the sequence $\{I^f(0, S, x_k, u_k)\}_{k=k_1}^{\infty}$ is bounded and that

$$\lim_{k \to \infty} |I^{g_k}(0, S, x_k, u_k) - I^f(0, S, x_k, u_k)| = 0. \tag{4.123}$$

In view of (4.116), (4.120), (4.122) and (4.123),

$$\lim_{k \to \infty} |I^f(0, S, x_k, u_k) - \sigma(f, y, x_k(S), 0, S)| = 0. \tag{4.124}$$

Since the sequence $\{\sigma(f, y, x_k(S), 0, S)\}_{k=k_1}^{\infty}$ is bounded, the sequence

$$\{I^f(0, S, x_k, u_k)\}_{k=k_1}^{\infty}$$

is bounded too for any integer $S \geq l_1$. By Proposition 3.25, extracting subsequences and re-indexing we may assume without loss of generality that there exists $(x, u) \in X(A, B, 0, \infty)$ such that for each integer $S \geq l_1$,

$$x_k(t) \to x(t) \text{ as } k \to \infty \text{ uniformly on } [0, S], \tag{4.125}$$

$$I^f(0, S, x, u) \leq \liminf_{k \to \infty} I^f(0, S, x_k, u_k). \tag{4.126}$$

By (4.125) and property (ii) the following property holds:

(iii) Let $\gamma > 0$ and let $l_\gamma > 0$ be as guaranteed by property (ii). Then

$$|x(t) - x_f| \leq \gamma \text{ for all } t \geq l_\gamma.$$

We show that the pair (x, u) is (f, A, B)-overtaking optimal. It view of property (iii) and Theorem 3.10, it is sufficient to show that the pair (x, u) is (f, A, B)-minimal.

Assume the contrary. Then there exist $\Delta > 0$, $Q_0 > l_1$ and $(\tilde{x}, \tilde{u}) \in X(A, B, 0, Q_0)$ such that

$$I^f(0, Q_0, \tilde{x}, \tilde{u}) < I^f(0, Q_0, x, u) - \Delta,$$

$$\tilde{x}(0) = y, \ \tilde{x}(Q_0) = x(Q_0). \tag{4.127}$$

By Proposition 3.29, there exists $\gamma > 0$ such that for each $z_1, z_2 \in R^n$ satisfying $|z_i - x_f| \leq 2\gamma$, $i = 1, 2$, we have

$$|\sigma(f, z_1, z_2, 1) - \mu(f)| \leq \Delta/16. \tag{4.128}$$

Let $l_\gamma > 0$, $k_\gamma \geq 1$ be as guaranteed by property (ii). By (4.124) and (4.126) there exists a natural number

$$k > 2k_\gamma + 2k_1 + 2l_\gamma + 2l_1 + 2Q_0 \tag{4.129}$$

such that

$$|I^f(0, Q_0 + l_\gamma + 1, x_k, u_k) - \sigma(f, y, x_k(Q_0 + l_\gamma + 1), Q_0 + l_\gamma + 1)| \leq \Delta/16, \tag{4.130}$$

$$I^f(0, Q_0 + l_\gamma, x, u) \leq I^f(0, Q_0 + l_\gamma, x_k, u_k) + \Delta/16. \tag{4.131}$$

It follows from (4.120) and property (ii) that

$$|x_k(t) - x_f| \leq \gamma, \ t \in [l_\gamma, T_k - l_\gamma]. \tag{4.132}$$

Property (iii) implies that

$$|x(t) - x_f| \leq \gamma, \ t \in [l_\gamma, \infty). \tag{4.133}$$

By (4.117) and (4.129),

$$[l_\gamma + Q_0, l_\gamma + Q_0 + 1] \subset [l_\gamma, T_k - l_\gamma]. \tag{4.134}$$

Proposition 3.27 and (4.127) imply that there exists

$$(\hat{x}, \hat{u}) \in X(A, B, 0, Q_0 + l_\gamma + 1)$$

such that

$$\hat{x}(t) = \tilde{x}(t), \ \hat{u}(t) = \tilde{u}(t), \ t \in [0, Q_0],$$

$$\hat{x}(t) = x(t), \ \hat{u}(t) = u(t), \ t \in [Q_0, Q_0 + l_\gamma],$$

$$\hat{x}(Q_0 + l_\gamma + 1) = x_k(Q_0 + l_\gamma + 1),$$

$$I^f(Q_0 + l_\gamma, Q_0 + l_\gamma + 1, \hat{x}, \hat{u}) = \sigma(f, x(Q_0 + l_\gamma), x_k(Q_0 + l_\gamma + 1), 1). \quad (4.135)$$

It follows from (4.127), (4.129), (4.130), and (4.135) that

$$I^f(0, Q_0 + l_\gamma + 1, \hat{x}, \hat{u})$$
$$\geq \sigma(f, y, x_k(Q_0 + l_\gamma + 1), Q_0 + l_\gamma + 1)$$
$$\geq I^f(0, Q_0 + l_\gamma + 1, x_k, u_k) - \Delta/16. \quad (4.136)$$

By (4.127), (4.129), (4.131)–(4.133), (4.135), and the choice of γ (see (4.128)),

$$I^f(0, Q_0 + l_\gamma + 1, \hat{x}, \hat{u}) - I^f(0, Q_0 + l_\gamma + 1, x_k, u_k)$$
$$I^f(0, Q_0 + l_\gamma, \hat{x}, \hat{u}) + I^f(Q_0 + l_\gamma, Q_0 + l_\gamma + 1, \hat{x}, \hat{u})$$
$$\quad - I^f(0, Q_0 + l_\gamma, x_k, u_k) - I^f(Q_0 + l_\gamma, Q_0 + l_\gamma + 1, x_k, u_k)$$
$$\leq I^f(0, Q_0 + l_\gamma, \hat{x}, \hat{u}) - I^f(0, Q_0 + l_\gamma, x_k, u_k)$$
$$\quad + \sigma(f, x(Q_0 + l_\gamma), x_k(Q_0 + l_\gamma + 1), 1) - \sigma(f, x_k(Q_0 + l_\gamma), x_k(Q_0 + l_\gamma + 1), 1)$$
$$\leq I^f(0, Q_0 + l_\gamma, \hat{x}, \hat{u}) - I^f(0, Q_0 + l_\gamma, x_k, u_k) + \Delta/8$$
$$= I^f(0, Q_0 + l_\gamma, \hat{x}, \hat{u}) - I^f(0, Q_0 + l_\gamma, x, u)$$
$$\quad + I^f(0, Q_0 + l_\gamma, x, u) - I^f(0, Q_0 + l_\gamma, x_k, u_k) + \Delta/8$$
$$\leq I^f(0, Q_0, \tilde{x}, \tilde{u}) - I^f(0, Q_0, x, u) + 3\Delta/16$$
$$< -\Delta + 3\Delta/16.$$

This contradicts (4.136). The contradiction we have reached proves that the pair (x, u) is (f, A, B)-overtaking optimal. By (4.125), there exists a natural number p_0 such that for all integers $p \geq p_0$,

$$|x_p(t) - x(t)| \leq \epsilon/2, \ t \in [0, L_0].$$

This contradicts property (i). The contradiction we have reached proves Theorem 4.14.

Chapter 5
Linear Control Systems with Discounting

In this chapter we extend the stability results obtained in Chap. 4 to the linear control systems with discounting. We establish the turnpike property of approximate solutions on large intervals and the convergence of approximate solutions in regions close to the endpoints of the time intervals. We also show that this convergence as well as the turnpike property are stable under small perturbations of objective functions (integrands).

5.1 Preliminaries and Main Results

We use the notation, definitions, and assumptions introduced in Sects. 3.1–3.3 of Chap. 3 and in Sects. 4.1 and 4.6 of Chap. 4.. Recall that $a_0 > 0$ and $\psi : [0, \infty) \to [0, \infty)$ is an increasing function such that

$$\lim_{t \to \infty} \psi(t) = \infty.$$

We continue to study the structure of optimal trajectories of the controllable linear control system

$$x' = Ax + Bu$$

where A and B are given matrices of the dimensions $n \times n$ and $n \times m$ and with the continuous integrand $f : R^n \times R^m \to R^1$ which satisfies assumptions (A1)–(A3) and (3.10).

© Springer International Publishing Switzerland 2015
A.J. Zaslavski, *Turnpike Theory of Continuous-Time Linear Optimal Control Problems*, Springer Optimization and Its Applications 104,
DOI 10.1007/978-3-319-19141-6_5

We continue to consider the set $\mathfrak{M}$ of all borelian functions $g : R^{n+m+1} \to R^1$ which satisfy the growth condition

$$g(t, x, u) \geq \max\{\psi(|x|),\ \psi(|u|),$$

$$\psi([|Ax + Bu| - a_0|x|]_+)[|Ax + Bu| - a_0|x|]_+\} - a_0 \qquad (5.1)$$

for each $(t, x, u) \in R^{n+m+1}$.

The set $\mathfrak{M}$ is equipped with the uniformity which is determined by the following base:

$$E(N, \epsilon, \lambda) = \{(f, g) \in \mathfrak{M} \times \mathfrak{M} : |f(t, x, u) - g(t, x, u)| \leq \epsilon$$

$$\text{for each } (t, x, u) \in R^{n+m+1} \text{ satisfying } |x|, |u| \leq N\}$$

$$\cap \{(f, g) \in \mathfrak{M} \times \mathfrak{M} : (|f(t, x, u)| + 1)(|g(t, x, u)| + 1)^{-1} \in [\lambda^{-1}, \lambda]$$

$$\text{for each } (t, x, u) \in R^{n+m+1} \text{ satisfying } |x| \leq N\}, \qquad (5.2)$$

where $N > 0$, $\epsilon > 0$ and $\lambda > 1$.

Remind that the uniform space $\mathfrak{M}$ is metrizable and complete.

Denote by $\mathfrak{M}_b$ the set of all functions $g \in \mathfrak{M}$ which are bounded on bounded subsets of R^{n+m+1}. Note that $\mathfrak{M}_b$ is a closed subset of $\mathfrak{M}$. We consider the topological subspace $\mathfrak{M}_b \subset \mathfrak{M}$ equipped with the relative topology.

Recall that $f : R^n \times R^m \to R^1$ is a continuous function which satisfies assumptions (3.10), (A1)–(A3). For each $(t, x, u) \in R^{n+m+1}$ set

$$F(t, x, u) = f(x, u). \qquad (5.3)$$

For each $g \in \mathfrak{M}$, each borelian function $\alpha : [0, \infty) \to R^1$ define a function $\alpha g : R^{n+m+1} \to R^1$ by

$$\alpha g(t, x, u) = \alpha(t)g(t, x, u),\ (t, x, u) \in [0, \infty) \times R^n \times R^m,$$

$$\alpha g(t, x, u) = 0,\ (t, x, u) \in (-\infty, 0) \times R^n \times R^m \qquad (5.4)$$

and for each $T_1 \in R^1$, each $T_2 > T_1$ and each $(x, u) \in X(A, B, T_1, T_2)$ set

$$I^{\alpha g}(T_1, T_2, x, u) = \int_{T_1}^{T_2} \alpha(t)g(t, x(t), u(t))dt.$$

In this chapter we prove the following two stability results. The first of them (Theorem 5.1) establishes the stability of the turnpike property of approximate solutions on large intervals for our linear control systems with discounting. The second one (Theorem 5.2) establishes the stability of the convergence of approximate solutions in regions close to the endpoints of the time intervals in the case of discounting.

Theorem 5.1. *Let $\epsilon, M > 0$. Then there exist $L > 0$, $\delta \in (0, \epsilon)$, $\lambda \in (0, 1)$, and a neighborhood $\mathcal{U}$ of F in $\mathfrak{M}_b$ such that for each $T > 2L$, each $g \in \mathcal{U}$, each borelian function $\alpha : [0, \infty) \to (0, 1]$ which satisfies*

$$\alpha(t_1)\alpha(t_2)^{-1} \geq \lambda \text{ for each } t_1, t_2 \in [0, T] \text{ satisfying } |t_1 - t_2| \leq L$$

and each $(x, u) \in X(A, B, 0, T)$ which satisfies at least one of the following conditions:

(a) $|x(0)| \leq M$, $I^{\alpha g}(0, T, x, u) \leq \sigma(\alpha g, x(0), 0, T) + \delta \inf\{\alpha(t) : t \in [0, T]\}$;

(b) $|x(0)|$, $|x(T)| \leq M$,

$$I^{\alpha g}(0, T, x, u) \leq \sigma(\alpha g, x(0), x(T), 0, T) + \delta \inf\{\alpha(t) : t \in [0, T]\}$$

there exist $p_1 \in [0, L]$, $p_2 \in [T - L, T]$ such that

$$|x(t) - x_f| \leq \epsilon \text{ for all } t \in [p_1, p_2].$$

Moreover if $|x(0) - x_f| \leq \delta$, then $p_1 = 0$ and if $|x(T) - x_f| \leq \delta$, then $p_2 = T$.

Theorem 5.2. *Let $\tau_0 \geq 1$, $\epsilon > 0$, $M > 0$. Then there exist $\delta \in (0, \epsilon)$, $\lambda \in (0, 1)$, $T_0 \geq \tau_0$ and a neighborhood $\mathcal{U}$ of F in $\mathfrak{M}_b$ such that for each $T \geq T_0$, each $g \in \mathcal{U}$, each borelian function $\alpha : [0, \infty) \to (0, 1]$ which satisfies*

$$\alpha(t_1)\alpha(t_2)^{-1} \geq \lambda \text{ for each } t_1, t_2 \in [0, T] \text{ satisfying } |t_1 - t_2| \leq T_0$$

and each $(x, u) \in X(A, B, 0, T)$ which satisfies

$$|x(0)| \leq M,$$

$$I^{\alpha g}(0, T, x, u) \leq \sigma(\alpha g, x(0), 0, T) + \delta \inf\{\alpha(t) : t \in [0, T]\}$$

there exists an $(f, -A, -B)$-overtaking optimal pair

$$(x_*, u_*) \in X(-A, -B, 0, \infty)$$

such that

$$\pi_-^f(x_*(0)) = \inf(\pi_-^f),$$

$$|x(T - t) - x_*(t)| \leq \epsilon \text{ for all } t \in [0, \tau_0].$$

This chapter is organized as follows. Section 5.2 contains auxiliary results. Theorem 5.1 is proved in Sect. 5.3. Section 5.4 contains the proof of Theorem 5.2.

5.2   Auxiliary Results for Theorems 5.1 and 5.2

Let $g \in \mathfrak{M}$, $L > 0$, $\lambda \in (0, 1)$ and let a borelian function $\alpha : [0, \infty) \to (0, 1]$ satisfy

$$\alpha(t_1)\alpha(t_2)^{-1} \geq \lambda \text{ for each } t_1, t_2 \geq 0 \text{ satisfying } |t_1 - t_2| \leq L.$$

Let

$$T \geq T_2 \geq T_1 \geq 0, \ T_2 - T_1 \leq L.$$

Define a function $g_{\alpha, T_1, T_2} : R^{n+m+1} \to R^1$ as follows. For every $(x, u) \in R^n \times R^m$ set

$$g_{\alpha, T_1, T_2}(t, x, u) = \alpha(t)g(t, x, u) \inf\{\alpha(s) : \ s \in [T_1, T_2]\}^{-1} + a_0(\lambda^{-1} - 1) \quad (5.5)$$

for all $t \in [T_1, T_2]$;

$$g_{\alpha, T_1, T_2}(t, x, u) = g_{\alpha, T_1, T_2}(T_1, x, u) \text{ for all } t \in (-\infty, T_1), \quad (5.6)$$

$$g_{\alpha, T_1, T_2}(t, x, u) = g_{\alpha, T_1, T_2}(T_2, x, u) \text{ for all } t \in (T_2, \infty). \quad (5.7)$$

Clearly, the function g_{α, T_1, T_2} is borelian and belongs to the space $\mathfrak{M}$ and if $g \in \mathfrak{M}_b$, then $g_{\alpha, T_1, T_2} \in \mathfrak{M}_b$.

Proposition 5.3. *Let $M, \tau > 0$. Then there exists $M_1 > 0$ such that for each $\tilde{y}, \tilde{z} \in R^n$ satisfying $|\tilde{y}|, |\tilde{z}| \leq M$ there is $(x, u) \in X(A, B, 0, \tau)$ such that $x(0) = \tilde{y}$, $x(\tau) = \tilde{z}$ and that*

$$|x(t)|, \ |u(t)| \leq M_1 \text{ for all } t \in [0, \tau].$$

Proof. It follows from Proposition 3.26 that for each $\tilde{y}, \tilde{z} \in R^n$ there exists a unique solution $x(\cdot), y(\cdot)$ of the following system

$$(x', y')^t = C((x, y)^t) \quad (5.8)$$

with the boundary constraints $x(0) = \tilde{y}$, $x(\tau) = \tilde{z}$ and

$$C((x, y)^t) = (Ax + BB^t y, x - A^t y)^t.$$

(Here B^t denotes the transpose of B.)

For any initial value $(x_0, y_0) \in R^n \times R^n$ there exists a unique solution of (5.8) $x(\cdot), y(\cdot)$ and

$$(x(s), y(s))^t = e^{sC}(x_0, y_0)^t, \ s \in R^1.$$

Clearly, for each $\tilde{y}, \tilde{z} \in R^n$ there exists a unique of vector $D(\tilde{y}, \tilde{z}) \in R^n$ such that the function

$$(x(s), y(s)) = (e^{sC}((\tilde{y}, D(\tilde{y}, \tilde{z}))^t))^t, \ s \in R^1$$

satisfies (5.8) with the boundary constraints

$$x(0) = \tilde{y}, \ x(\tau) = \tilde{z}.$$

It is easy to see that $D : R^n \times R^n \to R^n$ is a linear operator. Now the validity of the proposition follows from the boundedness of the set

$$\{|e^{sC}((\tilde{y}, D(\tilde{y}, \tilde{z}))^t)| : \ \tilde{y}, \tilde{z} \in R^n, \ |\tilde{y}|, |\tilde{z}| \leq M, \ s \in [0, \tau]\}.$$

We suppose that the sum over empty set is zero.

Lemma 5.4. *Let $M > 0$. Then there exist $L > 0$, $\lambda \in (0, 1)$, $M_0 > 0$, and a neighborhood $\mathcal{U}$ of F in $\mathfrak{M}_b$ such that for each $T > L$, each $g \in \mathcal{U}$, each borelian function $\alpha : [0, \infty) \to (0, 1]$ which satisfies*

$$\alpha(t_1)\alpha(t_2)^{-1} \geq \lambda \text{ for each } t_1, t_2 \in [0, T] \text{ satisfying } |t_1 - t_2| \leq L \qquad (5.9)$$

and each $(x, u) \in X(A, B, 0, T)$ which satisfies at least one of the following conditions:

(a) $|x(0)| \leq M$, $I^{\alpha g}(0, T, x, u) \leq \sigma(\alpha g, x(0), 0, T) + \inf\{\alpha(t) : \ t \in [0, T]\}$;

(b) $|x(0)|, \ |x(T)| \leq M$,

$$I^{\alpha g}(0, T, x, u) \leq \sigma(\alpha g, x(0), x(T), 0, T) + \inf\{\alpha(t) : \ t \in [0, T]\}$$

the inequality $|x(t)| \leq M_0$ holds for all $t \in [0, T]$.

Proof. We may assume that

$$M > |x_f| + |u_f| + 4. \qquad (5.10)$$

By Proposition 4.7 there exists $M_1 > M + 2$ such that the following property holds:

(i) for each $g \in \mathfrak{M}$, each $T_1 \in R^1$, each $T_2 \in [T_1 + 8^{-1}, T_1 + 8]$ and each $(x, u) \in X(A, B, T_1, T_2)$ which satisfies

$$I^g(T_1, T_2, x, u) \leq 64(|f(x_f, u_f)| + 4 + a_0)$$

we have

$$|x(t)| \leq M_1 \text{ for all } t \in [T_1, T_2].$$

By Proposition 5.3 there exists $M_2 > M_1$ such that the following property holds:

(ii) for each $z_1, z_2 \in R^n$ satisfying $|z_i| \leq M_1$, $i = 1, 2$ there is $(\xi, \eta) \in X(A, B, 0, 1)$ such that

$$\xi(0) = z_1, \quad \xi(1) = z_2,$$
$$|\xi(t)|, \ |\eta(t)| \leq M_2 \text{ for all } t \in [0, 1].$$

In view of the continuity of the function f, there exists $M_3 > 16$ such that

$$|f(z_1, z_2)| \leq M_3 \text{ for each } (z_1, z_2) \in R^{n+m} \text{ satisfying } |z_i| \leq M_2 + 2, \ i = 1, 2.$$
(5.11)

Choose numbers M_*, L, λ such that

$$M_* > 64(a_0 + 2) + 8 \cdot 128 M_3, \quad L > 8M_3 + 8, \tag{5.12}$$

$$\lambda \in (2^{-1}, 1). \tag{5.13}$$

By Proposition 4.7 there exists $M_0 > 0$ such that the following property holds:

(iii) for each $g \in \mathfrak{M}$, each $T_1 \in R^1$, each $T_2 \in [T_1 + 8^{-1}, T_1 + 8]$ and each $(x, u) \in X(A, B, T_1, T_2)$ which satisfies

$$I^g(T_1, T_2, x, u) \leq M_2 + M_1 + M + 64(a_0 + 2) + M_*$$

we have

$$|x(t)| \leq M_0 \text{ for all } t \in [T_1, T_2].$$

There exists a neighborhood $\mathcal{U}$ of F in $\mathfrak{M}_b$ such that

$$\mathcal{U} \subset \{g \in \mathfrak{M}_b : \ |F(t, z, v) - g(t, z, v)| \leq 1$$

$$\text{for all } (t, z, v) \in R^{n+m+1} \text{ satisfying } |z|, |v| \leq M_2 + 1\}. \tag{5.14}$$

Assume that

$$T \geq L, \ g \in \mathcal{U}, \tag{5.15}$$

a borelian function $\alpha : [0, \infty) \to (0, 1]$ satisfies (5.9), $(x, u) \in X(A, B, 0, T)$ satisfies at least one of the conditions (a), (b). We show that

$$|x(t)| \leq M_0, \ t \in [0, T]. \tag{5.16}$$

Assume the contrary. Then there is a number τ_0 such that

$$\tau_0 \in (0, T), \quad |x(\tau_0)| > M_0. \tag{5.17}$$

Clearly, there is $\tau_1 \in (0, T - 2)$ such that

$$\tau_1 < \tau_0 < \tau_1 + 2. \tag{5.18}$$

Set

$$S_1 = \max\{t \in [0, \tau_1] : \ |x(t) \le M_1\} \tag{5.19}$$

and set

$$S_2 = \min\{t \in [\tau_1 + 2, T] : \ |x(t)| \le M_1\} \tag{5.20}$$

if $\{t \in [\tau_1 + 2, T] : \ |x(t) \le M_1\} \ne \emptyset$; otherwise put $S_2 = T$.
  Recall that for any $z \in R^1$,

$$[z] = \min\{i \in R^1 : \ i \text{ is an integer and } i \le z\}.$$

Set

$$k_1 = [\tau_1 - S_1], \ k_2 = [S_2 - \tau_1 - 2]. \tag{5.21}$$

By (5.10), (5.19), (5.20), property (ii), and conditions (a) and (b), there exists $(\tilde{x}, \tilde{u}) \in X(A, B, 0, T)$ such that

$$\tilde{x}(t) = x(t), \ \tilde{u}(t) = u(t), \ t \in [0, S_1] \cup ([S_2, T] \setminus \{T\}), \tag{5.22}$$

$$\tilde{x}(S_1 + 1) = x_f, \ |\tilde{x}(t)|, \ |\tilde{u}(t)| \le M_2, \ t \in [S_1, S_1 + 1], \tag{5.23}$$

$$\tilde{x}(S_2 - 1) = x_f, \ |\tilde{x}(t)|, \ |\tilde{u}(t)| \le M_2, \ t \in [S_2 - 1, S_2], \tag{5.24}$$

$$\tilde{x}(S_2) = x(S_2) \text{ if } |x(S_2)| \le M_1, \ \text{ otherwise } \tilde{x}(S_2) = 0, \tag{5.25}$$

$$\tilde{x}(t) = x_f, \ \tilde{u}(t) = u_f, \ t \in [S_1 + 1, S_2 - 1]. \tag{5.26}$$

By (5.22)–(5.26) and conditions (a), (b),

$$\inf\{\alpha(t) : \ t \in [0, T]\} \ge I^{\alpha g}(0, T, x, u) - I^{\alpha g}(0, T, \tilde{x}, \tilde{u})$$
$$\ge I^{\alpha g}(S_1, S_2, x, u) - I^{\alpha g}(S_1, S_2, \tilde{x}, \tilde{u}). \tag{5.27}$$

It is easy to see that

$$I^{\alpha g}(S_1, S_2, x, u) = I^{\alpha g}(S_1 + k_1, S_2 - k_2, x, u)$$
$$+ \sum \{I^{\alpha g}(S_1 + i, S_1 + i + 1, x, u) : \; i \text{ is an integer}, 0 \le i < k_1\}$$
$$+ \sum \{I^{\alpha g}(S_2 - i - 1, S_2 - i, x, u) : \; i \text{ is an integer}, 0 \le i < k_2\}$$

$$(5.28)$$

and that

$$I^{\alpha g}(S_1, S_2, \tilde{x}, \tilde{u}) = I^{\alpha g}(S_1, S_1 + 1, \tilde{x}, \tilde{u})$$
$$+ I^{\alpha g}(S_2 - 1, S_2, \tilde{x}, \tilde{u}) + I^{\alpha g}(S_1 + 1, S_2 - 1, \tilde{x}, \tilde{u}). \quad (5.29)$$

We estimate $I^{\alpha g}(S_1 + k_1, S_2 - k_2, x, u)$. In view of (5.21),

$$(S_2 - k_2) - (S_1 + k_1) \ge 2, \quad (5.30)$$

$$(S_2 - k_2) - (S_1 + k_1) \le (S_2 - (S_2 - \tau_1 - 3)) - (S_1 + \tau_1 - S_1 - 1) \le 4. \quad (5.31)$$

It follows from (5.17), (5.18), and (5.21) that

$$\tau_0 \in [S_1 + k_1, S_2 - k_2], \; |x(\tau_0)| > M_0. \quad (5.32)$$

Let

$$\hat{g} = g_{\alpha, S_1 + k_1, S_2 - k_2}. \quad (5.33)$$

Clearly, $g \in \mathfrak{M}_b$. In view of (5.30)–(5.33) and property (iii),

$$I^{\hat{g}}(S_1 + k_1, S_2 - k_2, x, u) > M_*. \quad (5.34)$$

By (5.5), (5.33), and (5.34),

$$M_* < I^{\hat{g}}(S_1 + k_1, S_2 - k_2, x, u) = (S_2 - S_1 - k_2 - k_1)a_0(\lambda^{-1} - 1)$$
$$+ \inf\{\alpha(t) : \; t \in [S_1 + k_1, S_2 - k_2]\}^{-1}$$
$$I^{\alpha g}(S_1 + k_1, S_2 - k_2, x, u)$$

and

$$I^{\alpha g}(S_1 + k_1, S_2 - k_2, x, u)$$
$$> (M_* - (S_2 - S_1 - k_2 - k_1)a_0(\lambda^{-1} - 1)) \inf\{\alpha(t) : \; t \in [S_1 + k_1, S_2 - k_2]\}.$$

$$(5.35)$$

If $k_1 > 0$, then by (5.19) and (5.21),

$$|x(S_1 + i)| > M_1, \ i = 1, \ldots, S_1 + k_1. \tag{5.36}$$

Assume that

$$k_1 > 0, \ i \in \{0, \ldots, k_1 - 1\}.$$

We estimate

$$I^{\alpha g}(S_1 + i, S_1 + i + 1, x, u).$$

Let

$$\hat{g} = g_{\alpha, S_1 + i, S_1 + i + 1}. \tag{5.37}$$

In view of (5.36), (5.37), and property (i),

$$I^{\hat{g}}(S_1 + i, S_1 + i + 1, x, u) > 64(|f(x_f, u_f)| + 4 + 2a_0). \tag{5.38}$$

By (5.5), (5.33), and (5.38),

$$64(|f(x_f, u_f)| + 4 + 2a_0) < I^{\hat{g}}(S_1 + i, S_1 + i + 1, x, u)$$
$$= a_0(\lambda^{-1} - 1) + \inf\{\alpha(t) : t \in [S_1 + i, S_1 + i + 1]\}^{-1} I^{\alpha g}(S_1 + i, S_1 + i + 1, x, u)$$

and

$$I^{\alpha g}(S_1 + i, S_1 + i + 1, x, u)$$
$$> (64(|f(x_f, u_f)| + 4 + 2a_0) - a_0(\lambda^{-1} - 1)) \inf\{\alpha(t) : t \in [S_1 + i, S_1 + i + 1]\},$$
$$i = 0, \ldots, k_1 - 1.$$

Combined with (5.9) and (5.13) this implies that

$$I^{\alpha g}(S_1, S_1 + k_1, x, u)$$
$$\geq (64(|f(x_f, u_f)| + 4 + a_0)) \sum \{\alpha(S_1 + i)\lambda : i \text{ is an integer}, 0 \leq i < k_1\}. \tag{5.39}$$

If $k_2 > 0$, then by (5.20) and (5.21),

$$|x(S_2 - i)| > M_1, \ i = 1, \ldots, k_2. \tag{5.40}$$

Assume that

$$k_2 > 0, \ i \in \{0, \ldots, k_2 - 1\}.$$

We estimate

$$I^{\alpha g}(S_2 - i - 1, S_2 - i, x, u).$$

Let

$$\hat{g} = g_{\alpha, S_2 - i - 1, S_2 - i}. \qquad (5.41)$$

In view of (5.40), (5.41), and property (i),

$$I^{\hat{g}}(S_2 - i - 1, S_2 - i, x, u) > 64(|f(x_f, u_f)| + 4 + 2a_0).$$

Together with (5.5) this implies that

$$64(|f(x_f, u_f)| + 4 + 2a_0) < I^{\hat{g}}(S_2 - i - 1, S_2 - i, x, u)$$
$$= a_0(\lambda^{-1} - 1) + \inf\{\alpha(t) : t \in [S_2 - i - 1, S_2 - i]\}^{-1} I^{\alpha g}(S_2 - i - 1, S_2 - i, x, u)$$

and

$$I^{\alpha g}(S_2 - i - 1, S_2 - i, x, u)$$
$$> (64(|f(x_f, u_f)| + 4 + 2a_0) - a_0(\lambda^{-1} - 1)) \inf\{\alpha(t) : t \in [S_2 - i - 1, S_2 - i]\},$$
$$i = 0, \dots, k_2 - 1.$$

Combined with (5.9) and (5.13) this implies that

$$I^{\alpha g}(S_2 - k_2, S_2, x, u)$$
$$\geq 64(|f(x_f, u_f)| + 4 + a_0) \sum \{\alpha(S_2 - i)\lambda : i \text{ is an integer}, 0 \leq i < k_2\}. \quad (5.42)$$

It follows from (5.11), (5.13), (5.28), (5.31), (5.35), (5.39), and (5.42) that

$$I^{\alpha g}(S_1, S_2, x, u)$$
$$\geq 2^{-1} M_* \inf\{\lambda\alpha(t) : t \in [S_1 + k_1, S_2 - k_2]\}$$
$$+ 64(|f(x_f, u_f)| + 4 + a_0) \sum \{\alpha(S_1 + i)\lambda : i \text{ is an integer}, 0 \leq i < k_1\}$$
$$+ 64(|f(x_f, u_f)| + 4 + a_0) \sum \{\alpha(S_2 - i)\lambda : i \text{ is an integer}, 0 \leq i < k_2\}. \quad (5.43)$$

By (5.14), (5.15), (5.23)–(5.25), and the choice of M_3 (see (5.11)), for all $t \in [S_1, S_1 + 1] \cup [S_2 - 1, S_2]$,

$$|g(t, \tilde{x}(t), \tilde{u}(t))| \leq 1 + |F(t, \tilde{x}(t), \tilde{u}(t))| \leq M_3 + 1. \qquad (5.44)$$

In view of (5.10), (5.14), (5.15), and (5.26), for all $t \in [S_1 + 1, S_2 - 1]$,

$$|g(t, \tilde{x}(t), \tilde{u}(t))| \leq 1 + |F(t, \tilde{x}(t), \tilde{u}(t))| \leq |f(x_f, u_f)| + 1. \tag{5.45}$$

Relations (5.9), (5.12), (5.31), (5.44), and (5.45) imply that

$$I^{\alpha g}(S_1, S_2, \tilde{x}, \tilde{u})$$

$$= I^{\alpha g}(S_1, S_1 + 1, \tilde{x}, \tilde{u}) + I^{\alpha g}(S_2 - 1, S_2, \tilde{x}, \tilde{u}) + \int_{S_1+1}^{S_2-1} \alpha(t) g(t, \tilde{x}(t), \tilde{u}(t)) dt$$

$$\leq (M_3 + 1)\alpha(S_1)\lambda^{-1} + (M_3 + 1)\alpha(S_2 - 1)\lambda^{-1}$$

$$+ \int_{S_1+1}^{S_2-1} \alpha(t)|f(x_f, u_f)| + 1) dt$$

$$\leq (M_3 + 1)\lambda^{-1}(\alpha(S_1) + \alpha(S_2 - 1))$$

$$+ \lambda^{-1}(|f(x_f, u_f)| + 1)\left[\sum \{\alpha(S_1 + i) : \ i \text{ is an integer}, 0 \leq i < k_1\} \right.$$

$$\left. + \sum \{\alpha(S_2 - i) : \ i \text{ is an integer}, 0 \leq i < k_2\} \right]$$

$$+ (S_2 - S_1 - k_2 - k_1)\lambda^{-1} \inf\{\alpha(t) : \ t \in [S_1 + k_1, S_2 - k_2]\}.$$

By (5.9), (5.13), (5.27), (5.31), (5.43), and the relation above,

$$\inf\{\alpha(t) : \ t \in [0, T]\} \geq I^{\alpha g}(S_1, S_2, x, u) - I^{\alpha g}(S_1, S_2, \tilde{x}, \tilde{u})$$

$$\geq \inf\{\alpha(t) : \ t \in [S_1 + k_1, S_2 - k_2]\}(2^{-1}M_*\lambda - 4\lambda^{-1})$$

$$+ 16(|f(x_f, u_f)| + 4 + a_0) - 2(|f(x_f, u_f)| + 1)$$

$$\times \left(\sum \{\alpha(S_1 + i) : \ i \text{ is an integer}, 0 \leq i < k_1\} \right.$$

$$\left. + \sum \{\alpha(S_2 - i) : \ i \text{ is an integer}, 0 \leq i < k_2\} \right)$$

$$- (M_3 + 1)\lambda^{-1}(\alpha(S_1) + \alpha(S_2 - 1))$$

$$\geq 8^{-1}M_* \inf\{\alpha(t) : \ t \in [S_1 + k_1, S_2 - k_2]\}$$

$$+ 12 \sum \{\alpha(S_1 + i) : \ i \text{ is an integer}, 0 \leq i < k_1\}$$

$$+ 12 \sum \{\alpha(S_2 - i) : \ i \text{ is an integer}, 0 \leq i < k_2\}$$

$$- 4M_3(\alpha(S_1) + \alpha(S_2 - 1))$$

$$\geq 16^{-1}M_*\lambda\alpha(S_1 + k_1)$$
$$+12\sum\{\alpha(S_1 + i) : i \text{ is an integer}, 0 \leq i < k_1\} - 4M_3\alpha(S_1)$$
$$+16^{-1}M_*\lambda\alpha(S_2 - k_2)$$
$$+12\sum\{\alpha(S_2 - i) : i \text{ is an integer}, 0 \leq i < k_2\}$$
$$-4M_3\alpha(S_2 - 1). \tag{5.46}$$

There are two cases: $k_1 \leq L$; $k_1 > L$. If $k_1 \leq L$, then in view of (5.9),

$$\alpha(S_1 + k_1) \geq \lambda\alpha(S_1)$$

and by (5.11) and (5.13),

$$16^{-1}M_*\lambda\alpha(S_1 + k_1) - 4M_3\alpha(S_1)$$
$$\geq 16^{-1}M_*\lambda^2\alpha(S_1) - 4M_3\alpha(S_1)$$
$$\geq \alpha(S_1)(64^{-1}M_* - M_3) \geq 128^{-1}M_*\alpha(S_1) > 8M_3\alpha(S_1).$$

If $k_1 > L$, then (5.9), (5.12), and (5.13) imply that

$$12\sum\{\alpha(S_1 + i) : i \text{ is an integer}, 0 \leq i < k_1\} - 4M_3\alpha(S_1)$$
$$\geq 12\sum\{\alpha(S_1 + i) : i = 0,\ldots,L\} - 4M_3\alpha(S_1)$$
$$\geq 12L\lambda\alpha(S_1) - 4M_3\alpha(S_1) \geq (6L - 4M_3)\alpha(S_1) \geq 8\alpha(S_1).$$

In the both cases

$$16^{-1}M_*\lambda\alpha(S_1 + k_1)$$
$$+12\sum\{\alpha(S_1 + i) : i \text{ is an integer}, 0 \leq i < k_1\}$$
$$-4M_3\alpha(S_1) \geq 8\alpha(S_1). \tag{5.47}$$

There are two cases: $k_2 \leq L$; $k_2 > L$. If $k_2 \leq L$, then in view of (5.9),

$$\alpha(S_2 - k_2) \geq \lambda\alpha(S_2 - 1)$$

and by (5.11)–(5.13),

$$16^{-1}M_*\lambda\alpha(S_2 - k_2) - 4M_3\alpha(S_2 - 1)$$
$$\geq 16^{-1}M_*\lambda^2\alpha(S_2 - 1) - 4M_3\alpha(S_2 - 1)$$
$$\geq \alpha(S_2 - 1)(64^{-1}M_* - 4M_3) \geq 256^{-1}M_*\alpha(S_2 - 1) > 8\alpha(S_2 - 1).$$

If $k_2 > L$, then (5.9), (5.12), and (5.13) imply that

$$12 \sum \{\alpha(S_2 - i) : i \text{ is an integer}, 0 \leq i < k_2\} - 4M_3\alpha(S_2 - 1)$$
$$\geq 12 \sum \{\alpha(S_2 - i) : i = 0, \ldots, L\} - 4M_3\alpha(S_2 - 1)$$
$$\geq 12L\lambda\alpha(S_2 - 1) - 4M_3\alpha(S_2 - 1)$$
$$\geq (6L - 4M_3)\alpha(S_2 - 1) \geq 8\alpha(S_2 - 1).$$

In the both cases

$$16^{-1}M_*\lambda\alpha(S_2 - k_2)$$
$$+12 \sum \{\alpha(S_2 - i) : i \text{ is an integer}, 0 \leq i < k_2\}$$
$$-4M_3\alpha(S_2 - 1) \geq 8\alpha(S_2 - 1). \tag{5.48}$$

By (5.46)–(5.48),

$$\inf\{\alpha(t) : t \in [0, T]\} \geq 8\alpha(S_1) + 8\alpha(S_2 - 1),$$

a contradiction. The contradiction we have reached proves Lemma 5.4 itself.

5.3   Proof of Theorem 5.1

By Lemma 5.4, there exist $L_0 > 0$, $\lambda_0 \in (0, 1)$, $M_0 > 0$ and a neighborhood $\mathcal{U}_0$ of F in $\mathfrak{M}_b$ such that the following property holds:

(i) for each $T > L_0$, each $g \in \mathcal{U}_0$, each borelian function $\alpha : [0, \infty) \to (0, 1]$ which satisfies

$$\alpha(t_1)\alpha(t_2)^{-1} \geq \lambda_0 \text{ for each } t_1, t_2 \in [0, T] \text{ satisfying } |t_1 - t_2| \leq L_0$$

and each $(x, u) \in X(A, B, 0, T)$ which satisfies at least one of the conditions (a), (b) of Theorem 5.1 we have

$$|x(t)| \leq M_0, \ t \in [0, T].$$

By Proposition 3.33, there exists $\Delta_0 > 0$ such that

$$\sigma(f, y, z, T) \leq T\mu(f) + \Delta_0 \tag{5.49}$$

for all $y, z \in R^n$ satisfying $|y|, |z| \leq M_0 + 1$ and all $T \geq 1$.

By Theorem 4.1, there exist $L_1 \geq 1$ and $\delta_1 \in (0, 1/8)$ and a neighborhood $\mathcal{U}_1 \subset \mathcal{U}_0$ of F in $\mathfrak{M}_b$ such that the following property holds:

(ii) for each $T > 4L_1$, each $g \in \mathcal{U}_1$, each $(x, u) \in X(A, B, 0, T)$ and each finite sequence of numbers $\{S_i\}_{i=0}^{q}$ which satisfies

$$S_0 = 0, \ S_{i+1} - S_i \in [L_1, 2L_1], \ i = 0, \ldots, q-1, \ S_q \in [T - 2L_1, T],$$
$$I^g(S_i, S_{i+1}, x, u) \leq (S_{i+1} - S_i)\mu(f) + \Delta_0 + 8$$

for each integer $i \in [0, q-1]$,

$$I^g(S_i, S_{i+2}, x, u) \leq \sigma(g, x(S_i), x(S_{i+2}), S_i, S_{i+2}) + \delta_1$$

for each nonnegative integer $i \leq q - 2$ and

$$I^g(S_{q-2}, T, x, u) \leq \sigma(g, x(S_{q-2}), x(T), S_{q-2}, T) + \delta_1,$$

there exist $p_1, p_2 \in [0, T]$ such that $p_1 \leq p_2, p_1 \leq 2L_1, p_2 > T - 4L_1$ and that

$$|x(t) - x_f| \leq \epsilon \text{ for all } t \in [p_1, p_2].$$

Moreover if $|x(0) - x_f| \leq \delta_1$, then $p_1 = 0$ and if $|x(T) - x_f| \leq \delta_1$, then $p_2 = T$. Choose positive numbers L, δ such that

$$L > 8(L_0 + L_1 + 1), \delta < \min\{\epsilon, \delta_1, 1\}/64. \tag{5.50}$$

By Proposition 4.8, there exists a neighborhood $\mathcal{U}_2 \subset \mathcal{U}_1$ of F in $\mathfrak{M}_b$ such that the following property holds:

(iii) for each $g \in \mathcal{U}_2$, each $T_1 \in R^1$, each $T_2 \in [T_1 + 8^{-1}, T_1 + 8L]$ and each $(x, u) \in X(A, B, T_1, T_2)$ satisfying

$$\min\{I^f(T_1, T_2, x, u), I^g(T_1, T_2, x, u)\}$$
$$\leq 16\Delta_0 + 16 + 16L|\mu(f)| + 16L$$

we have

$$|I^f(T_1, T_2, x, u) - I^g(T_1, T_2, x, u)| \leq \delta_1/16.$$

By (5.2) there exist $\epsilon_1 \in (0, \min\{\epsilon, 1\}), \gamma_1 > 1$ and $N_1 > 1$ such that

$$\{g \in \mathfrak{M}_b : \ (F, g) \in E(N_1, \epsilon_1, \gamma_1)\} \subset \mathcal{U}_2. \tag{5.51}$$

Set

$$\Delta_1 = \sup\{|f(z,\xi)| : (z,\xi) \in R^{n+m}, \ |z|, |\xi| \le N_1\}, \tag{5.52}$$

$$\mathcal{U} = \{g \in \mathfrak{M}_b : (g, F) \in E(N_1, \epsilon_1/4, 2^{-1}(1 + \gamma_1))\}. \tag{5.53}$$

Choose a number λ such that

$$\lambda_0 < \lambda < 1, \ (\lambda^{-1} - 1)(\Delta_1 + 1 + a_0) < \epsilon_1/4, \tag{5.54}$$

$$2^{-1}(1 + \gamma_1)[\lambda^{-1} + |a_0 - 1|(\lambda^{-1} - 1)] < \gamma_1. \tag{5.55}$$

Assume that

$$T > 2L, \ g \in \mathcal{U}, \tag{5.56}$$

a borelian function $\alpha : [0, \infty) \to (0, 1]$ satisfies

$$\alpha(t_1)\alpha(t_2)^{-1} \ge \lambda \ \text{for each} \ t_1, t_2 \in [0, T] \ \text{satisfying} \ |t_1 - t_2| \le L, \tag{5.57}$$

$$(x, u) \in X(A, B, 0, T)$$

and at least one of the conditions (a), (b) of Theorem 5.1 holds.
By property (i), (5.50), (5.56), and (5.57),

$$|x(t)| \le M_0, \ t \in [0, T]. \tag{5.58}$$

Set

$$q = \lfloor T/L_1 \rfloor = \max\{i \in R^1 : i \le T/L_1 \ \text{is an integer}\}, \tag{5.59}$$

$$S_0 = 0, \ S_i = iL_1, \ i = 0, \dots, \lfloor T/L_1 \rfloor - 1, \ S_q = T. \tag{5.60}$$

Let

$$j, i \in \{0, \dots, q-1\}, \ j - i \in \{1, 2\}. \tag{5.61}$$

We estimate $I^f(S_i, S_j, x, u)$. By conditions (a) and (b),

$$I^{\alpha g}(S_i, S_j, x, u) \le \sigma(\alpha g, x(S_i), x(S_j), S_i, S_j) + \delta \inf\{\alpha(t) : t \in [0, T]\}. \tag{5.62}$$

Set

$$\hat{g} = g_{\alpha, S_i, S_j}. \tag{5.63}$$

Clearly, $\hat{g} \in \mathfrak{M}_b$. In view of (5.5)–(5.7) and (5.63), for each

$$(y, v) \in X(A, B, S_i, S_j)$$

we have

$$I^{\hat{g}}(S_i, S_j, y, v) = a_0(\lambda^{-1} - 1)(S_j - S_i)$$
$$+ \inf\{\alpha(t) : t \in [S_i, S_j]\}^{-1} I^{\alpha g}(S_i, S_j, y, v). \tag{5.64}$$

It follows from (5.62) and (5.64) that

$$I^{\hat{g}}(S_i, S_j, x, u) \leq \sigma(\hat{g}, x(S_i), x(S_j), S_i, S_j) + \delta. \tag{5.65}$$

We show that

$$(F, \hat{g}) \in E(N_1, \epsilon_1, \gamma_1), \ \hat{g} \in \mathcal{U}_2. \tag{5.66}$$

Let $(t, z, \xi) \in R^{n+m+1}$ satisfy

$$|z|, \ |\xi| \leq N_1. \tag{5.67}$$

We claim that

$$|\hat{g}(t, z, \xi) - f(z, \xi)| \leq \epsilon_1. \tag{5.68}$$

In view of (5.5)–(5.7) and (5.63), we may assume without loss of generality that

$$t \in [S_i, S_j]. \tag{5.69}$$

By (5.53), (5.56), and (5.67),

$$|g(t, z, \xi) - f(z, \xi)| \leq \epsilon_1/4. \tag{5.70}$$

Relations (5.52), (5.67), and (5.70) imply that

$$|g(t, z, \xi)| \leq \Delta_1 + \epsilon_1/4. \tag{5.71}$$

It follows from (5.5), (5.50), (5.54), (5.57), (5.61), (5.63), (5.69), and (5.71) that

$$|\hat{g}(t, z, \xi) - g(t, z, \xi)|$$
$$\leq a_0(\lambda^{-1} - 1) + |g(t, z, \xi)||\alpha(t) \inf\{\alpha(s) : s \in [S_i, S_j]\}^{-1} - 1|$$
$$\leq a_0(\lambda^{-1} - 1) + (\Delta_1 + 1)(\lambda^{-1} - 1) < \epsilon_1/4.$$

Together with (5.70) this implies (5.68).
    Let $(t, z, \xi) \in R^{n+m+1}$ satisfy

$$|z| \leq N_1. \tag{5.72}$$

We claim that

$$(|\hat{g}(t, z, \xi)| + 1)(|f(z, \xi)| + 1)^{-1} \in [\gamma^{-1}, \gamma_1]. \tag{5.73}$$

In view of (5.5)–(5.7) and (5.63), we may assume without loss of generality that $t \in [S_i, S_j]$. By (5.53), (5.56), and (5.72),

$$(|f(z, \xi)| + 1)(|g(t, z, \xi)| + 1)^{-1} \in [(2^{-1}(\gamma_1 + 1))^{-1}, 2^{-1}(\gamma_1 + 1)]. \tag{5.74}$$

In view of (5.5), (5.63), and the inclusion $t \in [S_i, S_j]$,

$$\hat{g}(t, z, \xi) = \alpha(t) \inf\{\alpha(s) : s \in [S_i, S_j]\}^{-1} g(t, z, \xi) + a_0(\lambda^{-1} - 1). \tag{5.75}$$

The inclusion $t \in [S_i, S_j]$, (5.5), (5.50), (5.57), and (5.63) imply that

$$|\hat{g}(t, z, \xi)| + 1$$
$$\leq \alpha(t) \inf\{\alpha(s) : s \in [S_i, S_j]\}^{-1} |g(t, z, \xi)| + a_0(\lambda^{-1} - 1) + 1$$
$$\leq \lambda^{-1}|g(t, z, \xi)| + a_0(\lambda^{-1} - 1) + 1$$
$$\leq \lambda^{-1}(|g(t, z, \xi)| + 1) + (a_0 - 1)(\lambda^{-1} - 1)$$

and

$$(|\hat{g}(t, z, \xi)| + 1)(|g(t, z, \xi)| + 1)^{-1} \leq \lambda^{-1} + |a_0 - 1|(\lambda^{-1} - 1). \tag{5.76}$$

It follows from (5.5), (5.63), and the inclusion $t \in [S_i, S_j]$ that

$$g(t, z, \xi) = (\hat{g}(t, z, \xi) - a_0(\lambda^{-1} - 1))\alpha(t)^{-1} \inf\{\alpha(s) : s \in [S_i, S_j]\}.$$

Together with (5.56) and (5.57) this implies that

$$|g(t, z, \xi)| + 1 \leq |\hat{g}(t, z, \xi)|\lambda^{-1} + a_0(\lambda^{-1} - 1)\lambda^{-1} + 1$$
$$\leq \lambda^{-1}(|\hat{g}(t, z, \xi)| + 1) + (\lambda^{-1} - 1)(a_0 - 1)$$

and

$$(|g(t, z, \xi)| + 1)(|\hat{g}(t, z, \xi)| + 1)^{-1} \leq \lambda^{-1} + |a_0 - 1|(\lambda^{-1} - 1). \tag{5.77}$$

By (5.55), (5.56), (5.74), (5.76), (5.77), and the inclusion $t \in [S_i, S_j]$,

$$(|f(z, \xi)| + 1)(|\hat{g}(t, z, \xi)| + 1)^{-1}, \ (|\hat{g}(t, z, \xi)| + 1)(|f(z, \xi)| + 1)^{-1}$$
$$\leq 2^{-1}(1 + \gamma_1)(\lambda^{-1} + |a_0 - 1|(\lambda^{-1} - 1)) < \gamma_1$$

and (5.73) holds.

It follows from (5.51), (5.68), and (5.73) that relation (5.66) holds. In view of (5.49), (5.58), (5.59), and (5.61),

$$\sigma(f, x(S_i), x(S_j), S_j - S_i) \leq (S_j - S_i)\mu(f) + \Delta_0. \tag{5.78}$$

Property (iii), (5.66), and (5.78) imply that

$$\sigma(\hat{g}, x(S_i), x(S_j), S_i, S_j) \leq (S_j - S_i)\mu(f) + \Delta_0 + \delta_1/16 \tag{5.79}$$

and

$$|\sigma(f, x(S_i), x(S_j), S_j - S_i) - \sigma(\hat{g}, x(S_i), x(S_j), S_i, S_j)| \leq \delta_1/16. \tag{5.80}$$

By (5.65), (5.79), and property (iii),

$$|I^{\hat{g}}(S_i, S_j, x, u) - I^f(S_i, S_j, x, u)| \leq \delta_1/16. \tag{5.81}$$

It follows from (5.65), (5.78), (5.80), and (5.81) that

$$I^f(S_i, S_j, x, u) \leq \sigma(f, x(S_i), x(S_j), S_j - S_i) + \delta_1/4$$
$$\leq (S_j - S_i)\mu(f) + \Delta_0 + 1. \tag{5.82}$$

Therefore (5.82) holds for all integers i, j satisfying (5.61). Together with (5.50), (5.56), (5.59), (5.60), (5.82), and property (ii) this implies that there exist $p_1, p_2 \in [0, T]$ such that $p_1 \leq p_2$, $p_1 \leq 2L_1$, $p_2 > T - 4L_1$ and that

$$|x(t) - x_f| \leq \epsilon \text{ for all } t \in [p_1, p_2].$$

Moreover if $|x(0) - x_f| \leq \delta_1$, then $p_1 = 0$ and if $|x(T) - x_f| \leq \delta_1$, then $p_2 = T$. Theorem 5.1 is proved.   □

5.4 Proof of Theorem 5.2

By Lemma 3.38 applied to the triplet $(f, -A - B)$ there exist

$$\delta_1 \in (0, \epsilon/4)$$

such that the following property holds:

(P1)   for each $(x, u) \in X(-A, -B, 0, \tau_0)$ which satisfies

$$\pi_-^f(x(0)) \leq \inf(\pi_-^f) + 64\delta_1,$$

$$f(0, \tau_0, x, u) - \tau_0 \mu(f) - \pi_-^f(x(0)) + \pi_-^f(x(\tau_0)) \le 64\delta_1$$

there exists an $(f, -A, -B)$-overtaking optimal pair

$$(x^*, u^*) \in X(-A, -B, 0, \infty)$$

such that

$$\pi_-^f(x^*(0)) = \inf(\pi_-^f),$$
$$|x(t) - x^*(t)| \le \epsilon \text{ for all } t \in [0, \tau_0].$$

In view of Propositions 3.13, 3.14, and 3.29, there exists $\delta_2 \in (0, \delta_1)$ such that: for each $z \in R^n$ satisfying $|z - x_f| \le 2\delta_2$,

$$|\pi_-^f(z)| = |\pi_-^f(z) - \pi_-^f(x_f)| \le \delta_1/8; \tag{5.83}$$

for each $y, z \in R^n$ satisfying

$$|y - x_f| \le 2\delta_2, \ |z - x_f| \le 2\delta_2$$

we have

$$|\sigma(f, y, z, 1) - \mu(f)| \le \delta_1/8. \tag{5.84}$$

By Theorem 5.1 and Lemma 5.4, there exist $l_0 > 0$, $\delta_3 \in (0, \delta_2/8)$, $\lambda_0 \in (0, 1)$ and a neighborhood $\mathcal{U}_0$ of F in $\mathfrak{M}_b$ such that the following property holds:

(P2)   for each $T > 2l_0$, each $g \in \mathcal{U}_0$, each borelian function $\alpha : [0, \infty) \to (0, 1]$ which satisfies

$$\alpha(t_1)\alpha(t_2)^{-1} \ge \lambda_0 \text{ for each } t_1, t_2 \in [0, T] \text{ satisfying } |t_1 - t_2| \le l_0$$

and each $(x, u) \in X(A, B, 0, T)$ which satisfies

$$|x(0)| \le M, I^{\alpha g}(0, T, x, u) \le \sigma(\alpha g, x(0), 0, T) + \delta_3 \inf\{\alpha(t) : t \in [0, T]\}$$

we have

$$|x(t) - x_f| \le \delta_2 \text{ for all } t \in [l_0, T - l_0],$$
$$|x(t)| \le M_0 \text{ for all } t \in [0, T].$$

By Theorem 3.8, there exists an $(f, -A, -B)$-overtaking optimal pair

$$(\bar{x}_*, \bar{u}_*) \in X(-A, -B, 0, \infty)$$

such that

$$\pi_-^f(\bar{x}_*(0)) = \inf(\pi_-^f). \tag{5.85}$$

(A3) implies that there exists $l_1 > 0$ such that

$$|\bar{x}_*(t) - x_f| \le \delta_2 \text{ for all } t \ge l_1. \tag{5.86}$$

Choose numbers

$$L > 8(l_0 + l_1 + \tau_0) + 8, \tag{5.87}$$

$$T_0 \ge 2l_1 + 2l_0 + 2\tau_0 + 8 + 8L. \tag{5.88}$$

By Proposition 4.8, there exists a neighborhood $\mathcal{U}_1 \subset \mathcal{U}_0$ of F in $\mathfrak{M}_b$ such that the following property holds:

(P3)   for each $g \in \mathcal{U}_1$, each $T_1 \in R^1$, each $T_2 \in [T_1 + 8^{-1}, T_1 + 8T_0]$ and each $(x, u) \in X(A, B, T_1, T_2)$ satisfying

$$\min\{I^f(T_1, T_2, x, u), I^g(T_1, T_2, x, u)\}$$
$$\le (|\mu(f)| + 2)8T_0 + 8 + |\pi_-^f(\bar{x}_*(0))|$$

we have

$$|I^f(T_1, T_2, x, u) - I^g(T_1, T_2, x, u)| \le \delta_3/8.$$

By (5.2), there exist $\epsilon_1 \in (0, \min\{\epsilon, 1\})$, $\gamma_1 > 1$ and $N_1 > 1$ such that

$$\{g \in \mathfrak{M}_b : (F, g) \in E(N_1, \epsilon_1, \gamma_1)\} \subset \mathcal{U}_1. \tag{5.89}$$

Set

$$\mathcal{U} = \{g \in \mathfrak{M}_b : (g, F) \in E(N_1, \epsilon_1/4, 2^{-1}(1 + \gamma_1))\}, \tag{5.90}$$

$$\Delta_1 = \sup\{|f(z, \xi)| : (z, \xi) \in R^{n+m}, |z|, |\xi| \le N_1\}. \tag{5.91}$$

Choose a number $\lambda \in (0, 1)$ such that

$$\lambda > 8/9, \ 2T_0 a_0(\lambda^{-1} - 1) < \delta_3/8, \tag{5.92}$$

$$\lambda_0 < \lambda < 1, \ 16(\lambda^{-2} - 1)(\Delta_1 + 1 + |\mu(f)| + a_0) < \epsilon_1/4, \tag{5.93}$$

$$2^{-1}(1 + \gamma_1)[\lambda^{-1} + |a_0 - 1|(\lambda^{-1} - 1) + |a_0\lambda^{-1} - 1|(\lambda^{-1} - 1)] < \gamma_1 \tag{5.94}$$

and a positive number

$$\delta < \min\{\delta_1, \delta_2, \delta_3\}/8. \tag{5.95}$$

Assume that

$$T \geq T_0, \ g \in \mathcal{U}, \tag{5.96}$$

a borelian function $\alpha : [0, \infty) \to (0, 1]$ satisfies

$$\alpha(t_1)\alpha(t_2)^{-1} \geq \lambda \text{ for each } t_1, t_2 \in [0, T] \text{ satisfying } |t_1 - t_2| \leq T_0 \tag{5.97}$$

and $(x, u) \in X(A, B, 0, T)$ satisfies

$$|x(0)| \leq M,$$

$$I^{\alpha g}(0, T, x, u) \leq \sigma(\alpha g, x(0), 0, T) + \delta \inf\{\alpha(t) : \ t \in [0, T]\}. \tag{5.98}$$

By property (P2), (5.88), (5.93), and (5.96)–(5.98),

$$|x(t)| \leq M_0 \text{ for all } t \in [0, T], \tag{5.99}$$

$$|x(t) - x_f| \leq \delta_2 \text{ for all } t \in [l_0, T - l_0]. \tag{5.100}$$

In view of (5.88),

$$[T - l_0 - l_1 - \tau_0 - 4, T - l_0 - l_1 - \tau_0] \subset [l_0, T - l_0 - l_1 - \tau_0]. \tag{5.101}$$

Relations (5.100) and (5.101) imply that

$$|x(t) - x_f| \leq \delta_2 \text{ for all } t \in [T - l_0 - l_1 - \tau_0 - 4, T - l_0 - l_1 - \tau_0]. \tag{5.102}$$

Proposition 3.27 implies that there exists $(x_1, u_1) \in X(A, B, 0, T)$ such that

$$x_1(t) = x(t), \ u_1(t) = u(t), \ t \in [0, T - l_0 - l_1 - \tau_0 - 4], \tag{5.103}$$

$$x_1(t) = \bar{x}_*(T - t), \ u_1(t) = \bar{u}_*(T - t), \ t \in [T - l_0 - l_1 - \tau_0 - 3, T], \tag{5.104}$$

$$I^f(T - l_0 - l_1 - \tau_0 - 4, T - l_0 - l_1 - \tau_0 - 3, x_1, u_1)$$
$$= \sigma(f, x(T - l_0 - l_1 - \tau_0 - 4), \bar{x}_*(l_0 + l_1 + \tau_0 + 3)). \tag{5.105}$$

By (5.103)–(5.105) and (5.98),

$$
\begin{aligned}
-\delta \inf\{\alpha(t) : \ t \in [0,T]\} &\le I^{\alpha g}(0,T,x_1,u_1) - I^{\alpha g}(0,T,x,u) \\
&= I^{\alpha g}(T - l_0 - l_1 - \tau_0 - 4, T - l_0 - l_1 - \tau_0 - 3, x_1, u_1) \\
&\quad + I^{\alpha g}(T - l_0 - l_1 - \tau_0 - 3, T, x_1, u_1) \\
&\quad - I^{\alpha g}(T - l_0 - l_1 - \tau_0 - 4, T - l_0 - l_1 - \tau_0 - 3, x, u) \\
&\quad - I^{\alpha g}(T - l_0 - l_1 - \tau_0 - 3, T, x, u). \tag{5.106}
\end{aligned}
$$

Let $S_1, S_2 \in [0,T]$ satisfy

$$
S_2 - S_1 \in [8^{-1}, L]. \tag{5.107}
$$

We show that

$$
(F, g_{\alpha,S_1,S_2}) \in E(N_1, \epsilon_1, \gamma_1), \ g_{\alpha,S_1,S_2} \in \mathcal{U}_1. \tag{5.108}
$$

Set

$$
\hat{g} = g_{\alpha,S_1,S_2}. \tag{5.109}
$$

Let $(t, z, \xi) \in R^{n+m+1}$ satisfy

$$
|z|, \ |\xi| \le N_1. \tag{5.110}
$$

We claim that

$$
|\hat{g}(t, z, \xi) - f(z, \xi)| \le \epsilon_1. \tag{5.111}
$$

In view of (5.5)–(5.7) and (5.109), we may assume without loss of generality that $t \in [S_1, S_2]$. By (5.90), (5.96), and (5.110),

$$
|g(t, z, \xi) - f(z, \xi)| \le \epsilon_1/4. \tag{5.112}
$$

Relations (5.91), (5.110), and (5.112) imply that

$$
|g(t, z, \xi)| \le \Delta_1 + \epsilon_1/4. \tag{5.113}
$$

It follows from (5.5), (5.88), (5.93), (5.97), (5.107), (5.109), and (5.113) that

$$
\begin{aligned}
&|\hat{g}(t, z, \xi) - g(t, z, \xi)| \\
&\le a_0(\lambda^{-1} - 1) + |g(t, z, \xi)||\alpha(t) \inf\{\alpha(s) : \ s \in [S_1, S_2]\}^{-1} - 1| \\
&\le a_0(\lambda^{-1} - 1) + (\Delta_1 + 1)(\lambda^{-1} - 1) < \epsilon_1/4.
\end{aligned}
$$

Together with (5.112) this implies (5.111).

Let $(t, z, \xi) \in R^{n+m+1}$ satisfy

$$|z| \le N_1. \tag{5.114}$$

We claim that

$$(|\hat{g}(t, z, \xi)| + 1)(|f(z, \xi)| + 1)^{-1} \in [\gamma^{-1}, \gamma_1]. \tag{5.115}$$

In view of (5.5)–(5.7) and (5.109), we may assume without loss of generality that $t \in [S_1, S_2]$. By (5.90), (5.96), and (5.114),

$$(|f(z, \xi)| + 1)(|g(t, z, \xi)| + 1)^{-1} \in [(2^{-1}(\gamma_1 + 1))^{-1}, 2^{-1}(\gamma_1 + 1)]. \tag{5.116}$$

In view of (5.5), (5.88), (5.97), (5.107), and (5.109),

$$\hat{g}(t, z, \xi) = \alpha(t) \inf\{\alpha(s) : s \in [S_1, S_2]\}^{-1} g(t, z, \xi) + a_0(\lambda^{-1} - 1), \tag{5.117}$$

$$|\hat{g}(t, z, \xi)| + 1$$
$$\le \alpha(t) \inf\{\alpha(s) : s \in [S_1, S_2]\}^{-1}|g(t, z, \xi)| + a_0(\lambda^{-1} - 1) + 1$$
$$\le \lambda^{-1}|g(t, z, \xi)| + a_0(\lambda^{-1} - 1) + 1$$
$$\le \lambda^{-1}(|g(t, z, \xi)| + 1) + (a_0 - 1)(\lambda^{-1} - 1)$$

and

$$(|\hat{g}(t, z, \xi)| + 1)(|g(t, z, \xi)| + 1)^{-1} \le \lambda^{-1} + |a_0 - 1|(\lambda^{-1} - 1). \tag{5.118}$$

It follows from (5.117) that

$$g(t, z, \xi) = (\hat{g}(t, z, \xi) - a_0(\lambda^{-1} - 1))\alpha(t)^{-1} \inf\{\alpha(s) : s \in [S_1, S_2]\}.$$

Together with (5.88), (5.97), and (5.107) this implies that

$$|g(t, z, \xi)| + 1 \le |\hat{g}(t, z, \xi)|\lambda^{-1} + a_0(\lambda^{-1} - 1)\lambda^{-1} + 1$$
$$\le \lambda^{-1}(|\hat{g}(t, z, \xi)| + 1) + (\lambda^{-1} - 1)|a_0\lambda^{-1} - 1|.$$

By the relation above,

$$(|g(t, z, \xi)| + 1)(|\hat{g}(t, z, \xi)| + 1)^{-1} \le \lambda^{-1} + |a_0\lambda^{-1} - 1|(\lambda^{-1} - 1).$$

Together with (5.94), (5.116), and (5.118) this implies that

$$(|f(z, \xi)| + 1)(|\hat{g}(t, z, \xi)| + 1)^{-1}, \ (|\hat{g}(t, z, \xi)| + 1)(|f(z, \xi)| + 1)^{-1}$$
$$\leq 2^{-1}(1 + \gamma_1)(\lambda^{-1} + |a_0 - 1|(\lambda^{-1} - 1) + |a_0/\lambda - 1|(\lambda^{-1} - 1)) \ < \gamma_1$$

and (5.115) holds.

Thus the following property holds:

(P4)   Relation (5.108) holds for all $S_1, S_2 \in [0, T]$ satisfying $S_2 - S_1 \in [1/8, L]$.

It follows from (5.84), (5.86), (5.102), and (5.105) that

$$I^f(T - l_0 - l_1 - \tau_0 - 4, T - l_0 - l_1 - \tau_0 - 3, x_1, u_1) \leq \mu(f) + \delta_1/8. \qquad (5.119)$$

Set

$$\hat{g} = g_{\alpha, T - l_0 - l_1 - \tau_0 - 4, T - l_0 - l_1 - \tau_0 - 3}. \qquad (5.120)$$

By (5.108), (5.119), (5.120), and properties (P3) and (P4),

$$I^{\hat{g}}(T - l_0 - l_1 - \tau_0 - 4, T - l_0 - l_1 - \tau_0 - 3, x_1, u_1) \leq \mu(f) + \delta_1/8 + \delta_3/8. \qquad (5.121)$$

It follows from (5.102) and the choice of δ_2 (see (5.84)) that

$$I^f(T - l_0 - l_1 - \tau_0 - 4, T - l_0 - l_1 - \tau_0 - 3, x, u) \geq \mu(f) - \delta_1/8. \qquad (5.122)$$

If

$$I^{\hat{g}}(T - l_0 - l_1 - \tau_0 - 4, T - l_0 - l_1 - \tau_0 - 3, x, u)$$
$$< \mu(f) - \delta_1/2,$$

then by (5.108), (5.120), and properties (P3) and (P4),

$$I^f(T - l_0 - l_1 - \tau_0 - 4, T - l_0 - l_1 - \tau_0 - 3, x, u)$$
$$< \mu(f) - \delta_1/2 + \delta_3/8 < \mu(f) - 3\delta_1/8$$

and this contradicts (5.122). Thus

$$I^{\hat{g}}(T - l_0 - l_1 - \tau_0 - 4, T - l_0 - l_1 - \tau_0 - 3, x, u) \geq \mu(f) - \delta_1/2. \qquad (5.123)$$

By (5.5), (5.120), and (5.121),

$$I^{\alpha g}(T - l_0 - l_1 - \tau_0 - 4, T - l_0 - l_1 - \tau_0 - 3, x_1, u_1)$$
$$\leq \inf\{\alpha(s) : \ s \in [T - l_0 - l_1 - \tau_0 - 4, T - l_0 - l_1 - \tau_0 - 3]\}$$

$$\times I^{\hat{g}}(T - l_0 - l_1 - \tau_0 - 4, T - l_0 - l_1 - \tau_0 - 3, x_1, u_1)$$

$$\leq (\mu(f) + \delta_1/8 + \delta_3/8) \inf\{\alpha(s) : s \in [T - l_0 - l_1 - \tau_0 - 4, T - l_0 - l_1 - \tau_0 - 3]\}. \tag{5.124}$$

In view of (5.5), (5.120), and (5.123),

$$I^{\alpha g}(T - l_0 - l_1 - \tau_0 - 4, T - l_0 - l_1 - \tau_0 - 3, x, u)$$

$$\geq \inf\{\alpha(s) : s \in [T - l_0 - l_1 - \tau_0 - 4, T - l_0 - l_1 - \tau_0 - 3]\}$$

$$\times (I^{\hat{g}}(T - l_0 - l_1 - \tau_0 - 4, T - l_0 - l_1 - \tau_0 - 3, x, u) - a_0(\lambda^{-1} - 1))$$

$$\geq (\mu(f) - \delta_1/2 - a_0(\lambda^{-1} - 1)) \inf\{\alpha(s) : s \in [T - l_0 - l_1 - \tau_0 - 4, T - l_0 - l_1 - \tau_0 - 3]\}. \tag{5.125}$$

It follows from (5.92), (5.124), and (5.125) that

$$I^{\alpha g}(T - l_0 - l_1 - \tau_0 - 4, T - l_0 - l_1 - \tau_0 - 3, x_1, u_1)$$

$$- I^{\alpha g}(T - l_0 - l_1 - \tau_0 - 4, T - l_0 - l_1 - \tau_0 - 3, x, u)$$

$$\leq \inf\{\alpha(s) : s \in [T - l_0 - l_1 - \tau_0 - 4, T - l_0 - l_1 - \tau_0 - 3]\}(5\delta_1/8 + \delta_3/8 + a_0(\lambda^{-1} - 1))$$

$$\leq \inf\{\alpha(s) : s \in [T - l_0 - l_1 - \tau_0 - 4, T - l_0 - l_1 - \tau_0 - 3]\}(5\delta_1/8 + \delta_3/4). \tag{5.126}$$

Relations (5.106) and (5.126) imply that

$$I^{\alpha g}(T - l_0 - l_1 - \tau_0 - 3, T, x_1, u_1) - I^{\alpha g}(T - l_0 - l_1 - \tau_0 - 3, T, x, u)$$

$$\geq -\delta \inf\{\alpha(t) : t \in [0, T]\}$$

$$- \inf\{\alpha(t) : t \in [T - l_0 - l_1 - \tau_0 - 4, T - l_0 - l_1 - \tau_0 - 3]\}(5\delta_1/8 + \delta_3/4)$$

$$\geq -\inf\{\alpha(t) : t \in [T - l_0 - l_1 - \tau_0 - 4, T - l_0 - l_1 - \tau_0 - 3]\}(5\delta_1/8 + 3\delta_3/8). \tag{5.127}$$

Since $(\bar{x}_*, \bar{u}_*)$ is an $(\bar{f}, -A, -B)$-overtaking optimal pair it follows from (5.104) and Proposition 3.12 that

$$I^f(T - l_0 - l_1 - \tau_0 - 3, T, x_1, u_1) = I^f(0, l_0 + l_1 + \tau_0 + 3, \bar{x}_*, \bar{u}_*)$$

$$= \mu(f)(l_0 + l_1 + \tau_0 + 3) + \pi_-^f(\bar{x}_*(0))$$

$$- \pi_-^f(\bar{x}_*(l_0 + l_1 + \tau_0 + 3)). \tag{5.128}$$

By (5.86) and the choice of δ_2 (see (5.83)),

$$\pi_-^f(\bar{x}_*(l_0 + l_1 + \tau_0 + 3)) \leq \delta_1/8.$$

By (5.128) and the relation above,

$$I^f(T - l_0 - l_1 - \tau_0 - 3, T, x_1, u_1) \leq \pi_-^f(\bar{x}_*(0)) + \mu(f)(l_0 + l_1 + \tau_0 + 3) + \delta_1/8. \tag{5.129}$$

Set

$$\hat{g} = g_{\alpha,T-l_0-l_1-\tau_0-3,T}.\tag{5.130}$$

By (5.87), (5.129), (5.130), and properties (P3) and (P4),

$$I^{\hat{g}}(T-l_0-l_1-\tau_0-3,T,x_1,u_1) \le \pi^f_-(\bar{x}_*(0))+\mu(f)(l_0+l_1+\tau_0+3)+\delta_1/8+\delta_3/8.\tag{5.131}$$

It follows from (5.5), (5.130), and (5.131) that

$$I^{\alpha g}(T-l_0-l_1-\tau_0-3,T,x_1,u_1)$$
$$\le \inf\{\alpha(s) : s \in [T-l_0-l_1-\tau_0-3,T]\}I^{\hat{g}}(T-l_0-l_1-\tau_0-3,T,x_1,u_1)$$
$$\le \inf\{\alpha(s) : s \in [T-l_0-l_1-\tau_0-3,T]\}$$
$$\times (\pi^f_-(\bar{x}_*(0)) + \mu(f)(l_0+l_1+\tau_0+3) + 3\delta_1/8).\tag{5.132}$$

In view of (5.127) and (5.132),

$$I^{\alpha g}(T-l_0-l_1-\tau_0-3,T,x,u)$$
$$\le \inf\{\alpha(t) : t \in [T-l_0-l_1-\tau_0-4, T-l_0-l_1-\tau_0-3]\}(5\delta_1/8+3\delta_3/8)$$
$$+\inf\{\alpha(s) : s \in [T-l_0-l_1-\tau_0-3,T]\}$$
$$\times (\pi^f_-(\bar{x}_*(0)) + \mu(f)(l_0+l_1+\tau_0+3) + 3\delta_1/8).\tag{5.133}$$

Relations (5.5), (5.87), (5.92), (5.97), (5.130), and (5.133) imply that

$$I^{\hat{g}}(T-l_0-l_1-\tau_0-3,T,x,u)$$
$$\le \inf\{\alpha(s) : s \in [T-l_0-l_1-\tau_0-3,T]\}^{-1}$$
$$\times I^{\alpha g}(T-l_0-l_1-\tau_0-3,T,x,u) + a_0(\lambda^{-1}-1)(l_0+l_1+\tau_0+3)$$
$$\le \inf\{\alpha(s) : s \in [T-l_0-l_1-\tau_0-3,T]\}^{-1}$$
$$\times \inf\{\alpha(t) : t \in [T-l_0-l_1-\tau_0-4, T-l_0-l_1-\tau_0-3]\}(5\delta_1/8+3\delta_3/8)$$
$$+\pi^f_-(\bar{x}_*(0)) + \mu(f)(l_0+l_1+\tau_0+3) + 3\delta_1/8$$
$$+ a_0(\lambda^{-1}-1)(l_0+l_1+\tau_0+3)$$
$$\le \lambda^{-1}(5\delta_1/8+3\delta_3/8) + a_0(\lambda^{-1}-1)(l_0+l_1+\tau_0+3)$$
$$+\pi^f_-(\bar{x}_*(0)) + \mu(f)(l_0+l_1+\tau_0+3) + 3\delta_1/8$$
$$\le \pi^f_-(\bar{x}_*(0)) + \mu(f)(l_0+l_1+\tau_0+3) + 9(5\delta_1+3\delta_3) + 3\delta_1/8 + \delta_3/8.\tag{5.134}$$

It follows from properties (P3) and (P4), (5.88), (5.130), and (5.134) that

$$I^f(T - l_0 - l_1 - \tau_0 - 3, T, x, u)$$
$$\leq \pi^f_-(\tilde{x}_*(0)) + \mu(f)(l_0 + l_1 + \tau_0 + 3)$$
$$+ 9(5\delta_1 + 3\delta_3) + 3\delta_1/8 + \delta_3/2. \tag{5.135}$$

Set

$$\tilde{x}(t) = x(T - t), \quad \tilde{u}(t) = u(T - t), \quad t \in [0, T]. \tag{5.136}$$

Clearly, $(\tilde{x}, \tilde{u}) \in X(-A, -B, 0, T)$ and in view of (5.135), (5.136),

$$I^f(0, l_0 + l_1 + \tau_0 + 3, \tilde{x}, \tilde{u}) = I^f(T - l_0 - l_1 - \tau_0 - 3, T, x, u)$$
$$\leq \pi^f_-(\tilde{x}_*(0)) + \mu(f)(l_0 + l_1 + \tau_0 + 3) + 60\delta_1. \tag{5.137}$$

It follows from (5.102) and (5.136) that

$$|\tilde{x}(l_0 + l_1 + \tau_0 + 3) - x_f| \leq \delta_2.$$

By the relation above and the choice of δ_2 (see (5.83)),

$$|\pi^f_-(\tilde{x}(l_0 + l_1 + \tau_0 + 3))| \leq \delta_1/8. \tag{5.138}$$

By Propositions 3.11, (5.137), and (5.138),

$$\pi^f_-(\tilde{x}(0)) - \pi^f_-(\tilde{x}_*(0))$$
$$+ I^f(0, \tau_0, \tilde{x}, \tilde{u}) - \tau_0\mu(f) - \pi^f_-(\tilde{x}(0)) + \pi^f_-(\tilde{x}(\tau_0))$$
$$\leq \pi^f_-(\tilde{x}(0)) - \pi^f_-(\tilde{x}_*(0)) + I^f(0, l_0 + l_1 + \tau_0 + 3, \tilde{x}, \tilde{u})$$
$$- \mu(f)(l_0 + l_1 + \tau_0 + 3) - \pi^f_-(\tilde{x}(0)) + \pi^f_-(\tilde{x}(l_0 + l_1 + \tau_0 + 3))$$
$$\leq \pi^f_-(\tilde{x}(0)) - \pi^f_-(\tilde{x}_*(0))$$
$$+ \mu(f)(l_0 + l_1 + \tau_0 + 3) + 60\delta_1$$
$$+ \pi^f_-(\tilde{x}_*(0)) - \mu(f)(l_0 + l_1 + \tau_0 + 3) - \pi^f_-(\tilde{x}(0)) + \delta_1/8$$
$$\leq 60\delta_1 + \delta_1/8 \leq 61\delta_1. \tag{5.139}$$

By (5.139), (5.85), and Proposition 3.11,

$$\pi^f_-(\tilde{x}(0)) \leq \inf(\pi^f_-) + 61\delta_1, \tag{5.140}$$
$$I^f(0, \tau_0, \tilde{x}, \tilde{u}) - \tau_0\mu(f) - \pi^f_-(\tilde{x}(0)) + \pi^f_-(\tilde{x}(\tau_0)) \leq 61\delta_1. \tag{5.141}$$

It follows from (5.140), (5.141), and property (P1) that there exists an $(f, -A, -B)$-overtaking optimal pair

$$(x_*, u_*) \in X(A, -B, 0, \infty)$$

such that

$$\pi_-^f(x_*(0)) = \inf(\pi_-^f),$$

$$|x(T - t) - x_*(t)| = |\tilde{x}(t) - x_*(t)| \leq \epsilon \text{ for all } t \in [0, \tau_0].$$

Theorem 5.2 is proved.   □

Chapter 6
Dynamic Zero-Sum Games with Linear Constraints

In this chapter we study the existence and turnpike properties of approximate solutions for a class of dynamic continuous-time two-player zero-sum games without using convexity-concavity assumptions. We describe the structure of approximate solutions which is independent of the length of the interval, for all sufficiently large intervals and show that approximate solutions are determined mainly by the objective function, and are essentially independent of the choice of interval and endpoint conditions.

6.1 Preliminaries and Main Results

We use the notation and definitions of Sects. 3.1 and 3.2 of Chap. 3.

Let n_1, n_2, m_1, m_2 be natural numbers, $f : R^{m_1} \times R^{m_1} \times R^{n_2} \times R^{m_2} \to R^1$ be a Borel measurable function, A_1, B_1, A_2, B_2 given matrices of dimensions $n_1 \times n_1, n_1 \times m_1, n_2 \times n_2, n_2 \times m_2$, respectively. Linear control controllable systems are described by

$$x'(t) = A_1 x(t) + B_1 u(t), \tag{6.1}$$

$$y'(t) = A_2 y(t) + B_2 v(t), \tag{6.2}$$

for almost every (a. e.) $t \in \mathcal{I}$, where $\mathcal{I}$ is either R^1 or $[T_1, \infty)$ or $[T_1, T_2]$ (here $-\infty < T_1 < T_2 < \infty$), $x : \mathcal{I} \to R^{n_1}, y : \mathcal{I} \to R^{n_2}$ are a. c. functions and the control functions $u : \mathcal{I} \to R^{m_1}, v : \mathcal{I} \to R^{m_2}$ are Lebesgue measurable functions.

Given $z_1, z_2 \in R^{n_1}$, $\xi_1, \xi_2 \in R^{n_2}$, and a positive number T we consider a continuous-time two-player zero-sum game over the interval $[0, T]$ denoted by $\Gamma(z_1, z_2, \xi_1, \xi_2, T)$. For this game the set of strategies for the first player is the set of all pairs $(x, u) \in X(A_1, B_1, 0, T)$ satisfying $x(0) = z_1$ and $x(T) = z_2$, the set

© Springer International Publishing Switzerland 2015
A.J. Zaslavski, *Turnpike Theory of Continuous-Time Linear Optimal Control Problems*, Springer Optimization and Its Applications 104,
DOI 10.1007/978-3-319-19141-6_6

of strategies for the second player is the set of all pairs $(y, v) \in X(A_2, B_2, 0, T)$ satisfying $y(0) = \xi_1$ and $y(T) = \xi_2$, and the objective function for the first player associated with the strategies

$$(x, u) \in X(A_1, B_1, 0, T) \text{ and } (y, v) \in X(A_2, B_2, 0, T)$$

is given by

$$\int_0^T f(x(t), u(t), y(t), v(t)) dt$$

if this integrand is well defined in the sense which is explained below.

We recall that $a_0 > 0$ and $\psi : [0, \infty) \to [0, \infty)$ be an increasing function such that

$$\lim_{t \to \infty} \psi(t) = \infty.$$

Suppose that

$$x_f \in R^{n_1}, \ u_f \in R^{m_1}, \ y_f \in R^{n_2}, \ v_f \in R^{m_2}$$

and

$$A_1 x_f + B_1 u_f = 0, \ A_2 y_f + B_2 v_f = 0. \tag{6.3}$$

We suppose that the following assumption holds.

(C1)

   (i) the function f is bounded on all bounded subsets of $R^{n_1} \times R^{m_1} \times R^{n_2} \times R^{m_2}$;
   (ii) the function $f(\cdot, \cdot, y_f, v_f) : R^{n_1} \times R^{m_1} \to R^1$ is continuous;
   (iii) for each $(x, u) \in R^{n_1} \times R^{m_1}$,

$$f(x, u, y_f, v_f) \geq \max\{\psi(|x|), \ \psi(|u|),$$
$$\psi([|A_1 x + B_1 u| - a_0|x|]_+)[|A_1 x + B_1 u| - a_0|x|]_+\} - a_0;$$

   (iv) the function $f(x_f, u_f, \cdot, \cdot) : R^{n_2} \times R^{m_2} \to R^1$ is continuous;
   (v) for each $(y, v) \in R^{n_2} \times R^{m_2}$,

$$f(x_f, u_f, y, v) \leq -\max\{\psi(|y|), \ \psi(|v|),$$
$$\psi([|A_2 y + B_2 v| - a_0|y|]_+)[|A_2 y + B_2 v| - a_0|y|]_+\} + a_0;$$

   (vi) for each $x \in R^{n_1}$ the function $f(x, \cdot, y_f, v_f) : R^{m_1} \to R^1$ is convex and for each $y \in R^{n_2}$ the function $f(x_f, u_f, y, \cdot) : R^{m_2} \to R^1$ is concave;

(vii) for each $M, \epsilon > 0$ there exist $\Gamma, \delta > 0$ such that

$$|f(x_1, u_1, y_f, v_f) - f(x_2, u_2, y_f, v_f)|$$
$$\leq \epsilon \max\{f(x_1, u_1, y_f, v_f), f(x_2, u_2, y_f, v_f)\}$$

for each $u_1, u_2 \in R^{m_1}$ and each $x_1, x_2 \in R^{n_1}$ which satisfy

$$|x_i| \leq M, \ |u_i| \geq \Gamma, \ i = 1, 2,$$
$$\max\{|x_1 - x_2|, |u_1 - u_2|\} \leq \delta;$$

(viii) for each $M, \epsilon > 0$ there exist $\Gamma, \delta > 0$ such that

$$|f(x_f, u_f, y_1, v_1) - f(x_f, u_f, y_2, v_2)|$$
$$\leq \epsilon \max\{-f(x_f, u_f, y_1, v_1), -f(x_f, u_f, y_2, v_2)\}$$

for each $y_1, y_2 \in R^{n_2}$ and each $v_1, v_2 \in R^{m_2}$ which satisfy

$$|y_i| \leq M, \ |v_i| \geq \Gamma, \ i = 1, 2,$$
$$\max\{|y_1 - y_2|, |v_1 - v_2|\} \leq \delta;$$

(ix) for each $K > 0$ there exist a constant $a_K > 0$ and an increasing function

$$\psi_K : [0, \infty) \to [0, \infty)$$

such that

$$\psi_K(t) \to \infty \text{ as } t \to \infty$$

and

$$f(x, u, y_f, u_f) \geq \psi_K(|u|)|u| - a_K$$

for each $u \in R^{m_1}$ and each $x \in R^{n_1}$ satisfying $|x| \leq K$, and

$$-f(x_f, u_f, y, v) \geq \psi_K(|v|)|v| - a_K$$

for each $v \in R^{m_2}$ and each $y \in R^{n_2}$ satisfying $|y| \leq K$;

(x) for each $M > 0$, there is a number $c_M > 0$ such that

$$f(x, u, y, v) \geq -c_M$$

for each $x \in R^{n_1}$, each $u \in R^{m_1}$, each $y \in R^{n_2}$, and each $v \in R^{m_2}$ satisfying $|y|, |v| \le M$ and

$$f(x, u, y, v) \le c_M$$

for each $x \in R^{n_1}$, each $u \in R^{m_1}$, each $y \in R^{n_2}$, and each $v \in R^{m_2}$ satisfying $|x|, |u| \le M$.

Define

$$f^{(1)}(x, u) = f(x, u, y_f, v_f), \quad (x, u) \in R^{n_1} \times R^{m_1}, \tag{6.4}$$

$$f^{(2)}(y, v) = -f(x_f, u_f, y, v), \quad (y, v) \in R^{n_2} \times R^{m_2}. \tag{6.5}$$

It is not difficult to see that (A1) holds for $f = f^{(i)}$, $i = 1, 2$. For $f = f^{(i)}$, $i = 1, 2$, we defined $\mu(f^{(i)})$ by (3.5):

$$\mu(f^{(i)}) = \inf\{\liminf_{T \to \infty} T^{-1} I^{f^{(i)}}(0, T, x, u) : (x, u) \in X(A_i, B_i, 0, \infty)\}. \tag{6.6}$$

Propositions 3.1 and 3.3 applied to the triplets $(f^{(i)}, A_i, B_i)$, $i = 1, 2$ imply the following results.

Proposition 6.1.

1.

$$\mu(f^{(1)}) = f(x_f, u_f, y_f, v_f)$$

if and only if there is $c > 0$ such that

$$I^{f^{(1)}}(0, T, x, u) \ge Tf(x_f, u_f, y_f, v_f) - c$$

for each $T > 0$ and each $(x, u) \in X(A_1, B_1, 0, \infty)$.

2.

$$\mu(f^{(2)}) = -f(x_f, u_f, y_f, v_f)$$

if and only if there is $c > 0$ such that

$$I^{f^{(2)}}(0, T, x, u) \ge -Tf(x_f, u_f, y_f, v_f) - c$$

for each $T > 0$ and each $(x, u) \in X(A_2, B_2, 0, \infty)$.

We suppose that the following assumption holds.

(C2) There is $c_* > 0$ such that for each $T > 0$ and each $(x, u) \in X(A_1, B_1, 0, \infty)$,

$$\int_0^T f(x(t), u(t), y_f, v_f)dt \geq Tf(x_f, u_f, y_f, v_f) - c_*$$

and for each $T > 0$ and each $(y, v) \in X(A_2, B_2, 0, \infty)$,

$$\int_0^T f(x_f, u_f, y(t), v(t))dt \leq Tf(x_f, u_f, y_f, v_f) + c_*.$$

By (C2) and Proposition 6.1,

$$\mu(f^{(1)}) = -\mu(f^{(2)}) = f(x_f, u_f, y_f, v_f). \tag{6.7}$$

We suppose that the following assumption holds.

(C3) If $(x, u) \in R^{n_1} \times R^{m_1}$ satisfies

$$A_1 x + B_1 u = 0, \ \mu(f^{(1)}) = f(x, u, y_f, u_f),$$

then $x = x_f$;
If $(x, u) \in R^{n_2} \times R^{m_2}$ satisfies

$$A_2 x + B_2 u = 0, \ \mu(f^{(2)}) = -f(x_f, u_f, x, u),$$

then $x = y_f$.

Clearly, (A2) holds for $f = f^{(i)}$, $i = 1, 2$.
Let $\bar{R} = R^1 \cup \{\infty, -\infty\}$. We suppose that for all real numbers λ,

$$\lambda + \infty = \infty, \ \lambda - \infty = -\infty, \ -\infty < \lambda < \infty,$$

$$\max\{\lambda, \infty\} = \infty, \ \max\{\lambda, -\infty\} = \lambda, \min\{\lambda, \infty\} = \lambda, \ \min\{\lambda, -\infty\} = -\infty.$$

For each $z \in R^1$ set

$$z^+ = \max\{z, 0\}, \ z^- = \max\{-z, 0\}.$$

Define

$$f^+(x_1, x_2, y_1, y_2) = \max\{f(x_1, x_2, y_1, y_2), 0\},$$
$$f^-(x_1, x_2, y_1, y_2) = \max\{-f(x_1, x_2, y_1, y_2), 0\}$$

for all $x_1 \in R^{m_1}$, $x_2 \in R^{m_1}$, $y_1 \in R^{n_2}$ and $y_2 \in R^{m_2}$.
Let $-\infty < T_1 < T_2 < \infty$,

$$(x, u) \in X(A_1, B_1, T_1, T_2), \ (y, v) \in X(A_2, B_2, T_1, T_2).$$

The pair $((x, u), (y, v))$ is called admissible if at least one of the integrals

$$\int_{T_1}^{T_2} f^+(x(t), u(t), y(t), v(t))dt, \quad \int_{T_1}^{T_2} f^-(x(t), u(t), y(t), v(t))dt$$

is finite. If $((x, u), (y, v)))$ is admissible, then we set

$$I^f(T_1, T_2, x, u, y, v) := \int_{T_1}^{T_2} f(x(t), u(t), y(t), v(t))dt$$

$$= \int_{T_1}^{T_2} f^+(x(t), u(t), y(t), v(t))dt$$

$$- \int_{T_1}^{T_2} f^-(x(t), u(t), y(t), v(t))dt.$$

We can apply the results of Chap. 3 for $f = f^{(i)}$, $i = 1, 2$.
Let us now define approximate solutions (saddle points) of games

$$\Gamma(z_1, z_2, \xi_1, \xi_2, T)$$

with $z_1, z_2 \in R^{n_1}$, $\xi_1, \xi_2 \in R^{n_2}$ and a positive constant T.

Let $M \geq 0$, $-\infty < T_1 < T_2 < \infty$, $(x, u) \in X(A_1, B_1, T_1, T_2)$, $(y, v) \in X(A_2, B_2, T_1, T_2)$ be such that the pair $((x, u), (y, v))$ is admissible. The pair $((x, u), (y, v))$ is called (M)-good if the integral $I^f(T_1, T_2, x, u, y, v)$ is finite and the following properties hold:

for each $(z, \xi) \in X(A_1, B_1, T_1, T_2)$ such that the pair $((z, \xi), (y, v))$ is admissible and $z(T_i) = x(T_i)$, $i = 1, 2$,

$$I^f(T_1, T_2, z, \xi, y, v) \geq I^f(T_1, T_2, x, u, y, v) - M;$$

for each $(z, \xi) \in X(A_2, B_2, T_1, T_2)$ such that the pair $((x, u), (z, \xi))$ is admissible and $z(T_i) = y(T_i)$, $i = 1, 2$,

$$I^f(T_1, T_2, x, u, z, \xi) \leq I^f(T_1, T_2, x, u, y, v) + M.$$

If $((x, u), (y, v))$ is (0)-good, then $((x, u), (y, v))$ is called a saddle point of the game $\Gamma(x(T_1), x(T_2), y(T_1), y(T_2), T_2 - T_1)$.

Note that the existence of a saddle point of the game $\Gamma(z_1, z_2, \xi_1, \xi_2, T)$ with $z_1, z_2 \in R^{n_1}$, $\xi_1, \xi_2 \in R^{n_2}$ and $T > 0$ is not guaranteed. Nevertheless, the next result which is proved in Sect. 6.2 holds.

Theorem 6.2. *Let $M > 0$. Then there exists $M_* > 0$ such that for each $T_1 \in R^1$, each $T_2 > T_1 + 2$, each $z_1, z_2 \in R^{n_1}$ satisfying $|z_i| \leq M$, $i = 1, 2$ and each $\xi_1, \xi_2 \in R^{n_2}$ satisfying $|\xi_i| \leq M$, $i = 1, 2$ there exists an (M_*)-good pair*

$$(x, u) \in X(A_1, B_1, T_1, T_2), \quad (y, v) \in X(A_2, B_2, T_1, T_2)$$

such that

$$x(T_i) = z_i, \ y(T_i) = \xi_i, \ i = 1, 2,$$

$$|x(t)|, |u(t)|, |y(t)|, |v(t)| \leq M_*, \ t \in [T_1, T_1 + 1] \cup [T_2 - 1, T_2],$$

$$x(t) = x_f, \ u(t) = u_f, y(t) = y_f, \ v(t) = v_f, \ t \in [T_1 + 1, T_2 - 1].$$

It should be mentioned that in Theorem 6.2 the constant M_* does not depend on the length of the interval $T_2 - T_1$.

In this chapter we establish a turnpike property of good solutions of our dynamic games which means that they spend most of the time in a small neighborhood of the pair (x_f, y_f). It is known in the optimal control theory that turnpike properties of approximately optimal solutions are deduced from an asymptotic turnpike property of solutions of corresponding infinite horizon optimal control problems [44, 53].

We say that f possesses the asymptotic turnpike property (or briefly ATP), if the following properties hold:
for each $(x, u) \in X(A_1, B_1, 0, \infty)$ such that

$$\sup \left\{ \int_0^T f(x(t), u(t), y_f, u_f) dt - Tf(x_f, u_f, y_f, v_f) : \ T \in (0, \infty) \right\} < \infty$$

we have $\lim_{t \to \infty} |x(t) - x_f| = 0$;
for each $(y, v) \in X(A_2, B_2, 0, \infty)$ such that

$$\inf \left\{ \int_0^T f(x_f, u_f, y(t), v(t)) dt - Tf(x_f, u_f, y_f, v_f) : \ T \in (0, \infty) \right\} > -\infty$$

we have $\lim_{t \to \infty} |y(t) - y_f| = 0$.

Clearly, f has (ATP) if and only if $f^{((i))}$, $i = 1, 2$ satisfy (A3).

The following theorem is the mail result of this chapter which will be proved in Sect. 6.3.

Theorem 6.3. *Let f possess (ATP) and $M, \epsilon > 0$. Then there exist $l > 0$ and an integer $Q \geq 1$ such that for each $T > Ql$ and each (M)-good admissible pair*

$$(x, u) \in X(A_1, B_1, 0, T), \ (y, v) \in X(A_2, B_2, 0, T)$$

such that

$$|x(0)|, \ |x(T)|, \ |y(0)|, \ |y(T)| \leq M$$

there exist a natural number $q \leq Q$ and a sequence of closed intervals $[a_i, b_i] \subset [0, T]$, $i = 1, \ldots, q$ such that

$$0 \leq b_i - a_i \leq l, \ i = 1, \ldots, q,$$
$$|x(t) - x_f| \leq \epsilon, \ |y(t) - y_f| \leq \epsilon$$

for all $t \in [0, T] \setminus \cup_{i=1}^{q}[a_i, b_i]$.

6.2 Proof of Theorem 6.2

Proposition 5.3 implies the following result.

Proposition 6.4. *Let $\tau, M > 0$. Then there exist $M_1 > 0$ such that for each $z_1, z_2 \in R^{n_1}$ satisfying $|z_i| \leq M$, $i = 1, 2$ and each $\xi_1, \xi_2 \in R^{n_2}$ satisfying $|\xi_i| \leq M$, $i = 1, 2$ there exist*

$$(x, u) \in X(A_1, B_1, 0, \tau), \ (y, v) \in X(A_2, B_2, 0, \tau)$$

such that

$$x(0) = z_1, \ x(\tau) = z_2, \ y(0) = \xi_1, \ y(\tau) = \xi_2,$$
$$|x(t)|, |u(t)|, |y(t)|, |v(t)| \leq M_1$$

for all $t \in [0, \tau]$.

Proof of Theorem 6.2. We may assume that

$$M > |x_f| + |u_f| + |y_f| + |v_f|. \tag{6.8}$$

By Proposition 6.4, there exist $M_1 > M$ such that the following property holds:

(P1)   for each $z_1, z_2 \in R^{n_1}$ satisfying $|z_i| \leq M$, $i = 1, 2$ and each $\xi_1, \xi_2 \in R^{n_2}$ satisfying $|\xi_i| \leq M$, $i = 1, 2$ there exist

$$(x, u) \in X(A_1, B_1, 0, 1), \ (y, v) \in X(A_2, B_2, 0, 1)$$

such that

$$x(0) = z_1, \ x(1) = z_2, \ y(0) = \xi_1, \ y(1) = \xi_2,$$
$$|x(t)|, |u(t)|, |y(t)|, |v(t)| \leq M_1$$

for all $t \in [0, 1]$.

By (C1)(i) and (C1)(x) there exists $M_2 > 0$ such that

$$|f(z_1, z_2, \xi_1, \xi_2)| \le M_2$$

for each $z_1 \in R^{n_1}$, $z_2 \in R^{m_1}$, each $\xi_1 \in R^{n_2}$, $\xi_2 \in R^{m_2}$

$$\text{satisfying } |z_i|, |\xi_i| \le M_1, \ i = 1, 2, \tag{6.9}$$

$$f(z_1, z_2, \xi_1, \xi_2) \ge -M_2$$

for each pair $z_1 \in R^{n_1}$, $z_2 \in R^{m_1}$, each pair $\xi_1 \in R^{n_2}$, $\xi_2 \in R^{m_2}$

$$\text{satisfying } |\xi_i| \le M_1, \ i = 1, 2, \tag{6.10}$$

$$f(z_1, z_2, \xi_1, \xi_2) \le M_2$$

for each pair $z_1 \in R^{n_1}$, $z_2 \in R^{m_1}$, each pair $\xi_1 \in R^{n_2}$, $\xi_2 \in R^{m_2}$

$$\text{satisfying } |z_i| \le M_1, \ i = 1, 2. \tag{6.11}$$

Set

$$M_* = 8M_2 + c_* + M_1. \tag{6.12}$$

Let

$$T_1 \in R^1, \ T_2 > T_1 + 2, \ z_1, z_2 \in R^{n_1}, \ \xi_1, \xi_2 \in R^{n_2},$$

$$|z_i|, \ |\xi_i| \le M, \ i = 1, 2. \tag{6.13}$$

By property (P1), (6.8), and (6.13), there exist

$$(x, u) \in X(A_1, B_1, T_1, T_2), \ (y, v) \in X(A_2, B_2, T_1, T_2)$$

such that

$$x(T_1) = z_1, x(t) = x_f, \ u(t) = u_f, \ t \in [T_1 + 1, T_2 - 1], \ x(T_2) = z_2, \tag{6.14}$$

$$|x(t)|, |u(t)| \le M_1, \ t \in [T_1, T_1 + 1] \cup [T_2 - 1, T_2], \tag{6.15}$$

$$y(T_1) = \xi_1, y(t) = y_f, \ v(t) = v_f, \ t \in [T_1 + 1, T_2 - 1], \ y(T_2) = \xi_2, \tag{6.16}$$

$$|y(t)|, |v(t)| \le M_1, \ t \in [T_1, T_1 + 1] \cup [T_2 - 1, T_2], \tag{6.17}$$

It follows from (6.8), (6.9), and (6.14)–(6.17) that

$$|f(x(t), u(t), y(t), v(t))| \leq M_2, \ t \in [T_1, T_2]. \tag{6.18}$$

By (6.14)–(6.18),

$$\left| \int_{T_1}^{T_2} f(x(t), u(t), y(t), v(t))dt - (T_2 - T_1)f(x_f, u_f, y_f, v_f) \right|$$

$$\leq \left| \int_{T_1}^{T_1+1} f(x(t), u(t), y(t), v(t))dt \right.$$

$$+ \left. \int_{T_2-1}^{T_2} f(x(t), u(t), y(t), v(t))dt - 2f(x_f, u_f, y_f, v_f) \right| \leq 4M_2. \tag{6.19}$$

We show that the pair $((x, u), (y, v))$ is M_*-good. Let

$$(\xi, \eta) \in X(A_1, B_1, T_1, T_2),$$

$$\xi(T_i) = x(T_i), \ i = 1, 2.$$

In view of (6.16), (6.17), and (C1)(x), the pair $((\xi, \eta), (y, v))$ is admissible. It follows from (6.8)–(6.10), (6.16), (6.17), (6.19), and (C2) that

$$\int_{T_1}^{T_2} f(\xi(t), \eta(t), y(t), v(t))dt$$

$$= \int_{T_1}^{T_1+1} f(\xi(t), \eta(t), y(t), v(t))dt + \int_{T_1+1}^{T_2-1} f(\xi(t), \eta(t), y(t), v(t))dt$$

$$+ \int_{T_2-1}^{T_2} f(\xi(t), \eta(t), y(t), v(t))dt$$

$$\geq -2M_2 + \int_{T_1+1}^{T_2-1} f(\xi(t), \eta(t), y_f, v_f)dt$$

$$\geq -2M_2 + (T_2 - T_1 - 2)f(x_f, u_f, y_f, v_f) - c_*$$

$$\geq \int_{T_1}^{T_2} f(x(t), u(t), y(t), v(t))dt - 6M_2 - c_* - 2|f(x_f, u_f, y_f, v_f)|$$

$$\geq \int_{T_1}^{T_2} f(x(t), u(t), y(t), v(t))dt - M_*.$$

Let

$$(\xi, \eta) \in X(A_2, B_2, T_1, T_2),$$

$$\xi(T_i) = y(T_i), \ i = 1, 2.$$

In view of (6.14), (6.15), and (C1)(x), the pair $((x, u), (\xi, \eta))$ is admissible. It follows from (6.8), (6.11)–(6.15), and (6.19), and (C2) that

$$
\int_{T_1}^{T_2} f(x(t), u(t), \xi(t), \eta(t))dt
$$

$$
= \int_{T_1}^{T_1+1} f(x(t), u(t), \xi(t), \eta(t))dt + \int_{T_1+1}^{T_2-1} f(x(t), u(t), \xi(t), \eta(t))dt
$$

$$
+ \int_{T_2-1}^{T_2} f(x(t), u(t), \xi(t), \eta(t))dt
$$

$$
\leq 2M_2 + \int_{T_1+1}^{T_2-1} f(x_f, u_f, \xi(t), \eta(t))dt
$$

$$
\leq 2M_2 + (T_2 - T_1 - 2)f(x_f, u_f, y_f, v_f) + c_*
$$

$$
\leq \int_{T_1}^{T_2} f(x(t), u(t), y(t), v(t))dt + 6M_2 + c_*
$$

$$
\leq \int_{T_1}^{T_2} f(x(t), u(t), y(t), v(t))dt + M_*.
$$

Theorem 6.2 is proved. □

6.3 Proof of Theorem 6.3

By Theorem 6.2, there exists $M_1 > 0$ such that the following property holds:

(P2)   for each $T_1 \in R^1$, each $T_2 > T_1 + 2$, each $z_1, z_2 \in R^n$ satisfying $|z_i| \leq M$, $i = 1, 2$ and each $\xi_1, \xi_2 \in R^{n_2}$ satisfying $|\xi_i| \leq M$, $i = 1, 2$ there exists an (M_1)-good pair

$$
(\tilde{x}, \tilde{u}) \in X(A_1, B_1, T_1, T_2), \quad (\tilde{y}, \tilde{v}) \in X(A_2, B_2, T_1, T_2)
$$

such that

$$
\tilde{x}(T_i) = z_i, \ \tilde{y}(T_i) = \xi_i, \ i = 1, 2,
$$

$$
|\tilde{x}(t)|, |\tilde{u}(t)|, |\tilde{y}(t)|, |\tilde{v}(t)| \leq M_1, \ t \in [T_1, T_1 + 1] \cup [T_2 - 1, T_2],
$$

$$
\tilde{x}(t) = x_f, \ \tilde{u}(t) = u_f, \tilde{y}(t) = y_f, \ \tilde{v}(t) = v_f, \ t \in [T_1 + 1, T_2 - 1].
$$

By (C1)(x), there exists $M_2 > 0$ such that

$$
f(z_1, z_2, \xi_1, \xi_2) \geq -M_2
$$

for each $z_1 \in R^{n_1}$, $z_2 \in R^{m_1}$, each $\xi_1 \in R^{n_2}$, $\xi_2 \in R^{m_2}$

$$\text{satisfying } |\xi_i| \leq M_1, \ i = 1, 2, \tag{6.20}$$

$$f(z_1, z_2, \xi_1, \xi_2) \leq M_2$$

$$\text{for each } z_1 \in R^{n_1}, \ z_2 \in R^{m_1}, \text{ each } \xi_1 \in R^{n_2}, \ \xi_2 \in R^{m_2}$$

$$\text{satisfying } |z_i| \leq M_1, \ i = 1, 2. \tag{6.21}$$

By Theorem 4.3 and Proposition 3.2, there exist $l_1, l_2 > 0$ and integers $Q_1, Q_2 \geq 1$ such that the following properties hold:

(P3)   for each $T > l_1 Q_1$ and each $(x, u) \in X(A_1, B_1, 0, T)$ which satisfies

$$\int_0^T f(x(t), u(t), y_f, v_f) dt \leq 2M + 4M_2 + c_* + Tf(x_f, u_f, y_f, v_f)$$

there exist strictly increasing sequences of numbers $\{a_i\}_{i=1}^q, \{b_i\}_{i=1}^q \subset [0, T]$ such that $q \leq Q_1$, for all $i = 1, \dots, q$,

$$0 \leq b_i - a_i \leq l_1,$$

$b_i \leq a_{i+1}$ for all integers i satisfying $1 \leq i < q$ and that

$$|x(t) - x_f| \leq \epsilon \text{ for all } t \in [0, T] \setminus \cup_{i=1}^q [a_i, b_i];$$

(P4)   for each $T > l_2 Q_2$ and each $(y, v) \in X(A_2, B_2, 0, T)$ which satisfies

$$\int_0^T f(x_f, u_f, y(t), v(t))) dt \geq -2M - 4M_2 - c_* + Tf(x_f, u_f, y_f, v_f)$$

there exist strictly increasing sequences of numbers $\{a_i\}_{i=1}^q, \{b_i\}_{i=1}^q \subset [0, T]$ such that $q \leq Q_2$, for all $i = 1, \dots, q$,

$$0 \leq b_i - a_i \leq l_2,$$

$b_i \leq a_{i+1}$ for all integers i satisfying $1 \leq i < q$ and that

$$|y(t) - x_f| \leq \epsilon \text{ for all } t \in [0, T] \setminus \cup_{i=1}^q [a_i, b_i].$$

Set

$$Q = Q_1 + Q_2 + 4,$$

$$l = \max\{l_1, l_2\} + 4. \tag{6.22}$$

Assume that $T > Ql$ and

$$(x, u) \in X(A_1, B_1, 0, T), \quad (y, v) \in X(A_2, B_2, 0, T)$$

is an (M)-good admissible pair such that

$$|x(0)|, \ |x(T)|, \ |y(0)|, \ |y(T)| \leq M. \tag{6.23}$$

By property (P2) and (6.23) there exists an (M_1)-good pair $((\tilde{x}, \tilde{u}), (\tilde{y}, \tilde{v}))$ such that

$$(\tilde{x}, \tilde{u}) \in X(A_1, B_1, 0, T), \ \tilde{x}(0) = x(0), \ \tilde{x}(T) = x(T), \tag{6.24}$$

$$(\tilde{y}, \tilde{v}) \in X(A_2, B_2, 0, T), \ \tilde{y}(0) = y(0), \ \tilde{y}(T) = y(T), \tag{6.25}$$

$$|\tilde{x}(t)|, |\tilde{u}(t)|, |\tilde{y}(t)|, |\tilde{v}(t)| \leq M_1, \ t \in [0, 1] \cup [T - 1, T], \tag{6.26}$$

$$\tilde{x}(t) = x_f, \ \tilde{u}(t) = u_f, \tilde{y}(t) = y_f, \ \tilde{v}(t) = v_f, \ t \in [1, T - 1]. \tag{6.27}$$

In view of (C2),

$$\int_1^{T-1} f(x(t), u(t), y_f, v_f) dt \geq (T - 2) f(x_f, u_f, y_f, v_f) - c_*, \tag{6.28}$$

$$\int_1^{T-1} f(x_f, u_f, y(t), v(t)) dt \leq (T - 2) f(x_f, u_f, y_f, v_f) + c_*. \tag{6.29}$$

Since the pair $((x, u), (y, v))$ is (M)-good, it follows from (6.24)–(6.27) that

$$-M + \int_0^1 f(x(t), u(t), \tilde{y}(t), \tilde{v}(t)) dt$$

$$+ \int_1^{T-1} f(x(t), u(t), \tilde{y}(t), \tilde{v}(t)) dt + \int_{T-1}^T f(x(t), u(t), \tilde{y}(t), \tilde{v}(t)) dt$$

$$= -M + \int_0^T f(x(t), u(t), \tilde{y}(t), \tilde{v}(t)) dt$$

$$\leq \int_0^T f(x(t), u(t), y(t), v(t)) dt$$

$$\leq M + \int_0^T f(\tilde{x}(t), \tilde{u}(t), y(t), v(t)) dt$$

$$= M + \int_0^1 f(\tilde{x}(t), \tilde{u}(t), y(t), v(t)) dt + \int_1^{T-1} f(\tilde{x}(t), \tilde{u}(t), y(t), v(t)) dt$$

$$+ \int_{T-1}^T f(\tilde{x}(t), \tilde{u}(t), y(t), v(t)) dt. \tag{6.30}$$

By (6.21), (6.20), and (6.26)–(6.30),

$$-M - 2M_2 + (T-2)f(x_f, u_f, y_f, v_f) - c_*$$

$$\leq -M - 2M_2 + \int_1^{T-1} f(x(t), u(t), y_f, v_f))dt$$

$$\leq -M + \int_0^1 f(x(t), u(t), \tilde{y}(t), \tilde{v}(t))dt$$

$$+ \int_{T-1}^T f(x(t), u(t), \tilde{y}(t), \tilde{v}(t))dt + \int_1^{T-1} f(x(t), u(t), y_f, v_f)dt$$

$$\leq M + 2M_2 + \int_1^{T-1} f(x_f, u_f, y(t), v(t))dt$$

$$\leq M + 2M_2 + c_* + (T-2)f(x_f, u_f, y_f, v_f). \tag{6.31}$$

It follows from (6.28) and (6.31) that

$$\int_1^{T-1} f(x(t), u(t), y_f, v_f)dt \leq 2M + 4M_2 + c_* + (T-2)f(x_f, u_f, y_f, v_f),$$

$$\int_1^{T-1} f(x_f, u_f, y(t), v(t))dt \geq -2M - 4M_2 - c_* + (T-2)f(x_f, u_f, y_f, v_f).$$

By the inequalities above, the relation $T > Ql$, and properties (P3) and (P4), there exist

$$\{a_i^{(1)}\}_{i=1}^{q_1}, \ \{b_i^{(1)}\}_{i=1}^{q_1} \subset [1, T-1],$$

$$\{a_i^{(2)}\}_{i=1}^{q_2}, \ \{b_i^{(2)}\}_{i=1}^{q_2} \subset [1, T-1],$$

such that $q_i \leq Q_i$, $i = 1, 2$, for all $i = 1, \ldots, q_1$,

$$0 \leq b_i^{(1)} - a_i^{(1)} \leq l_1,$$

for all $i = 1, \ldots, q_2$,

$$0 \leq b_i^{(2)} - a_i^{(2)} \leq l_2,$$

$$|x(t) - x_f| \leq \epsilon \text{ for all } t \in [1, T-1] \setminus \cup_{i=1}^{q_1}[a_i^{(1)}, b_i^{(1)}],$$

$$|y(t) - y_f| \leq \epsilon \text{ for all } t \in [1, T-1] \setminus \cup_{i=1}^{q_2}[a_i^{(2)}, b_i^{(2)}].$$

This completes the proof of Theorem 6.3. □

6.4 Examples

We use the notation and definitions of Sect. 6.1.

Example 6.5. Assume that $f : R^{n_1} \times R^{m_1} \times R^{n_2} \times R^{m_2} \to R^1$ is a Borel measurable function which is bounded on all bounded subsets of $R^{n_1} \times R^{m_1} \times R^{n_2} \times R^{m_2}$ and satisfies (C1)(x). Let

$$x_f \in R^{n_1}, \; u_f \in R^{m_1}, \; y_f \in R^{n_2}, \; v_f \in R^{m_2}$$

satisfy (6.3).

Assume that $a_1 > 0, l_1 \in R^{m_1}, l_2 \in R^{m_2}, \psi_0 : [0, \infty) \to [0, \infty)$ is an increasing function such that

$$\lim_{t \to \infty} \psi_0(t) = \infty$$

and that $L_1 : R^{n_1} \times R^{m_1} \to [0, \infty)$ and $L_2 : R^{n_2} \times R^{m_2} \to [0, \infty)$ are continuous functions such that for all $(x, u) \in R^{n_1} \times R^{m_1}$,

$$L_1(x, u) = 0 \text{ if and only if } x = x_f, \; u = u_f,$$

$$L_1(x, u) \geq \max\{\psi_0(|x|), \; \psi_0(|u|)|u|\} - a_1 + |l_1||A_1 x + B_1 u|,$$

for all $(y, v) \in R^{n_2} \times R^{m_2}$,

$$L_2(y, v) = 0 \text{ if and only if } y = y_f, \; v = v_f,$$

$$L_2(y, v) \geq \max\{\psi_0(|y|), \; \psi_0(|v|)|v|\} - a_1 + |l_2||A_2 y + B_2 v|,$$

for each $M, \epsilon > 0$ there exist $\Gamma, \delta > 0$ such that for $i = 1, 2$,

$$|L_i(x_1, u_1) - L_i(x_2, u_2)| \leq \epsilon \max\{L_i(x_1, u_1), \; L_i(x_2, u_2)\}$$

for each $x_1, x_2 \in R^{n_i}$ and each $u_1, u_2 \in R^{m_i}$ which satisfy

$$|x_1|, |x_2| \leq M, \; |u_1|, |u_2| \geq \Gamma, \; |x_1 - x_2|, \; |u_1 - u_2| \leq \delta$$

and that for $i = 1, 2$ and each $x \in R^{n_i}$, the function $L_i(x, \cdot) : R^{m_i} \to R^1$ is convex.

Assume that for each $(x, u) \in R^{n_1} \times R^{m_1}$,

$$f(x, u, y_f, v_f) = f(x_f, u_f, y_f, v_f) + L_1(x, u) + \langle l_1, A_1 x + B_1 u \rangle$$

and that for each $(y, v) \in R^{n_2} \times R^{m_2}$,

$$f(x_f, u_f, y, v) = f(x_f, u_f, y_f, v_f) - L_2(y, v) - \langle l_2, A_2 y + B_2 v \rangle.$$

We show that assumptions (C1)–(C3) and (ATP) hold for the integrand f. It is clear that (C1)(i), (C1)(ii), (C1)(iv), (C1)(vi), (C1)(vii), (C1)(viii), and (C1)(ix) hold. It is not difficult to see that (C1)(iii) and (C1)(v) hold under the appropriate choice of $a_0 > 0$, ψ. Thus assumption (C1) holds.

Set

$$f^{(1)}(x, u) = f(x, u, y_f, v_f) = f(x_f, u_f, y_f, v_f) + L_1(x, u) + \langle l_1, A_1 x + B_1 u \rangle$$

for each $(x, u) \in R^{n_1} \times R^{m_1}$,

$$f^{(2)}(y, v) = -f(x_f, u_f, y, v) = -f(x_f, u_f, y_f, v_f) + L_2(y, v) + \langle l_2, A_2 y + B_2 v \rangle$$

for each $(y, v) \in R^{n_2} \times R^{m_2}$.

In Sect. 3.1 (see also Example 1.12) it was explained that for the triplets (f_i, A_i, B_i), $i = 1, 2$ assumptions (A1) holds under the appropriate choice of $a_0 > 0$, ψ and in view of Proposition 3.6, (A2) holds for the triplets (f_i, A_i, B_i), $i = 1, 2$,

$$\mu(f^{(1)}) = f(x_f, u_f, y_f, v_f), \quad \mu(f^{(2)}) = -f(x_f, u_f, y_f, v_f)$$

and the following properties hold:

for every (f_1, A_1, B_1)-good trajectory-control pair $(x, u) \in X(A_1, B_1, 0, \infty)$,

$$\lim_{t \to \infty} x(t) = x_f;$$

for every (f_2, A_2, B_2)-good trajectory-control pair $(y, v) \in X(A_2, B_2, 0, \infty)$,

$$\lim_{t \to \infty} y(t) = y_f.$$

Combined with Proposition 6.1 this implies that (C2) and (C3) hold and that the integrand f possesses (ATP). Therefore Theorems 6.2 and 6.3 hold for the integrand f.

Example 6.6. Assume that $f : R^{n_1} \times R^{m_1} \times R^{n_2} \times R^{m_2} \to R^1$ is a Borel measurable function which is bounded on all bounded subsets of $R^{n_1} \times R^{m_1} \times R^{n_2} \times R^{m_2}$ and satisfies (C1)(x). Let

$$x_f \in R^{n_1}, \ u_f \in R^{m_1}, \ y_f \in R^{n_2}, \ v_f \in R^{m_2}$$

satisfy (6.3).

Set

$$f^{(1)}(x, u) = f(x, u, y_f, v_f)$$

for each $(x, u) \in R^{n_1} \times R^{m_1}$,

$$f^{(2)}(y, v) = -f(x_f, u_f, y, v)$$

for each $(y, v) \in R^{n_2} \times R^{m_2}$.

Suppose that $f^{(1)}, f^{(2)}$ are continuous strictly convex functions, assumption (A1) holds for triplets (f_i, A_i, B_i), $i = 1, 2$ and for $i = 1, 2$,

$$f^{(i)}(x, u)/|u| \to \infty \text{ as } |u| \to \infty \text{ uniformly in } x \in R^{n_i}.$$

As it was mentioned in Sect. 3.1 (see also Example 1.11) Corollary 2.11 of Chap. 2 implies that assumptions (A2) and (A3) hold for the triplets (f_i, A_i, B_i), $i = 1, 2$, Combined with Proposition 6.1 this implies that (C1)–(C3) and (ATP) hold for the integrand f. Therefore Theorems 6.2 and 6.3 hold for the integrand f.

Chapter 7
Genericity Results

In this chapter we continue to study the class of optimal control problems studied in Chaps. 3 and 4. There we established the convergence of approximate solutions on large intervals in the regions close to the end points. In this chapter we show that for a typical (in the sense of Baire category) integrand the values of approximate solutions at the end points converge to the limit which is a unique solution of the corresponding minimization problem associated with the integrand.

7.1 Preliminaries and Main Results

We use the notation, definitions, and assumptions introduced in Sects. 3.1–3.3 of Chap. 3 and in Sects. 4.1 and 4.6 of Chap. 4. Recall that $a_0 > 0$ and $\psi : [0, \infty) \to [0, \infty)$ is an increasing function such that

$$\lim_{t \to \infty} \psi(t) = \infty.$$

We continue to study the structure of optimal trajectories of the controllable linear control system

$$x' = Ax + Bu$$

where A and B are given matrices of dimensions $n \times n$ and $n \times m$.

We consider the space $\mathfrak{M}$ of all borelian functions $g : R^{n+m+1} \to R^1$ which satisfy the growth assumption (5.1). The space $\mathfrak{M}$ is equipped with the uniformity which is determined by the base (5.2). This uniform space is metrizable and complete. In Chap. 5 we considered the set $\mathfrak{M}_b$ of all functions $g \in \mathfrak{M}$ which are

© Springer International Publishing Switzerland 2015
A.J. Zaslavski, *Turnpike Theory of Continuous-Time Linear Optimal Control Problems*, Springer Optimization and Its Applications 104,
DOI 10.1007/978-3-319-19141-6_7

bounded on bounded subsets of R^{n+m+1}. Note that $\mathfrak{M}_b$ is a closed subset of $\mathfrak{M}$. We consider the topological subspace $\mathfrak{M}_b \subset \mathfrak{M}$ equipped with the relative topology.

Denote by $\mathfrak{M}_c$ the set of all continuous functions $g \in \mathfrak{M}$ which satisfy assumption (A1) and which do not depend on the variable t. By Proposition 4.10 of [52], $\mathfrak{M}_c$ is a closed subset of $\mathfrak{M}$. We consider the topological subspace $\mathfrak{M}_c \subset \mathfrak{M}$ equipped with the relative topology. In this chapter any element $g \in \mathfrak{M}_c$ will be considered as a function $g : R^{n+m} \to R^1$.

Let $f \in \mathfrak{M}_c$. By Proposition 4.9 of [52], there exists an (A, B)-trajectory-control pair

$$\tilde{x}_f : R^1 \to R^n, \quad \tilde{u}_f : R^1 \to R^m$$

whose restriction to any finite interval $[T_1, T_2]$ belongs to $X(A, B, T_1, T_2)$ and a number $\tilde{b}_f > 0$ such that the following assumption holds:

(B1)   (i)

$$\sigma(f, \tilde{x}_f(T_1), \tilde{x}_f(T_2), T_1, T_2) = I^f(T_1, T_2, \tilde{x}_f, \tilde{u}_f)$$

for each $T_1 \in R^1$ and each $T_2 > T_1$;
   (ii)

$$\sup\{I^f(j, j+1, \tilde{x}_f, \tilde{u}_f) : j = 0, \pm 1, \pm 2, \ldots\} < \infty;$$

(iii) for each $S_1 > 0$ there exist $S_2 > 0$ and $c > 0$ such that

$$I^f(T_1, T_2, \tilde{x}_f, \tilde{u}_f) \leq I^f(T_1, T_2, x, u) + S_2$$

for each $T_1 \in R^1$, each $T_2 \geq T_1 + c$ and each $(x, u) \in X(A, B, T_1, T_2)$ which satisfies $|x(T_1)|, \; |x(T_2)| \leq S_1$;
(iv) for each $\epsilon > 0$ there exists $\delta > 0$ such that for each $(T, z) \in R^{n+1}$ which satisfies

$$|z - \tilde{x}_f(T)| \leq \delta$$

there are

$$\tau_1 \in (T, T + \tilde{b}_f] \text{ and } \tau_2 \in [T - \tilde{b}_f, T),$$

and trajectory-control pairs

$$(x_1, u_1) \in X(A, B, T, \tau_1), \; (x_2, u_2) \in X(A, B, \tau_2, T)$$

which satisfy

$$x_1(T) = x_2(T) = z,$$

$$x_i(\tau_i) = \tilde{x}_f(\tau_i), \ i = 1, 2,$$

$$|x_1(t) - \tilde{x}_f(t)| \le \epsilon \text{ for all } t \in [T, \tau_1],$$

$$|x_2(t) - \tilde{x}_f(t)| \le \epsilon \text{ for all } t \in [\tau_2, T],$$

$$I^f(T, \tau_1, x_1, u_1) \le I^f(T, \tau_1, \tilde{x}_f, \tilde{u}_f) + \epsilon,$$

$$I^f(\tau_2, T, x_2, u_2) \le I^f(\tau_2, T, \tilde{x}_f, \tilde{u}_f) + \epsilon.$$

Note that assumption (B1) means that the trajectory-control pair

$$\tilde{x}_f : R^1 \to R^n, \ \tilde{u}_f : R^1 \to R^m$$

is a solution of the corresponding infinite horizon optimal control problem associated with the integrand f and that certain controllability properties hold near this trajectory-control pair.

Let $f \in \mathfrak{M}_c$. Set

$$\mu(f) := \inf\{\liminf_{T \to \infty} T^{-1} I^f(0, T, x, u) \in X(A, B, 0, \infty)\}. \tag{7.1}$$

By (5.1), the value $\mu(f)$ is well defined and finite and

$$f(0, 0) \ge \mu(f) - a_0.$$

Denote by $\mathfrak{M}_*$ the set of all $f \in \mathfrak{M}_c$ for which there exists $(x_f, u_f) \in R^n \times R^m$ such that

$$Ax_f + Bu_f = 0, \ \mu(f) = f(x_f, u_f). \tag{7.2}$$

In Sect. 7.2 we prove the following result.

Proposition 7.1. $\mathfrak{M}_*$ *is a closed subset of the space* $\mathfrak{M}_c$.

We consider the topological subspace $\mathfrak{M}_* \subset \mathfrak{M}_c$ equipped with the relative topology.

For every $f \in \mathfrak{M}_*$ let $(x_f, u_f) \in R^n \times R^m$ satisfy (7.2).

Let $f \in \mathfrak{M}_*$. By (7.2), Propositions 3.1 and 3.2 (see Chap. 3), and Proposition 4.6 of [52] (see also Proposition 7.7), assumption (B1) holds with the constant trajectory-control pair

$$\tilde{x}_f(t) = x_f, \ \tilde{u}_f(t) = u_f \text{ for all } t \in R^1.$$

For each $r \in (0, 1)$ set

$$f_r(x, u) = f(x, u) + r \min\{|x - x_f|, 1\}, \quad (x, u) \in R^n \times R^m. \tag{7.3}$$

It is not difficult to see that $f_r \in \mathfrak{M}_*$ for all $r \in (0, 1)$.

Theorem 2.3 and Corollary 2.4 of [52] imply the following result.

Theorem 7.2. *There exists a set $\mathcal{F}_0 \subset \mathfrak{M}_*$ which is a countable intersection of open everywhere dense subsets of $\mathfrak{M}_*$ such that for each $f \in \mathcal{F}_0$ the following property holds.*

For each $S, \epsilon > 0$ there exist real numbers $\Delta > 0$, $\delta \in (0, \epsilon)$ such that for each $T_1 \in R^1$, each $T_2 \geq T_1 + 2\Delta$ and each $(x, u) \in X(A, B, T_1, T_2)$ which satisfies

$$I^f(T_1, T_2, x, u) \leq \sigma(f, T_2 - T_1) + S$$

and

$$I^f(T_1, T_2, x, u) \leq \sigma(f, x(T_1), x(T_2), T_2 - T_1) + \delta$$

the inequality

$$|x(t) - x_f| \leq \epsilon \text{ holds for all } t \in [T_1 + \Delta, T_2 - \Delta].$$

Let $\mathcal{F}_0 \subset \mathfrak{M}_*$ be as guaranteed by Theorem 7.2. Propositions 3.1, 3.2, and Theorem 7.2 imply the following result.

Proposition 7.3. *Every $f \in \mathcal{F}_0$ satisfies assumptions (A2) and (A3).*

Denote by $\tilde{\mathcal{F}}$ the set of all $f \in \mathfrak{M}_*$ which satisfy assumptions (A2) and (A3).

The following theorem is the main result of this chapter.

Theorem 7.4. *There exists a set $\mathcal{F} \subset \tilde{\mathcal{F}}$ which is a countable intersection of open everywhere dense subsets of $\mathfrak{M}_*$ such that for each $f \in \mathcal{F}$ the following properties hold:.*

(i) there exists a unique point $x_+^f \in R^n$ such that $\pi^f(x_+^f) = \inf(\pi^f)$;
(ii) there exists a unique point $x_-^f \in R^n$ such that $\pi_-^f(x_-^f) = \inf(\pi_-^f)$.

Theorem 7.4 follows from Propositions 3.21 and 3.22 and the next result which is proved in Sect. 7.4.

Theorem 7.5. *There exists a set $\mathcal{F} \subset \tilde{\mathcal{F}}$ which is a countable intersection of open everywhere dense subsets of $\mathfrak{M}_*$ such that for each $f \in \mathcal{F}$ there exists a unique point $x^f \in R^n$ such that $\pi^f(x^f) = \inf(\pi^f)$.*

Let $f \in \mathfrak{M}_*$. Denote by $\mathcal{L}(f)$ the set of all pairs $(x, u) \in R^n \times R^m$ such that

$$Ax + Bu = 0,$$

$$\mu(f) = f(x, u).$$

Clearly, $\mathcal{L}(f)$ is a nonempty set which is not necessarily a singleton and if $f \in \tilde{\mathcal{F}}$, then by (A2),

$$\{x \in R^n : \text{ there is } u \in R^m \text{ such that } (x, u) \in \mathcal{L}(f)\}$$

is a singleton denoted by $\{x_f\}$, where $x_f \in R^n$.

Chapter 7 is organized as follows. Proposition 7.1 is proved in Sect. 7.2. Section 7.3 contains auxiliary results for Theorem 7.5 which is proved in Sect. 7.4.

7.2 Proof of Proposition 7.1

Assumptions (A1) and (B1) and Proposition 2.7 of [52] imply the following result.

Lemma 7.6. *Let $f \in \mathfrak{M}_c$ and $\tilde{x}_f(\cdot)$, $\tilde{u}_f(\cdot)$ be as guaranteed by (B1). Then the following assertions hold.*

1.

$$\sup\{|\tilde{x}(t)| : t \in R^1\} < \infty.$$

2. Let $S_0 > 0$. Then there exist $S > 0$, $c \geq 1$ such that for each $T_1 \in R^1$, each $T_2 \geq T_1 + c$ and each trajectory-control pair

$$(x, u) \in X(A, B, T_1, T_2)$$

satisfying $|x(T_1)| \leq S_0$ the inequality

$$I^f(T_1, T_2, \tilde{x}_f, \tilde{u}_f) \leq I^f(T_1, T_2, x, u) + S$$

holds.

Proposition 7.7 (Proposition 4.6 of [52]). *Let $f \in \mathfrak{M}_c$, $M, \tau, \epsilon > 0$. Then there exists a number $\delta > 0$ such that:*

1. for each $T \in R^1$, each $y_1, y_2, z_1, z_2 \in R^n$ satisfying

$$|y_i|, |z_i| \leq M, \ i = 1, 2, \ |y_1 - y_2|, \ |z_1 - z_2| \leq \delta$$

the following relation holds:

$$|\sigma(f, y_1, z_1, T, T + \tau) - \sigma(f, y_2, z_2, T, T + \tau)| \leq \epsilon.$$

2. for each $T \in R^1$, each $y_1, y_2, z_1.z_2 \in R^n$ satisfying

$$|y_i|, |z_i| \leq M, \ i = 1, 2, \ |y_1 - y_2|, \ |z_1 - z_2| \leq \delta$$

and each trajectory-control pair

$$(x_1, u_1) \in X(A, B, T, T + \tau)$$

which satisfies

$$x_1(T) = y_1, \; x_1(T + \tau) = z_1,$$
$$I^f(T, T + \tau, x_1, u_1) = \sigma(f, y_1, z_1, T, T + \tau)$$

there exists a trajectory-control pair

$$(x_2, u_2) \in X(A, B, T, T + \tau)$$

such that

$$x_2(T) = y_2, \; x_2(T + \tau) = z_2,$$
$$|I^f(T, T + \tau, x_2, u_2) - I^f(T, T + \tau, x_1, u_1)| \le \epsilon,$$
$$|x_1(t) - x_2(t)| \le \epsilon, \; t \in [T, T + \tau].$$

Proof of Proposition 7.1. Let

$$\{f_i\}_{i=1}^{\infty} \subset \mathfrak{M}_*, \; \lim_{i \to \infty} f_i = f \tag{7.4}$$

in $\mathfrak{M}_c$. We show that $f \in \mathfrak{M}_*$.

Let an (A, B)-trajectory-control pair $(\tilde{x}_f, \tilde{u}_f)$ be as guaranteed by (B1). Clearly, for each integer $i \ge 1$ there exists $(x_{f_i}, u_{f_i}) \in R^n \times R^m$ such that

$$Ax_{f_i} + Bu_{f_i} = 0, \; \mu(f_i) = f_i(x_{f_i}, u_{f_i}). \tag{7.5}$$

Lemma 7.6 implies that

$$\mu(f) = \liminf_{T \to \infty} T^{-1} I^f(0, T, \tilde{x}_f, \tilde{u}_f). \tag{7.6}$$

Let $\epsilon \in (0, 1)$. By Lemma 7.6,

$$\sup\{|\tilde{x}_f(t)| : t \in R^1\} < \infty.$$

Proposition 7.7 implies that there exists a number

$$M_0 > \sup\{|\sigma(f, z_1, z_2, 0, 1)| :$$
$$z_1, z_2 \in R^n, \; |z_1|, |z_2| \le \sup\{|\tilde{x}_f(t)| : t \in R^1\} + 4\}. \tag{7.7}$$

In view of (7.6), there exists a number T_0 such that

$$T_0 \geq 40 + (2M_0 + 2)(4\epsilon)^{-1}, \quad 2T_0^{-1}|\mu(f)| < \epsilon/4, \tag{7.8}$$

$$|\mu(f) - T_0^{-1}I^f(0, T_0, \tilde{x}_f, \tilde{u}_f)| \leq \epsilon T_0^{-1}. \tag{7.9}$$

Clearly, there exists $(x, u) \in X(A, B, 0, T_0 + 2)$ such that

$$x(0) = 0, \ x(T_0 + 2) = 0,$$

$$x(t) = \tilde{x}_f(t - 1), \ u(t) = \tilde{u}_f(t - 1), \ t \in [1, T_0 + 1],$$

$$I^f(0, 1, x, u) \leq \sigma(f, 0, \tilde{x}_f(0), 0, 1) + 1,$$

$$I^f(T_0 + 1, T_0 + 2, x, u) \leq \sigma(f, \tilde{x}_f(T_0), 0, T_0 + 1, T_0 + 2) + 1. \tag{7.10}$$

By (7.7), (7.9), and (7.10),

$$I^f(0, T_0 + 2, x, u) \leq 2M_0 + I^f(0, T_0, \tilde{x}_f, \tilde{u}_f) \leq 2M_0 + T_0\mu(f) + \epsilon. \tag{7.11}$$

It follows from (7.4), (7.11), and Proposition 4.8 that there exists a natural number k_0 such that for each integer $k \geq k_0$,

$$I^{f_k}(0, T_0 + 2, x, u) \leq 2M_0 + T_0\mu(f) + 2. \tag{7.12}$$

In view of (7.4), (7.8), (7.10), and (7.12), for each integer $k \geq k_0$,

$$f_k(x_{f_k}, u_{f_k}) = \mu(f_k) \leq (T_0 + 2)^{-1}I^{f_k}(0, T_0 + 2, x, u)$$

$$\leq (T_0 + 2)^{-1}(2M_0 + 2) + T_0(T_0 + 2)^{-1}\mu(f) \leq \mu(f) + \epsilon/2. \tag{7.13}$$

(A1) and (7.13) imply that the sequence $\{(x_{f_k}, u_{f_k})\}_{k=1}^{\infty}$ is bounded. Extracting a subsequence and re-indexing we may assume that there exists

$$\xi = \lim_{k \to \infty} x_{f_k}, \ \eta = \lim_{k \to \infty} u_{f_k}. \tag{7.14}$$

By (7.4), (7.13), and (7.14),

$$f(\xi, \eta) = \lim_{k \to \infty} f_k(x_{f_k}, u_{f_k}) = \lim_{k \to \infty} \mu(f_k) \leq \mu(f) + \epsilon/2.$$

Since ϵ is any number belonging to the interval $(0, 1)$ we conclude that

$$f(\xi, \eta) \leq \mu(f).$$

Evidently, $A\xi + B\eta = 0$. This implies that $\mu(f) = f(\xi, \eta), f \in \mathfrak{M}_*$ and completes the proof of Proposition 7.1.   □

7.3   Auxiliary Results for Theorem 7.5

In the sequel we need the following result (see Proposition 3.7.1 of [44]).

Lemma 7.8. *Let Ω be a closed subset of R^s. Then there exists a bounded nonnegative function $\phi \in C^\infty(R^s)$ such that $\Omega = \{x \in R^s : \phi(x) = 0\}$ and for each sequence of nonnegative integers $p_1, \ldots, p_s$, the function $\partial^{|p|}\phi/\partial x_1^{p_1} \ldots \partial x_s^{p_s} : R^s \to R^1$ is bounded, where $|p| = \sum_{i=1}^s p_i$.*

Lemma 7.9. *Let $f \in \tilde{\mathcal{F}}$, $\{f_k\}_{k=1}^\infty \subset \mathfrak{M}_*$,*

$$\lim_{k \to \infty} f_k = f,$$

$$(x_k, u_k) \in \mathcal{L}(f_k), \ k = 1, 2, \ldots.$$

Then the sequence $\{(x_k, u_k)\}_{k=1}^\infty$ is bounded, any its limit point belongs to $\mathcal{L}(f)$, there exists $\lim_{k \to \infty} x_k$ and for every $(x, u) \in \mathcal{L}(f)$,

$$f(x, u) = \lim_{k \to \infty} f(x_k, u_k) = \lim_{k \to \infty} f_k(x_k, u_k).$$

Proof. Let $(x, u) \in \mathcal{L}(f)$. It is clear that for all integers $k \geq 1$,

$$f_k(x_k, u_k) \leq f_k(x, u) \to f(x, u) \text{ as } k \to \infty. \tag{7.15}$$

By (A1) and (7.15) the sequence $\{(x_k, u_k)\}_{k=1}^\infty$ is bounded.

Let (ξ, η) be its limit point. Clearly,

$$A\xi + Bu = 0.$$

In view of (7.15),

$$f(\xi, \eta) \leq \limsup_{k \to \infty} f(x_k, u_k) = \limsup_{k \to \infty} f_k(x_k, u_k) = f(x, u) = \mu(f).$$

This implies that $f(\xi, \eta) = \mu(f)$ and $\xi = x$. This completes the proof of Lemma 7.9.

Lemma 7.10. *Let $f \in \tilde{\mathcal{F}}$ and $\epsilon \in (0, 1)$. Then there exist $\delta > 0$ and a neighborhood $\mathcal{U}$ of f in $\mathfrak{M}_*$ such that for each $g \in \mathcal{U} \cap \tilde{\mathcal{F}}$ and each $z \in R^n$ satisfying $|z - x_f| \leq \delta$,*

$$|\pi^g(z)| \leq \epsilon.$$

Proof. By Lemma 7.9, there exists a neighborhood $\mathcal{U}_0$ of f in $\mathfrak{M}_*$ such that for each $g \in \mathcal{U}_0 \cap \mathfrak{M}_*$,

$$|x_f - x_g| \leq 4^{-1}. \tag{7.16}$$

By Proposition 7.7, there exists $\gamma \in (0, \epsilon/2)$ such that

$$|\sigma(f, z_1, z_2, 0, 1) - \mu(f)| \leq \epsilon/16$$

for each $z_1, z_2 \in R^n$ satisfying $|z_i - x_f| \leq 2\gamma$, $i = 1, 2$.   (7.17)

By Proposition 4.8, there exists a neighborhood $\mathcal{U}_1$ of f in $\mathfrak{M}_*$ such that $\mathcal{U}_1 \subset \mathcal{U}_0$ and the following property holds:

(i) for each $g \in \mathcal{U}_1$, each $S \in R^1$, each and each trajectory-control pair $(x, u) \in X(A, B, S, S + 1)$ which satisfies

$$\min\{I^g(S, S + 1, x, u),\ I^f(S, S + 1, x, u)\} \leq |\mu(f)| + 4$$

the inequality

$$|I^g(S, S + 1, x, u) - I^f(S, S + 1, x, u)| \leq \epsilon/16$$

holds.

By Theorem 4.2, there exist $\delta \in (0, \gamma/4)$, $l_0 > 0$ and a neighborhood $\mathcal{U}_2$ of f in $\mathcal{M}_*$ such that $\mathcal{U}_2 \subset \mathcal{U}_1$ and the following property holds:

(ii) for each $T > 2l_0$, each $g \in \mathcal{U}_2$ and each $(x, u) \in X(A, B, 0, T)$ which satisfies

$$|x(0) - x_f| \leq \delta,\ |x(T)| \leq |x_f| + 1,$$
$$I^g(0, T, x, u) \leq \sigma(g, x(0), x(T), 0, T) + \delta$$

we have

$$|x(t) - x_f| \leq \gamma \text{ for all } t \in [0, T - l_0].$$

By Lemma 7.9, there exists a neighborhood $\mathcal{U}$ of f in $\mathfrak{M}_*$ such that $\mathcal{U} \subset \mathcal{U}_2$ and the following property holds:

(iii) for each $g \in \mathcal{U} \cap \tilde{\mathcal{F}}$,

$$|x_f - x_g| \leq \delta.$$

Assume that

$$g \in \mathcal{U} \cap \tilde{\mathcal{F}},\ z_1, z_2 \in R^n,\ |z_i - x_f| \leq \delta,\ i = 1, 2.$$ (7.18)

Theorem 3.8 and (7.18) imply that there exists an (g, A, B)-overtaking optimal pair $(x, u) \in X(A, B, 0, \infty)$ satisfying

$$x(0) = z_1.$$ (7.19)

In view of (A3) and (7.18),

$$\lim_{t \to \infty} x(t) = x_g. \qquad (7.20)$$

It is clear that (7.16) holds. By (7.16) and (7.20), for all sufficiently large numbers t,

$$|x(t)| \le |x_f| + 1.$$

Together with (7.18) and property (ii) this implies that

$$|x(t) - x_f| \le \gamma \text{ for all } t \in [0, \infty). \qquad (7.21)$$

In view of (7.19),

$$\pi^g(z_1) = \liminf_{T \to \infty}[I^g(0, T, x, u) - T\mu(g)]$$
$$= I^g(0, 1, x, u) - \mu(g) + \liminf_{T \to \infty}[I^g(1, T, x, u) - (T - 1)\mu(g)]. \qquad (7.22)$$

By Proposition 3.27, there exists $(x_1, u_1) \in X(A, B, 0, \infty)$ such that

$$x_1(0) = z_2, \ x_1(t) = x(t), \ u_1(t) = u(t) \text{ for all } t \ge 1,$$

$$I^g(0, 1, x_1, u_1) = \sigma(g, z_2, x(1), 0, 1). \qquad (7.23)$$

Proposition 3.11, (7.19), (7.22), and (7.23) imply that

$$\pi^g(z_2) - \pi^g(z_1) \le \sigma(g, z_2, x(1), 0, 1) - \sigma(g, z_1, x(1), 0, 1). \qquad (7.24)$$

It follows from (7.17), (7.18), and (7.21) that

$$|\sigma(f, z_i, x(1), 0, 1) - \mu(f)| \le \epsilon/16, \ i = 1, 2. \qquad (7.25)$$

In view of (7.25) and property (i), for $i = 1, 2$,

$$|\sigma(f, z_i, x(1), 0, 1) - \sigma(g, z_i, x(1), 0, 1)| \le \epsilon/16. \qquad (7.26)$$

By (7.25) and (7.26),

$$|\sigma(g, z_1, x(1), 0, 1) - \sigma(g, z_2, x(1), 0, 1)| \le \epsilon/4.$$

Together with (7.24) this implies that

$$\pi^g(z_2) - \pi^g(z_1) \le \epsilon/4.$$

Thus we have shown that

$$|\pi^g(z_2) - \pi^g(z_1)| \le \epsilon/4$$

for all $z_1, z_2 \in R^n$ satisfying $|z_i - x_f| \le \delta$, $i = 1, 2$. Combined with property (iii) and the equality $\pi^g(x_g) = 0$ this completes the proof of Lemma 7.10.

Lemma 7.11. *Let $f \in \tilde{\mathcal{F}}$ and $M_0 > 0$. Then there exist $M_1 > 0$ and a neighborhood $\mathcal{U}$ of f in $\mathfrak{M}_*$ such that for each $g \in \mathcal{U} \cap \tilde{\mathcal{F}}$ and each $z \in R^n$ satisfying $|z| \le M_0$,*

$$\pi^g(z) \le M_1.$$

Proof. By Lemma 7.9, there exists a neighborhood $\mathcal{U}_0$ of f in $\mathfrak{M}_*$ such that for each $g \in \mathcal{U}_0 \cap \tilde{\mathcal{F}}_*$,

$$|x_f - x_g| \le 1, \ |\mu(f) - \mu(g)| \le 1. \tag{7.27}$$

By Proposition 3.28, there exists $M_2 > 0$ such that

$$|\sigma(f, z_1, z_2, 1)| \le M_2$$

for each $z_1, z_2 \in R^n$ satisfying $|z_i| \le |x_f| + M_0 + 1$, $i = 1, 2$. \tag{7.28}

By Proposition 4.8, there exists a neighborhood $\mathcal{U}$ of f in $\mathfrak{M}_*$ such that $\mathcal{U} \subset \mathcal{U}_0$ and the following property holds:

(iv) for each $g \in \mathcal{U}$ and each trajectory-control pair $(x, u) \in X(A, B, 0, 1)$ which satisfies

$$\min\{I^g(0, 1, x, u), I^f(0, 1, x, u)\} \le M_2 + 1$$

the inequality

$$|I^g(0, 1, x, u) - I^f(0, 1, x, u)| \le 1$$

holds.
Set

$$M_1 = M_2 + 2 + |\mu(f)|. \tag{7.29}$$

Assume that

$$g \in \mathcal{U} \cap \tilde{\mathcal{F}}, \; z \in R^n, \; |z| \le M_0. \tag{7.30}$$

In view of (7.30) and the choice of $\mathcal{U}_0$ (see (7.27)), (7.27) is true.

By Proposition 3.27, there exists $(x, u) \in X(A, B, 0, \infty)$ such that

$$x(0) = z, \; x(t) = x_g, \; u(t) = u_g \text{ for all } t \ge 1,$$

$$I^f(0, 1, x, u) = \sigma(f, z, x_g, 1). \tag{7.31}$$

In view of (7.31) and Proposition 3.11,

$$\pi^g(z) \le \liminf_{T \to \infty}[I^g(0, T, x, u) - T\mu(g)]$$

$$= I^g(0, 1, x, u) - \mu(g). \tag{7.32}$$

It follows from (7.27), (7.28), (7.30), and (7.31) that

$$I^f(0, 1, x, u) = \sigma(f, z, x_g, 0, 1) \le M_2.$$

Together with (7.30) and property (iv) this implies that

$$I^g(0, 1, x, u) \le I^f(0, 1, x, u) + 1 \le M_2 + 1.$$

Combined with (7.27) and (7.32) this implies that

$$\pi^g(z) \le M_2 + 2 + |\mu(f)| = M_1.$$

Lemma 7.11 is proved.

Lemma 7.12. *Let $f \in \tilde{\mathcal{F}}$ and $M_0 > 0$. Then there exist $M_1 > 0$ and a neighborhood $\mathcal{U}$ of f in $\mathfrak{M}_*$ such that for each $g \in \mathcal{U} \cap \tilde{\mathcal{F}}$ and each $z \in R^n$ satisfying*

$$\pi^g(z) \le \inf(\pi^g) + M_0$$

the inequality $|z| \le M_1$ holds.

Proof. By Lemma 7.9, there exists a neighborhood $\mathcal{U}$ of f in $\mathfrak{M}_*$ such that for each $g \in \mathcal{U} \cap \tilde{\mathcal{F}}$,

$$|\mu(f) - \mu(g)| \le 1. \tag{7.33}$$

By Proposition 4.7, there exists $M_1 > 0$ such that for each $g \in \mathfrak{M}_*$ and each trajectory-control pair $(x, u) \in X(A, B, 0, 1)$ which satisfies

$$I^g(0, 1, x, u) \le M_0 + 2 + |\mu(f)|$$

we have

$$|x(t)| \le M_1, \ t \in [0, 1]. \tag{7.34}$$

Assume that

$$g \in \mathcal{U} \cap \tilde{\mathcal{F}}, \ z \in R^n, \ \pi^g(z) \le \inf(\pi^g) + M_0. \tag{7.35}$$

Evidently, (7.33) holds. By Theorem 3.8 and (7.35), there exists (g, A, B)-overtaking optimal pair $(x, u) \in X(A, B, 0, \infty)$ such that

$$x(0) = z.$$

Proposition 3.12 implies that

$$\pi^g(z) = I^g(0, 1, x, u) - \mu(g) + \pi^g(x(1)). \tag{7.36}$$

By (7.35) and (7.36),

$$I^g(0, 1, x, u) - \mu(g) = \pi^g(z) - \pi^g(x(1))$$
$$\le \pi^g(z) - \inf(\pi^g) \le M_0. \tag{7.37}$$

In view of (7.33) and (7.37),

$$I^f(0, 1, x, u) \le M_0 + 2 + |\mu(f)|.$$

By the relation above and the choice of M_1 (see (7.34)),

$$|z| = |x(0)| \le M_1.$$

Lemma 7.12 is proved.

Lemma 7.13. *Let* $f \in \tilde{\mathcal{F}}$, $\epsilon \in (0, 1)$, $L_0 > 0$. *Then there exist a neighborhood* $\mathcal{U}$ *of* f *in* $\mathfrak{M}_*$ *and* $\delta > 0$ *such that for each* $g \in \mathcal{U} \cap \tilde{\mathcal{F}}$ *and each* (g, A, B)-*overtaking optimal pair* $(x, u) \in X(A, B, 0, \infty)$ *satisfying*

$$\pi^g(x(0)) \le \inf(\pi^g) + \delta$$

there exists an (f, A, B)-*overtaking optimal pair* $(\xi, \eta) \in X(A, B, 0, \infty)$ *such that*

$$\pi^f(\xi(0)) = \inf(\pi^f),$$
$$|\xi(t) - x(t)| \le \epsilon \text{ for all } t \in [0, L_0].$$

Proof. By Lemma 3.38, there exists $\gamma_0 \in (0, \epsilon/2)$ such that the following property holds:

(v) for each $(x, u) \in X(A, B, 0, L_0)$ which satisfies

$$\pi^f(x(0)) \le \inf(\pi^f) + \gamma_0,$$

$$I^f(0, L_0, x, u) - L_0 \mu(f) - \pi^f(x(0)) + \pi^f(x(L_0)) \le \gamma_0$$

there exists an (f, A, B)-overtaking optimal pair $(\xi, \eta) \in X(A, B, 0, \infty)$ such that

$$\pi^f(\xi(0)) = \inf(\pi^f),$$

$$|x(t) - \xi(t)| \le \epsilon \text{ for all } t \in [0, L_0].$$

By Proposition 3.29, there exists $\gamma_1 \in (0, \gamma_0)$ such that

$$|\sigma(f, z_1, z_2, 1) - \mu(f)| \le \gamma_0/64$$

for each $z_1, z_2 \in R^n$ satisfying $|z_i - x_f| \le \gamma_1$, $i = 1, 2$.   (7.38)

By Lemmas 7.9, 7.11, and 7.12, exist a neighborhood $\mathcal{U}_0$ of f in $\mathfrak{M}_*$, $M_0 > 0$, $M_1 > 0$ such that the following properties hold:

(vi) for each $g \in \mathcal{U}_0 \cap \tilde{\mathcal{F}}$,

$$|x_f - x_g| \le 1, \ |\mu(f) - \mu(g)| \le 1;$$

(vii) for each $g \in \mathcal{U}_0 \cap \tilde{\mathcal{F}}$ and each $z \in R^n$ satisfying $\pi^g(z) \le \inf(\pi^g) + 4$, we have

$$|z| \le M_0;$$

(viii) for each $g \in \mathcal{U}_0 \cap \tilde{\mathcal{F}}$ and each $z \in R^n$ satisfying $|z| \le M_0 + 1$ we have

$$\pi^g(z) \le M_1.$$

We may assume without loss of generality that

$$M_0 > |x_f| + 2.   (7.39)$$

Properties (vii) and (viii) imply that for each $g \in \mathcal{U}_0 \cap \tilde{\mathcal{F}}$,

$$\inf(\pi^g) \le M_1.   (7.40)$$

It follows from Lemma 7.10 that there exist $\delta \in (0, \gamma_1/4)$ and a neighborhood $\mathcal{U}_1$ of f in $\mathfrak{M}_*$ such that the following property holds:

(ix) for each $g \in \mathcal{U}_1 \cap \tilde{\mathcal{F}}$ and each $z \in R^n$ satisfying $|z - x_f| \leq \delta$,

$$|\pi^g(z)| \leq \gamma_0/32.$$

By Lemma 7.9, there exists a neighborhood $\mathcal{U}_2$ of f in $\mathfrak{M}_*$ such that $\mathcal{U}_2 \subset \mathcal{U}_1$ and for each $g \in \mathcal{U}_2 \cap \tilde{\mathcal{F}}$,

$$|x_f - x_g| \leq \delta. \tag{7.41}$$

By Theorem 4.2, there exist $\delta_* \in (0, \delta)$, $L_1 > 4 + L_0$ and a neighborhood $\mathcal{U}_3$ of f in $\mathcal{M}_*$ such that $\mathcal{U}_3 \subset \mathcal{U}_2$ and the following property holds:

(x) for each $T > 2L_1$, each $g \in \mathcal{U}_3$ and each $(x, u) \in X(A, B, 0, T)$ which satisfies

$$|x(0)|, \ |x(T)| \leq M_0 + 1,$$
$$I^g(0, T, x, u) \leq \sigma(g, x(0), x(T), 0, T) + \delta_*$$

we have

$$|x(t) - x_f| \leq \delta \text{ for all } t \in [L_1, T - L_1].$$

By Proposition 4.8, there exists a neighborhood $\mathcal{U}_4$ of f in $\mathfrak{M}_*$ such that and the following property holds:

(xi) for each $g \in \mathcal{U}_4$, each $T \in [1, L_1 + 2]$ each and each trajectory-control pair $(x, u) \in X(A, B, 0, T)$ which satisfies

$$\min\{I^g(0, T, x, u), \ I^f(0, T, x, u)\} \leq 4 + M_1 + 2L_1(|\mu(f)| + 2)$$

the inequality

$$|I^g(0, T, x, u) - I^f(0, T, x, u)| \leq \gamma_0/64$$

holds.

By Lemma 7.9, there exists a neighborhood $\mathcal{U}_5$ of f in $\mathfrak{M}_*$ such that for each $g \in \mathcal{U} \cap \tilde{\mathcal{F}}$,

$$|\mu(g) - \mu(f)| \leq (L_1 + 2)^{-1}\gamma_0/64. \tag{7.42}$$

Set

$$\mathcal{U} = \cap_{i=1}^{5}\mathcal{U}_i. \tag{7.43}$$

Assume that

$$g_1, g_2 \in \mathcal{U} \cap \tilde{\mathcal{F}}, \tag{7.44}$$

$$z \in R^n, \ \pi^{g_1}(z) \le \inf(\pi^{g_1}) + 1 \tag{7.45}$$

and $(x, u) \in X(A, B, 0, \infty)$ is a (g_1, A, B)-overtaking optimal pair satisfying

$$x(0) = z. \tag{7.46}$$

Proposition 3.15 and (7.46) imply that

$$\pi^{g_1}(z) = \lim_{T \to \infty} [I^{g_1}(0, T, x, u) - T\mu(g_1)]. \tag{7.47}$$

By (7.43)–(7.45),

$$|z| \le M_0. \tag{7.48}$$

In view of (7.44) and (A3),

$$\lim_{t \to \infty} x(t) = x_{g_1}. \tag{7.49}$$

It follows from (7.43), (7.44), and property (vi) that

$$|x_f - x_{g_1}| \le 1. \tag{7.50}$$

By (7.39), (7.49), and (7.50), for all sufficiently large numbers t,

$$|x(t) - x_f| \le |x(t) - x_{g_1}| + |x_{g_1} - x_f| \le 5/4,$$

$$|x(t)| \le |x_f| + 2 < M_0. \tag{7.51}$$

Since the pair (x, u) is (g_1, A, B)-overtaking optimal it follows from (7.44), (7.46), (7.48), (7.51), and property (x) that

$$|x(t) - x_f| \le \delta \text{ for all } t \ge L_1. \tag{7.52}$$

Since the pair (x, u) is (g_1, A, B)-overtaking optimal it follows from (7.46) and Proposition 3.12 that

$$\pi^{g_1}(z) = I^{g_1}(0, L_1, x, u) - L_1\mu(g_1) + \pi^{g_1}(x(L_1)). \tag{7.53}$$

In view of (7.43), (7.44), (7.48), and property (viii),

$$\pi^{g_1}(z) \le M_1. \tag{7.54}$$

Property (ix), (7.43), (7.44), and (7.52) imply that

$$|\pi^{g_1}(x(L_1))| \le 32^{-1}\gamma_0. \tag{7.55}$$

By (7.53)–(7.55) and property (vi),

$$I^{g_1}(0, L_1, x, u) \le 2 + M_1 + L_1|\mu(g_1)|$$
$$\le 2 + M_1 + L_1(|\mu(f)| + 1). \tag{7.56}$$

It follows from (7.43), (7.44), (7.56), and property (xi) that

$$|I^{g_1}(0, L_1, x, u) - I^f(0, L_1, x, u)| \le \gamma_0/64,$$

$$|I^f(0, L_1, x, u) - I^{g_2}(0, L_1, x, u)| \le \gamma_0/64. \tag{7.57}$$

In view of (7.57),

$$|I^{g_1}(0, L_1, x, u) - I^{g_2}(0, L_1, x, u)| \le \gamma_0/32. \tag{7.58}$$

Proposition 3.27 and (7.44) imply that there exists $(x_1, u_1) \in X(A, B, 0, \infty)$ such that

$$x_1(t) = x(t), \ u_1(t) = u(t) \text{ for all } t \in [0, L_1],$$
$$x_1(t) = x_{g_2}, \ u_1(t) = u_{g_2} \text{ for all } t \in [L_1 + 1, \infty),$$

$$I^{g_2}(L_1, L_1 + 1, x_1, u_1) = \sigma(g_2, x(L_1), x_{g_2}, 0, 1). \tag{7.59}$$

By (7.46), (7.59), and Proposition 3.11,

$$\pi^{g_2}(z) \le \liminf_{T \to \infty}[I^{g_2}(0, T, x_1, u_1) - T\mu(g_2)]$$
$$= I^{g_2}(0, L_1, x, u) - L_1\mu(g_2) + \sigma(g_2, x(L_1), x_{g_2}, 0, 1) - \mu(g_2). \tag{7.60}$$

In view of (7.42)–(7.44),

$$|L_1\mu(g_2) - L_1\mu(g_1)| \le \gamma_0/32. \tag{7.61}$$

It follows from (7.38), (7.41), (7.43), (7.44), and (7.52) that

$$|\sigma(f, x(L_1), x_{g_2}, 0, 1) - \mu(f)| \le \gamma_0/64. \tag{7.62}$$

Property (xi) and (7.43), (7.44), (7.62) imply that

$$|\sigma(f, x(L_1), x_{g_2}, 0, 1) - \sigma(g_i, x(L_1), x_{g_2}, 0, 1)| \le \gamma_0/64, \ i = 1, 2. \tag{7.63}$$

It follows from (7.53), (7.55), and (7.58)–(7.63) that

$$\pi^{g_2}(z) \le I^{g_1}(0, L_1, x, u) + \gamma_0/32 - L_1\mu(g_1) + \gamma_0/32 + \gamma_0/8$$
$$\le \pi^{g_1}(z) + 7\gamma_0/32. \tag{7.64}$$

By (7.43), (7.44), (7.52), (7.58), (7.61), and property (ix),

$$|[I^{g_2}(0, L_1, x, u) - L_1\mu(g_2) - \pi^{g_2}(x(L_1))]$$
$$- [I^{g_1}(0, L_1, x, u) - L_1\mu(g_1) - \pi^{g_1}(x(L_1))]|$$
$$\le \gamma_0/16 + \gamma_0/16.$$

Therefore, in view of (7.44) and the relation above, we have shown that the following property holds:

(C)   for each

$$g_1, g_2 \in \mathcal{U} \cap \tilde{\mathcal{F}},$$

each $z \in R^n$ satisfying

$$\pi^{g_1}(z) \le \inf(\pi^{g_1}) + 1$$

and each (g_1, A, B)-overtaking optimal pair $(x, u) \in X(A, B, 0, \infty)$ satisfying $x(0) = z$ we have

$$\pi^{g_2}(z) \le \pi^{g_1}(z) + \gamma_0/4,$$
$$|[I^{g_2}(0, L_1, x, u) - L_1\mu(g_2) - \pi^{g_2}(x(L_1))]$$
$$- [I^{g_1}(0, L_1, x, u) - L_1\mu(g_1) - \pi^{g_1}(x(L_1))]| \le \gamma_0/8.$$

Property (C) implies that for each

$$g_1, g_2 \in \mathcal{U} \cap \tilde{\mathcal{F}},$$

we have

$$|\inf(\pi^{g_2}) - \inf(\pi^{g_1})| \le \gamma_0/4. \tag{7.65}$$

Assume that

$$g \in \mathcal{U} \cap \tilde{\mathcal{F}}, \tag{7.66}$$

and $(x, u) \in X(A, B, 0, \infty)$ is a (g, A, B)-overtaking optimal pair satisfying

$$\pi^g(x(0)) \le \inf(\pi^g) + \delta. \tag{7.67}$$

By Property (C) and (7.65)–(7.67),

$$\pi^f(x(0)) \le \pi^g(x(0)) + \gamma_0/4 \le \inf(\pi^g) + \delta + \gamma_0/4 \le \inf(\pi^f) + \delta + \gamma_0/2. \quad (7.68)$$

Since $(x, u) \in X(A, B, 0, \infty)$ is a (g, A, B)-overtaking optimal pair it follows from (7.65)–(7.77), Proposition 3.12, and property (C) that

$$
\begin{aligned}
& I^f(0, L_1, x, u) - L_1\mu(f) - \pi^f(x(L_1)) + \pi^f(x(0)) \\
& \quad \le \gamma_0/8 + [I^g(0, L_1, x, u) - L_1\mu(g) - \pi^g(x(L_1))] \\
& \qquad + \pi^g(x(0)) - \pi^g(x(0)) + \pi^f(x(0)) \\
& \quad \le \gamma_0/8 - \pi^g(x(0)) + \pi^f(x(0)) \le \gamma_0/8 + \gamma_0/4. \quad (7.69)
\end{aligned}
$$

By (7.68), (7.69), the relation $L_1 > L_0$, property (v), and Proposition 3.11, there exists an (f, A, B)-overtaking optimal pair $(\xi, \eta) \in X(A, B, 0, \infty)$ such that

$$\pi^f(\xi(0)) = \inf(\pi^f),$$

$$|x(t) - \xi(t)| \le \epsilon \text{ for all } t \in [0, L_0].$$

Lemma 7.13 is proved.

7.4 Proof of Theorem 7.5

Denote by E the set of all $f \in \tilde{\mathcal{F}}$ for which there exists a unique point $z_f \in R^n$ satisfying

$$\pi^f(z_f) = \inf(\pi^f).$$

Lemma 7.14. *The set E is an everywhere dense subset of $\mathfrak{M}_*$.*

Proof. Let $f \in \tilde{\mathcal{F}}$. In order to prove the lemma it is sufficient to show that for every neighborhood V of f in $\mathfrak{M}_*$ we have $V \cap E \ne \emptyset$. There are two cases:

$$\pi^f(x_f) > \inf(\pi^f); \quad (7.70)$$

$$\pi^f(x_f) = \inf(\pi^f). \quad (7.71)$$

Assume that (7.70) holds. There exists $z_0 \in R^n$ such that

$$\pi^f(z_0) = \inf(\pi^f). \quad (7.72)$$

By Theorem 3.8, there exists an (f, A, B)-overtaking optimal pair $(y, v) \in X(A, B, 0, \infty)$ such that

$$y(0) = z_0. \quad (7.73)$$

(A3) implies that

$$\lim_{t\to\infty} y(t) = x_f. \tag{7.74}$$

Together with (7.70) and (7.72) this implies that there exists $\epsilon > 0$ such that for all sufficiently large numbers t,

$$\pi^f(z_0) + \epsilon < \pi^f(y(t)).$$

Therefore there exists a number $\tau_0 \geq 0$ such that

$$\pi^f(y(\tau_0)) = \pi^f(z_0),$$

$$\pi^f(y(t)) > \pi^f(z_0) \text{ for all } t > \tau_0.$$

We may assume without loss of generality that $\tau_0 = 0$. Then

$$\pi^f(y(t)) > \pi^f(z_0) \text{ for all } t > 0. \tag{7.75}$$

Since $(y, v) \in X(A, B, 0, \infty)$ is a (f, A, B)-overtaking optimal pair it follows from Propositions 3.12 and 3.15 that

$$\pi^f(z_0) = \lim_{T\to\infty} [I^f(0, T, y, v) - T\mu(f)] = \lim_{T\to\infty} [\pi^f(y(0)) - \pi^f(y(T))]. \tag{7.76}$$

By Lemma 7.8 and (7.74), there exists a bounded nonnegative function $\phi \in C^\infty(R^n)$ such that for each sequence of nonnegative integers $p_1, \ldots, p_n$, the function $\partial^{|p|}\phi/\partial x_1^{p_1} \ldots \partial x_n^{p_n} : R^n \to R^1$ is bounded, where $|p| = \sum_{i=1}^n p_i$ and

$$\{x \in R^n : \phi(x) = 0\} = \{x_f\} \cup \{y(t) : t \in [0, \infty)\}. \tag{7.77}$$

For any $r \in (0, 1)$ define a function $f_r : R^n \times R^m \to R^1$ by

$$f_r(x, y) = f(x, y) + r\phi(x), \ (x, y) \in R^{n+m}. \tag{7.78}$$

Let $r \in (0, 1)$. Clearly, $f_r \in \mathfrak{M}_c$. In view of (7.78),

$$\mu(f_r) \geq \mu(f).$$

Since $(y, v) \in X(A, B, 0, \infty)$ is a (f, A, B)-good pair it follows from (7.77) and (7.78) that

$$\mu(f_r) \leq \liminf_{T\to\infty} T^{-1}I^{f_r}(0, L_1, y, v)$$

$$= \liminf_{T\to\infty} T^{-1}I^f(0, L_1, y, v) = \mu(f).$$

Thus

$$\mu(f_r) = \mu(f) = f(x_f, u_f) = f_r(x_f, u_f) \tag{7.79}$$

where $(x_f, u_f) \in \mathcal{L}(f)$.

It is easy now to see that f_r satisfies assumption (A2). Proposition 3.4, (A3) which holds for the function f, and (7.79) imply that if $(x, u) \in X(A, B, 0, \infty)$ is an (f_r, A, B)-good pair, then it is also an (f, A, B)-good pair and $\lim_{t \to \infty} x(t) = x_f$. Thus f_r satisfies assumption (A3). Therefore we have shown that for each $r \in (0, 1)$,

$$\mu(f_r) = \mu(f), \ \mathcal{L}(f_r) = \mathcal{L}(f), \ f_r \in \tilde{\mathcal{F}}. \tag{7.80}$$

Let $r \in (0, 1)$. We show that $f_r \in E$. By Proposition 3.11, (3.12), (7.73), (7.77), (7.78), and (7.80),

$$\pi^{f_r}(z_0) \leq \liminf_{T \to \infty}[I^{f_r}(0, T, y, v) - T\mu(f_r)]$$
$$= \liminf_{T \to \infty}[I^f(0, T, y, v) - T\mu(f)] = \pi^f(z_0). \tag{7.81}$$

Proposition 3.11, (3.12), (7.78), (7.80), and (7.81) imply that

$$\pi^{f_r}(z) \geq \pi^f(z) \text{ for all } z \in R^n, \ \pi^{f_r}(z_0) = \pi^f(z_0). \tag{7.82}$$

Let $z \in R^n \setminus \{z_0\}$. We show that $\pi^{f_r}(z) > \pi^{f_r}(z_0)$. If

$$z \in \{y(t) : \ t \in [0, \infty)\} \cup \{x_f\},$$

then in view of (7.70), (7.72), (7.75), and (7.82),

$$\pi^{f_r}(z_0) = \pi^f(z_0) < \pi^f(z) \leq \pi^{f_r}(z).$$

Assume that

$$z \notin \{y(t) : \ t \in [0, \infty)\} \cup \{x_f\}. \tag{7.83}$$

By Theorem 3.8 and (7.80), there exists an (f_r, A, B)-overtaking optimal pair $(x, u) \in X(A, B, 0, \infty)$ such that

$$x(0) = z. \tag{7.84}$$

Since (x, u) is an (f_r, A, B)-overtaking optimal pair it follows from Propositions 3.11 and 3.15, (7.72), (7.77), (7.78), (7.80), (7.82), (7.83), and (7.84) that

$$\pi^{fr}(z) = \lim_{T \to \infty} [I^{fr}(0, T, x, u) - T\mu(f)]$$

$$= \lim_{T \to \infty} \left[I^f(0, T, x, u) - T\mu(f) + r \int_0^T \phi(u(t))dt \right]$$

$$\geq \pi^f(z) + r \int_0^1 \phi(u(t))dt > \pi^f(z) \geq \pi^{fr}(z_0).$$

Thus we have shown that

$$\pi^{fr}(z) > \pi^{fr}(z_0) \text{ for all } z \in R^n \setminus \{z_0\},$$

$f_r \in E$ for all $r \in (0, 1)$ and for any neighborhood V of f in $\mathfrak{M}_*$ we have $V \cap E \neq \emptyset$.

Assume that (7.71) holds. By Lemma 7.8, there exists a bounded nonnegative function $\phi \in C^\infty(R^n)$ such that for each sequence of nonnegative integers $p_1, \ldots, p_n$, the function $\partial^{|p|}\phi/\partial x_1^{p_1} \ldots \partial x_n^{p_n} : R^n \to R^1$ is bounded, where $|p| = \sum_{i=1}^n p_i$ and

$$\{x \in R^n : \phi(x) = 0\} = \{x_f\}. \tag{7.85}$$

For any $r \in (0, 1)$ define a function $f_r : R^n \times R^m \to R^1$ by

$$f_r(x, y) = f(x, y) + r\phi(x), \quad (x, y) \in R^{n+m}. \tag{7.86}$$

Let $r \in (0, 1)$. Clearly, $f_r \in \mathfrak{M}_c$ and

$$\mu(f_r) = \mu(f) = f(x_f, u_f) = f_r(x_f, u_f) \tag{7.87}$$

where $(x_f, u_f) \in \mathcal{L}(f) = \mathcal{L}(f_r)$.

Assume that $(x, u) \in X(A, B, 0, \infty)$ is an (f_r, A, B)-good pair. Then by (7.87), Proposition 3.4, and (A3) which holds for f, (x, u) is also an (f, A, B)-good pair and $\lim_{t \to \infty} x(t) = x_f$. Thus f_r satisfies assumptions (A2) and (A3) and $f_r \in \bar{\mathcal{F}}$. It follows from Propositions 3.11 and 3.13, (3.12) and (7.86) that

$$\pi^{fr}(x_f) = \pi^f(x_f) = 0, \ \pi^{fr}(z) \geq \pi^f(z) \text{ for all } z \in R^n. \tag{7.88}$$

Let $z \in R^n \setminus \{x_f\}$. By Theorem 3.8, there exists an (f_r, A, B)-overtaking optimal pair $(x, u) \in X(A, B, 0, \infty)$ such that

$$x(0) = z.$$

It follows from the relation above, Propositions 3.11 and 3.15, (7.71), and (7.85)–(7.87) that

$$\pi^{fr}(z) = \lim_{T\to\infty} [I^{fr}(0,T,x,u) - T\mu(f)]$$

$$= \lim_{T\to\infty} \left[I^f(0,T,x,u) - T\mu(f) + r \int_0^T \phi(u(t))dt \right]$$

$$\geq \pi^f(z) + r \int_0^1 \phi(u(t))dt > \pi^f(z) \geq \pi^{fr}(x_f).$$

Thus we have shown that $f_r \in E$ for all $r \in (0,1)$ and for any neighborhood V of f in $\mathfrak{M}_*$ we have $V \cap E \neq \emptyset$. This completes the proof of Lemma 7.14.

Completion of the Proof of Theorem 7.5. By Lemma 7.14, the set E is everywhere dense. For each $f \in E$ there exist $x_f, z_f \in R^n$, $u_f \in R^m$ such that

$$Ax_f + Bu_f = 0, \quad f(x_f, u_f) = \mu(f),$$

$$\pi^f(z_f) = \inf(\pi^f), \quad \{z_f\} = \{z \in R^n : \pi^f(z) = \inf(\pi^f)\}. \qquad (7.89)$$

Let $f \in E$ and $k \geq 1$ be an integer. By (7.13), there exists an open neighborhood $\mathcal{U}(f,k)$ of f in $\mathfrak{M}_*$ such that the following property holds:

(a) for each $g \in \mathcal{U}(f,k) \cap \tilde{F}$ and each $z \in R^n$ satisfying

$$\pi^g(z) = \inf(\pi^g)$$

we have $|z - z_f| \leq 2^{-k}$.

Set

$$\mathcal{F} = \cap_{k=1}^\infty \cup \{\mathcal{U}(f,k) : f \in E\} \cap \tilde{F}. \qquad (7.90)$$

Clearly, $\mathcal{F}$ is a countable intersection of open everywhere dense subsets of $\mathfrak{M}_*$ and $\mathcal{F} \subset \tilde{F}$.

Let $g \in \mathcal{F}$, $z_1, z_2 \in R^n$ and

$$\pi^g(z_i) = \inf(\pi^g), \quad i = 1, 2. \qquad (7.91)$$

Let $k \geq 1$ be an integer. In view of (7.90), there exist $f_k \in E$ such that

$$g \in \mathcal{U}(f_k, k).$$

By the relation above, (7.91), and property (a),

$$|z_i - z_{f_k}| \leq 2^{-k}, \quad i = 1, 2$$

and $|z_1 - z_2| \leq 2^{-k+1}$ for any integer $k \geq 1$. This implies that $z_1 = z_2$. Theorem 7.5 is proved. $\qquad \qquad \square$

Chapter 8
Variational Problems with Extended-Valued Integrands

In this chapter we study the structure of approximate solutions of autonomous variational problems with a lower semicontinuous extended-valued integrand. In our recent research we showed that approximate solutions are determined mainly by the integrand, and are essentially independent of the choice of time interval and data, except in regions close to the endpoints of the time interval. In this chapter our goal is to study the structure of approximate solutions in regions close to the endpoints of the time intervals.

8.1 Existence of Solutions and Their Turnpike Properties

In this chapter we consider the following variational problems

$$\int_0^T f(v(t), v'(t))dt \to \min, \qquad (P_1)$$

$v : [0, T] \to R^n$ is an absolutely continuous (a.c.) function such that

$$v(0) = x, \ v(T) = y,$$

$$\int_0^T f(v(t), v'(t))dt \to \min, \qquad (P_2)$$

$v : [0, T] \to R^n$ is an a. c. function such that $v(0) = x$

© Springer International Publishing Switzerland 2015
A.J. Zaslavski, *Turnpike Theory of Continuous-Time Linear Optimal Control Problems*, Springer Optimization and Its Applications 104,
DOI 10.1007/978-3-319-19141-6_8

and

$$\int_0^T f(v(t), v'(t))dt \to \min, \tag{P_3}$$

$v : [0, T] \to R^n$ is an a. c. function,

where $x, y \in R^n$. Here R^n is the n-dimensional Euclidean space with the Euclidean norm $| \cdot |$ and $f : R^n \times R^n \to R^1 \cup \{\infty\}$ is an extended-valued integrand.

In [46, 49, 53] we studied the problems (P_1) and (P_2) and showed under certain assumptions that the turnpike property holds and that the turnpike $\bar{x}$ is a unique solution of the minimization problem $f(x, 0) \to \min, x \in R^n$.

In this chapter which is based on [55] we study the structure of approximate solutions of the problems (P_2) and (P_3) in regions close to the endpoints of the time intervals. We show that in regions close to the right endpoint T of the time interval these approximate solutions are determined only by the integrand, and are essentially independent of the choice of interval and endpoint value x. For the problems (P_3), approximate solutions are determined only by the integrand function also in regions close to the left endpoint 0 of the time interval.

More precisely, we define $\bar{f}(x, y) = f(x, -y)$ for all $x, y \in R^n$ and consider the set $\mathcal{P}(\bar{f})$ of all solutions of a corresponding infinite horizon variational problem associated with the integrand $\bar{f}$. For a given pair of real positive numbers ϵ, τ, we show that if T is large enough and $v : [0, T] \to R^n$ is an approximate solution of the problem (P_2), then $|v(T - t) - y(t)| \le \epsilon$ for all $t \in [0, \tau]$, where $y(\cdot) \in \mathcal{P}(\bar{f})$. The prototype of our result was obtained in [47] under an additional assumption that the function $f : R^n \times R^n \to R^1 \cup \{\infty\}$ is strictly convex. This assumption played a crucial role there. In this chapter the integrand f is a nonconvex function.

We denote by $\text{mes}(E)$ the Lebesgue measure of a Lebesgue measurable set $E \subset R^1$, denote by $| \cdot |$ the Euclidean norm of the space R^n and by $\langle \cdot, \cdot \rangle$ the inner product of R^n. For each function $f : X \to R^1 \cup \{\infty\}$, where X is a nonempty, set

$$\text{dom}(f) = \{x \in X : f(x) < \infty\}.$$

Let a be a real positive number, $\psi : [0, \infty) \to [0, \infty)$ be an increasing function such that

$$\lim_{t \to \infty} \psi(t) = \infty \tag{8.1}$$

and let $f : R^n \times R^n \to R^1 \cup \{\infty\}$ be a lower semicontinuous function such that the set

$$\text{dom}(f) = \{(x, y) \in R^n \times R^n : f(x, y) < \infty\} \tag{8.2}$$

is nonempty, convex, and closed and that

$$f(x, y) \geq \max\{\psi(|x|), \ \psi(|y|)|y|\} - a \text{ for each } x, y \in R^n. \tag{8.3}$$

Recall that a function v defined on an infinite subinterval of R^1 with values in R^n is called absolutely continuous (a. c.) if v is absolutely continuous on any finite subinterval of its domain.

For each $x, y \in R^n$ and each $T > 0$ define

$$\sigma(f, T, x) = \inf\left\{ \int_0^T f(v(t), v'(t))dt : \ v : [0, T] \to R^n \right.$$

$$\left. \text{is an a. c. function satisfying } v(0) = x \right\}, \tag{8.4}$$

$$\sigma(f, T, x, y) = \inf\left\{ \int_0^T f(v(t), v'(t))dt : \ v : [0, T] \to R^n \right.$$

$$\left. \text{is an a. c. function satisfying } v(0) = x, \ v(T) = y \right\}, \tag{8.5}$$

$$\sigma(f, T) = \inf\left\{ \int_0^T f(v(t), v'(t))dt : \ v : [0, T] \to R^n \right.$$

$$\left. \text{is an a.c. function} \right\}, \tag{8.6}$$

$$\hat{\sigma}(f, T, y) = \inf\left\{ \int_0^T f(v(t), v'(t))dt : \ v : [0, T] \to R^n \right.$$

$$\left. \text{is an a. c. function satisfying } v(T) = y \right\}. \tag{8.7}$$

(Here we assume that infimum over an empty set is infinity.)

We suppose that there exists a point $\bar{x} \in R^n$ such that

$$f(\bar{x}, 0) \leq f(x, 0) \text{ for each } x \in R^n \tag{8.8}$$

and that the following assumptions hold:

(A1) $(\bar{x}, 0)$ is an interior point of the set dom(f) and the function f is continuous at the point $(\bar{x}, 0)$;

(A2) for each $M > 0$ there exists $c_M > 0$ such that

$$\sigma(f, T, x) \geq Tf(\bar{x}, 0) - c_M$$

for each $x \in R^n$ satisfying $|x| \leq M$ and each real number $T > 0$;

(A3)   for each $x \in R^n$ the function $f(x, \cdot) : R^n \to R^1 \cup \{\infty\}$ is convex.

Assumption (A2) implies that for each a.c. function $v : [0, \infty) \to R^n$ the function

$$T \to \int_0^T f(v(t), v'(t))dt - Tf(\bar{x}, 0), \quad T \in (0, \infty)$$

is bounded from below.

It should be mentioned that inequality (8.8) and assumptions (A1)–(A3) are common in the literature and hold for many infinite horizon optimal control problems. In particular, we need inequality (8.8) and assumption (A2) in the cases when the problems (P_1) and (P_2) possess the turnpike property and the point $\bar{x}$ is its turnpike. Assumption (A2) means that the constant function $\bar{v}(t) = \bar{x}, t \in [0, \infty)$ is an approximate solution of the infinite horizon variational problem with the integrand f related to the problems (P_1) and (P_2).

We say that an a. c. function $v : [0, \infty) \to R^n$ is (f)-good [44, 51] if

$$\sup \left\{ \left| \int_0^T f(v(t), v'(t))dt - Tf(\bar{x}, 0) \right| : T \in (0, \infty) \right\} < \infty.$$

The following result was obtained in [46].

Proposition 8.1. *Let $v : [0, \infty) \to R^n$ be an a.c. function. Then either the function v is (f)-good or*

$$\int_0^T f(v(t), v'(t))dt - Tf(\bar{x}, 0) \to \infty \text{ as } T \to \infty.$$

Moreover, if the function v is (f)-good, then $\sup\{|v(t)| : t \in [0, \infty)\} < \infty$.

For each pair of numbers $T_1 \in R^1$, $T_2 > T_1$ and each a.c. function $v : [T_1, T_2] \to R^n$ put

$$I^f(T_1, T_2, v) = \int_{T_1}^{T_2} f(v(t), v'(t))dt \tag{8.9}$$

and for any $T \in [T_1, T_2]$ set $I^f(T, T, v) = 0$.

For each $M > 0$ denote by $X_{M,f}$ the set of all $x \in R^n$ such that $|x| \leq M$ and there exists an a.c. function $v : [0, \infty) \to R^n$ which satisfies

$$v(0) = x, \; I^f(0, T, v) - Tf(\bar{x}, 0) \leq M \text{ for each } T \in (0, \infty). \tag{8.10}$$

It is clear that $\cup\{X_{M,f} : M \in (0, \infty)\}$ is the set of all points $x \in X$ for which there exists an (f)-good function $v : [0, \infty) \to R^n$ such that $v(0) = x$.

We suppose that the following assumption holds:

(A4) (the asymptotic turnpike property) for each (f)-good function $v : [0, \infty) \to R^n$, $\lim_{t \to \infty} |v(t) - \bar{x}| = 0$.

Examples of integrands f which satisfy assumptions (A1)–(A4) are considered in [46, 53].

The following turnpike result for the problem (P_2) was established in [46].

Theorem 8.2. *Let ϵ, M be positive numbers. Then there exist an integer $L \geq 1$ and a real number $\delta > 0$ such that for each real number $T > 2L$ and each a.c. function $v : [0, T] \to R^n$ which satisfies*

$$v(0) \in X_{M,f} \text{ and } I^f(0, T, v) \leq \sigma(f, T, v(0)) + \delta$$

there exist a pair of numbers $\tau_1 \in [0, L]$ and $\tau_2 \in [T - L, T]$ such that

$$|v(t) - \bar{x}| \leq \epsilon \text{ for all } t \in [\tau_1, \tau_2]$$

and if $|v(0) - \bar{x}| \leq \delta$, then $\tau_1 = 0$.

Let $M > 0$. Denote by $Y_{M,f}$ the set of all points $x \in R^n$ for which there exist a number $T \in (0, M]$ and an a. c. function $v : [0, T] \to R^n$ such that $v(0) = \bar{x}$, $v(T) = x$ and $I^f(0, T, v) \leq M$.

The following turnpike results for the problems (P_1) were established in [49].

Theorem 8.3. *Let $\epsilon, M_0, M_1 > 0$. Then there exist numbers $L, \delta > 0$ such that for each number $T > 2L$, each point $z_0 \in X_{M_0,f}$ and each point $z_1 \in Y_{M_1,f}$, the value $\sigma(f, T, z_0, z_1)$ is finite and for each a.c. function $v : [0, T] \to R^n$ which satisfies*

$$v(0) = z_0, \ v(T) = z_1, \ I^f(0, T, v) \leq \sigma(f, T, z_0, z_1) + \delta$$

there exists a pair of numbers $\tau_1 \in [0, L], \tau_2 \in [T - L, T]$ such that

$$|v(t) - \bar{x}| \leq \epsilon, \ t \in [\tau_1, \tau_2].$$

Moreover if $|v(0) - \bar{x}| \leq \delta$, then $\tau_1 = 0$ and if $|v(T) - \bar{x}| \leq \delta$, then $\tau_2 = T$.

In the sequel we use a notion of an overtaking optimal function [44, 53].

An a.c. function $v : [0, \infty) \to R^n$ is called (f)-overtaking optimal if for each a.c. function $u : [0, \infty) \to R^n$ satisfying $u(0) = v(0)$ the inequality

$$\limsup_{T \to \infty} [I^f(0, T, v) - I^f(0, T, u)] \leq 0$$

holds.

The following result which establishes the existence of an overtaking optimal function was obtained in [46].

Theorem 8.4. *Assume that $x \in R^n$ and that there exists an (f)-good function $v :$ $[0, \infty) \to R^n$ satisfying $v(0) = x$. Then there exists an (f)-overtaking optimal function $u_* : [0, \infty) \to R^n$ such that $u_*(0) = x$.*

Assumption (A1) implies that there exists a number $\bar{r} \in (0, 1)$ such that:

$$\Omega_0 := \{(x, y) \in R^n \times R^n : |x - \bar{x}| \le \bar{r} \text{ and } |y| \le \bar{r}\} \subset \text{dom}(f); \qquad (8.11)$$

$$\Delta_0 := \sup\{|f(z_1, z_2)| : (z_1, z_2) \in \Omega_0\} < \infty. \qquad (8.12)$$

It is easy to see that the value $\sigma(f, T, x, y)$ is finite for each number $T \ge 1$ and each pair of points $x, y \in R^n$ such that $|x - \bar{x}|, |y - \bar{x}| \le \bar{r}/2$.

Let $M > 0$. Denote by $\bar{Y}_{Mf}$ the set of all points $x \in R^n$ such that $|x| \le M$ and for which there exist a number $T \in (0, M]$ and an a. c. function $v : [0, T] \to R^n$ such that $v(0) = x$, $v(T) = \bar{x}$ and $I^f(0, T, v) \le M$.

It is easy to see that the following result holds.

Proposition 8.5. *For each $M > 0$ there exists $M_0 > 0$ such that $\bar{Y}_{Mf} \subset X_{M_0f}$.*

The next result follows from Lemma 8.27 and (A1).

Proposition 8.6. *For each $M > 0$ there exists $M_0 > 0$ such that $X_{Mf} \subset \bar{Y}_{M_0f}$.*

An a. c. function $v : [0, \infty) \to R^n$ is called (f)-minimal [5, 44] if for each $T_1 \ge 0$, each $T_2 > T_1$ and each a.c. function $u : [T_1, T_2] \to R^n$ satisfying $u(T_i) = v(T_i)$, $i = 1, 2$, we have

$$\int_{T_1}^{T_2} f(v(t), v'(t))dt \le \int_{T_1}^{T_2} f(u(t), u'(t))dt.$$

Following theorem obtained in [48] shows that the optimality notions introduced above are equivalent.

Theorem 8.7. *Assume that $x \in R^n$ and that there exists an (f)-good function $\tilde{v} :$ $[0, \infty) \to R^n$ satisfying $\tilde{v}(0) = x$. Let $v : [0, \infty) \to R^n$ be an a.c. function such that $v(0) = x$. Then the following conditions are equivalent:*

(i) *the function v is (f)-overtaking optimal;* (ii) *the function v is (f)-good and (f)-minimal;* (iii) *the function v is (f)-minimal and $\lim_{t \to \infty} v(t) = \bar{x}$;* (iv) *the function v is (f)-minimal and $\lim\inf_{t \to \infty} |v(t) - \bar{x}| = 0$.*

The following two theorems obtained in [48] describe the asymptotic behavior of overtaking optimal functions.

Theorem 8.8. *Let ϵ be a positive number. Then there exists a positive number δ such that:*

(i) *For each point $x \in R^n$ satisfying $|x - \bar{x}| \le \delta$ there exists an (f)-overtaking optimal and (f)-good function $v : [0, \infty) \to R^n$ such that $v(0) = x$.*

(ii) For each (f)-overtaking optimal function $v : [0, \infty) \to R^n$ satisfying $|v(0) - \bar{x}| \le \delta$, the inequality $|v(t) - \bar{x}| \le \epsilon$ holds for all numbers $t \in [0, \infty)$.

Theorem 8.9. *Let $\epsilon, M > 0$. Then there exists $L > 0$ such that for each $x \in X_{Mf}$ and each (f)-overtaking optimal function $v : [0, \infty) \to R^n$ satisfying $v(0) = x$ the following inequality holds:*

$$|v(t) - \bar{x}| \le \epsilon \text{ for all } t \in [L, \infty).$$

In Sect. 8.5 we prove the following turnpike result for approximate solutions of the problems of the type (P_3).

Theorem 8.10. *Let $\epsilon > 0$. Then there exist numbers $L, \delta > 0$ such that for each number $T > 2L$ and each a. c. function $v : [0, T] \to R^n$ which satisfies*

$$I^f(0, T, v) \le \sigma(f, T) + \delta$$

there exists a pair of numbers $\tau_1 \in [0, L], \tau_2 \in [T - L, T]$ such that

$$|v(t) - \bar{x}| \le \epsilon, \ t \in [\tau_1, \tau_2].$$

Moreover if $|v(0) - \bar{x}| \le \delta$, then $\tau_1 = 0$ and if $|v(T) - \bar{x}| \le \delta$, then $\tau_2 = T$.

8.2 Preliminaries

We use the notation, definitions, and assumptions introduced in Sect. 8.1. We define a function $\pi^f(x)$, $x \in R^n$ which plays an important role in our study. For all $x \in R^n \setminus \cup\{X_{Mf} : M \in (0, \infty)\}$ set

$$\pi^f(x) = \infty.$$

Let

$$x \in \cup\{X_{Mf} : M \in (0, \infty)\}. \tag{8.13}$$

Denote by $\Lambda(f, x)$ the set of all (f)-overtaking optimal functions $v : [0, \infty) : R^n$ satisfying $v(0) = x$. By (8.13) and Theorem 8.4, the set $\Lambda(f, x)$ is nonempty. In view of (8.13), any element of $\Lambda(f, x)$ is an (f)-good function. Define

$$\pi^f(x) = \liminf_{T \to \infty}[I^f(0, T, v) - Tf(\bar{x}, 0)], \tag{8.14}$$

where $v \in \Lambda(f, x)$. Clearly, $\pi^f(x)$ does not depend on the choice of v. In view of (A2) and (8.13), $\pi^f(x)$ is finite. Definition (8.14) and the definition of overtaking optimal functions imply the following result.

Proposition 8.11. *1. Let $v : [0, \infty) \to R^n$ be an (f)-good function. Then*

$$\pi^f(v(0)) \leq \liminf_{T \to \infty} [I^f(0, T, v) - Tf(\bar{x}, 0)]$$

and for each $T \geq 0$ and each $S > T$,

$$\pi^f(v(T)) \leq I^f(T, S, v) - (S - T)f(\bar{x}, 0) + \pi^f(v(S)). \tag{8.15}$$

2. Let $S > T \geq 0$ and $v : [0, S] \to R^n$ be an a. c. function such that $\pi^f(v(T)), \pi^f(v(S)) < \infty$. Then (8.15) holds.

The next result follows from definition (8.14).

Proposition 8.12. *Let $v : [0, \infty) \to R^n$ be an (f)-overtaking optimal and (f)-good function. Then for each $T \geq 0$ and each $S > T$,*

$$\pi^f(v(T)) = I^f(T, S, v) - (S - T)f(\bar{x}, 0) + \pi^f(v(S)).$$

Proposition 8.13. *$\pi^f(\bar{x}) = 0$.*

Proof. Set $v(t) = \bar{x}$ for all $t \geq 0$. By Theorem 8.7 and (A2), the function v is a (f)-overtaking optimal. In view of (8.14), $\pi^f(\bar{x}) = 0$.

The following two results are proved in Sect. 8.6.

Proposition 8.14. *The function π^f is finite in a neighborhood of $\bar{x}$ and continuous at $\bar{x}$.*

Proposition 8.15. *For each $M > 0$ the set $\{x \in R^n : \pi^f(x) \leq M\}$ is bounded.*

(Here we assume that an empty set is bounded.)
Set

$$\inf(\pi^f) = \inf\{\pi^f(z) : z \in R^n\}. \tag{8.16}$$

By (A2) and Proposition 8.15, $\inf(\pi^f)$ is finite. Set

$$X_f = \{x \in R^n : \pi^f(x) \leq \inf(\pi^f) + 1\}. \tag{8.17}$$

Proposition 8.16. *Assume that $x \in \cup\{X_{M,f} : M \in (0, \infty)\}$ and $v \in \Lambda(f, x)$. Then*

$$\pi^f(x) = \lim_{T \to \infty} [I^f(0, T, v) - Tf(\bar{x}, 0)].$$

Proof. It follows from (A4) and Propositions 8.12–8.14 that

$$\pi^f(x) = \lim_{T \to \infty} (\pi^f(v(0)) - \pi^f(v(T))) = \lim_{T \to \infty} [I^f(0, T, v) - Tf(\bar{x}, 0)].$$

Proposition 8.16 is proved.

The next result is proved in Sect. 8.6.

Proposition 8.17. *There exists $M > 0$ such that $X_f \subset X_{M,f}$.*

Propositions 8.6 and 8.17 imply the following result.

Proposition 8.18. *There exists $L > 0$ such that $X_f \subset \bar{Y}_{L,f}$.*

The following result is proved in Sect. 8.6.

Proposition 8.19. *The function $\pi^f : R^n \to R^1 \cup \{\infty\}$ is lower semicontinuous.*

Set

$$\mathcal{D}(f) = \{x \in R^n : \pi^f(x) = \inf(\pi^f)\}. \tag{8.18}$$

By Propositions 8.15 and 8.19, the set $\mathcal{D}(f)$ is nonempty bounded and closed subset of R^n. The following proposition is proved in Sect. 8.6.

Proposition 8.20. *Let $v : [0, \infty) \to R^n$ be an (f)-good function such that for all $T > 0$,*

$$I^f(0, T, v) - Tf(\bar{x}, 0) = \pi^f(v(0)) - \pi^f(v(T)). \tag{8.19}$$

Then v is an (f)-overtaking optimal function.

The next result easily follows from (8.17), (8.18), Proposition 8.17, and Theorem 8.9.

Proposition 8.21. *For each $\epsilon > 0$ there exists $T_\epsilon > 0$ such that for each $z \in \mathcal{D}(f)$ and each $v \in \Lambda(f, z)$ the inequality $|v(t) - \bar{x}| \leq \epsilon$ holds for all $t \geq T_\epsilon$.*

In order to study the structure of solutions of the problems (P_2) and (P_3) we introduce the following notation and definitions.

Define a function $\bar{f} : R^n \times R^n \to R^1 \cup \{\infty\}$ by

$$\bar{f}(x, y) = f(x, -y) \text{ for all } x, y \in R^n. \tag{8.20}$$

Clearly,

$$\text{dom}(\bar{f}) = \{(x, y) \in R^n \times R^n : (x, -y) \in \text{dom}(f)\}, \tag{8.21}$$

dom($\bar{f}$) is a nonempty closed convex set, $\bar{f}$ is a lower semicontinuous function satisfying

$$\bar{f}(x, y) \geq \max\{\psi(|x|), \ \psi(|y|)|y|\} - a \text{ for each } x, y \in R^n. \tag{8.22}$$

The notation introduced for the function f is also used for the function $\bar{f}$. Namely, for each pair of numbers $T_1 \in R^1$, $T_2 > T_1$, and each a.c. function $v : [T_1, T_2] \to R^n$ put

$$I^{\bar{f}}(T_1, T_2, v) = \int_{T_1}^{T_2} \bar{f}(v(t), v'(t))dt$$

and for each $x, y \in R^n$ and each $T > 0$ define

$$\sigma(\bar{f}, T, x) = \inf\{I^{\bar{f}}(0, T, v) : \ v : [0, T] \to R^n$$
$$\text{is an a. c. function satisfying } v(0) = x\},$$

$$\sigma(\bar{f}, T, x, y) = \inf\{I^{\bar{f}}(0, T, v) : \ v : [0, T] \to R^n$$
$$\text{is an a. c. function satisfying } v(0) = x, \ v(T) = y\},$$

$$\sigma(\bar{f}, T) = \inf\{I^{\bar{f}}(0, T, v) : \ v : [0, T] \to R^n \text{ is an a.c. function}\},$$

$$\hat{\sigma}(\bar{f}, T, y) = \inf\{I^{\bar{f}}(0, T, v) : \ v : [0, T] \to R^n$$
$$\text{is an a. c. function satisfying } v(T) = y\}.$$

Let $v : [0, T] \to R^n$ be an a.c. function. Set

$$\bar{v}(t) = v(T - t), \ t \in [0, T].$$

It is easy to see that

$$\int_0^T \bar{f}(\bar{v}(t), \bar{v}'(t))dt = \int_0^T f(\bar{v}(t), -\bar{v}'(t))dt$$

$$= \int_0^T f(v(T - t), v'(T - t))dt$$

$$= \int_0^T f(v(t), v'(t))dt. \tag{8.23}$$

Clearly, for all $x \in R^n$,

$$\bar{f}(\bar{x}, 0) = f(\bar{x}, 0) \leq f(x, 0) = \bar{f}(x, 0), \tag{8.24}$$

$(\bar{x}, 0)$ is an interior point of the set dom$(\bar{f})$ and the function $\bar{f}$ is continuous at the point $(\bar{x}, 0)$. Thus (A1) holds for the function $\bar{f}$. Clearly, for each $x \in R^n$ the function $\bar{f}(x, 0) : R^n \to R^1 \cup \{\infty\}$ is convex. Thus (A3) holds for $\bar{f}$. The next result easily follows from (8.23).

Proposition 8.22. *Let $T > 0$, $M \geq 0$, and $v_i : [0, T] \to R^n$, $i = 1, 2$ be a. c. functions. Then*

$$I^f(0, T, v_1) \geq I^f(0, T, v_2) - M$$

if and only if

$$I^{\bar{f}}(0, T, \bar{v}_1) \geq I^{\bar{f}}(0, T, \bar{v}_2) - M.$$

Proposition 8.22 implies the following result.

Proposition 8.23. *Let $T > 0$ and $v : [0, T] \to R^n$ be an a. c. function. Then the following assertions hold:*

$$I^f(0, T, v) \leq \sigma(f, T) + M \text{ if and only if}$$

$$I^{\bar{f}}(0, T, \bar{v}) \leq \sigma(\bar{f}, T) + M;$$

$$I^f(0, T, v) \leq \sigma(f, T, v(0), v(T)) + M \text{ if and only if}$$

$$I^{\bar{f}}(0, T, \bar{v}) \leq \sigma(\bar{f}, T, \bar{v}(0), \bar{v}(T)) + M;$$

$$I^f(0, T, v) \leq \hat{\sigma}(f, T, v(T)) + M \text{ if and only if } I^{\bar{f}}(0, T, \bar{v}) \leq \sigma(\bar{f}, T, \bar{v}(0)) + M;$$

$$I^f(0, T, v) \leq \sigma(f, T, v(0)) + M \text{ if and only if } I^{\bar{f}}(0, T, \bar{v}) \leq \hat{\sigma}(\bar{f}, T, \bar{v}(T)) + M.$$

The next result is proved in Sect. 8.6.

Proposition 8.24. *1. For each $M > 0$ there exists $c_M > 0$ such that $\sigma(\bar{f}, T, x) \geq T\bar{f}(\bar{x}, 0) - c_M$ for each $x \in R^n$ satisfying $|x| \leq M$ and each $T > 0$.*
2. For each $(\bar{f})$-good function $v : [0, \infty) \to R^n$, $\lim_{t \to \infty} v(t) = \bar{x}$.

In view of Proposition 8.24 the function $\bar{f}$ satisfies assumptions (A2) and (A4). We have already mentioned that assumptions (A1) and (A3) hold for the function $\bar{f}$. Therefore all the results stated above for the function f are also true for the function $\bar{f}$.

8.3 The Main Results

We use the notation, definitions, and assumptions introduced in Sects. 8.1 and 8.2. Recall that $f : R^n \times R^n \to R^1 \cup \{\infty\}$ is a lower semicontinuous function with a nonempty closed convex dom(f) satisfying (8.3), $\bar{x} \in R^n$ satisfies (8.8), and that (A1)–(A4) hold.

The following two theorems which describe the structure of approximate solutions of the problems (P_1) and (P_2) are proved in Sects. 8.8 and 8.9, respectively.

Theorem 8.25. *Let $L_0 > 0$, $\tau_0 > 0$ and $\epsilon > 0$. Then there exist $\delta > 0$ and $T_0 > \tau_0$ such that for each $T \geq T_0$ and each a. c. function $v : [0, T] \to R^n$ which satisfies*

$$v(0) \in \bar{Y}_{L_0, f}, \; I^f(0, T, v) \leq \sigma(f, T, v(0)) + \delta$$

there exists an $(\bar{f})$-overtaking optimal function $v^* : [0, \infty) \to R^n$ such that $v^*(0) \in \mathcal{D}(\bar{f})$ and $|v(T - t) - v^*(t)| \leq \epsilon$ for all $t \in [0, \tau_0]$.

Theorem 8.26. *Let $\tau_0 > 0$ and $\epsilon > 0$. Then there exist $\delta > 0$ and $T_0 > \tau_0$ such that for each $T \geq T_0$ and each a. c. function $v : [0, T] \to R^n$ which satisfies*

$$I^f(0, T, v) \leq \sigma(f, T) + \delta$$

there exist an (f)-overtaking optimal function $u^* : [0, \infty) \to R^n$ and an $(\bar{f})$-overtaking optimal function $v^* : [0, \infty) \to R^n$ such that $u^*(0) \in \mathcal{D}(f)$, $v^*(0) \in \mathcal{D}(\bar{f})$ and that for all $t \in [0, \tau_0]$,

$$|v(t) - u^*(t)| \leq \epsilon \text{ and } |v(T - t) - v^*(t)| \leq \epsilon.$$

The results of this section were obtained in [55]. Note that in [47] analogous results were obtained for strictly convex integrands.

8.4 Auxiliary Results

This section contains several auxiliary results which will be used in the paper.

Lemma 8.27 ([46]). *Let $M, \epsilon > 0$. Then there exists $L_0 > 0$ such that for each number $T \geq L_0$, each a.c. function $v : [0, T] \to R^n$ satisfying*

$$|v(0)| \leq M, \; I^f(0, T, v) \leq Tf(\bar{x}, 0) + M$$

and each number $s \in [0, T - L_0]$ the inequality

$$\min\{|v(t) - \bar{x}| : \, t \in [s, s + L_0]\} \leq \epsilon$$

holds.

Lemma 8.28 ([46]). *Let $\epsilon > 0$. Then there exists a number $\delta \in (0, \bar{r}/2)$ such that for each number $T \geq 2$ and each a.c. function $v : [0, T] \to R^n$ which satisfies*

$$|v(0) - \bar{x}|, \; |v(T) - \bar{x}| \leq \delta,$$

$$I^f(0, T, v) \leq \sigma(f, T, v(0), v(T)) + \delta$$

the inequality $|v(t) - \bar{x}| \leq \epsilon$ is true for all numbers $t \in [0, T]$.

Lemma 8.29 ([46]). *Let $M_0, M_1 > 0$. Then there exists $M_2 > 0$ such that for each $T > 0$ and each a.c. function $v : [0, T] \to R^n$ which satisfies*

$$|v(0)| \leq M_0, \; I^f(0, T, v) \leq Tf(\bar{x}, 0) + M_1$$

the following inequality holds:

$$|v(t)| \leq M_2 \text{ for all } t \in [0, T].$$

Proposition 8.30 ([46]). *Let $T > 0$ and let $v_k : [0, T] \to R^n$, $k = 1, 2, \ldots$ be a sequence of a.c. functions such that the sequence $\{I^f(0, T, v_k)\}_{k=1}^{\infty}$ is bounded and that the sequence $\{v_k(0)\}_{k=1}^{\infty}$ is bounded. Then there exist a strictly increasing sequence of natural numbers $\{k_i\}_{i=1}^{\infty}$ and an a.c. function $v : [0, T] \to R^n$ such that*

$$v_{k_i}(t) \to v(t) \text{ as } i \to \infty \text{ uniformly on } [0, T],$$

$$I^f(0, T, v) \leq \liminf_{i \to \infty} I^f(0, T, v_{k_i}).$$

Lemma 8.31 ([46]). *Let $\epsilon > 0$. Then there exists $\delta > 0$ such that for each a.c. function $v : [0, 1] \to R^n$ satisfying $|v(0) - \bar{x}|, |v(1) - \bar{x}| \leq \delta$,*

$$I^f(0, 1, v) \geq f(\bar{x}, 0) - \epsilon.$$

Lemma 8.32. *Let $0 < L_0 < L_1$ and $M_0 > 0$. Then there exists $M_1 > 0$ such that for each $T \in [L_0, L_1]$ and each a. c. function $v : [0, T] \to R^n$ satisfying $I^f(0, T, v) \leq M_0$ the inequality $|v(t)| \leq M_1$ holds for all $t \in [0, T]$.*

Proof. By (8.1), there exists $c_0 > 0$ such that $\psi(t) \geq 4$ for all $t \geq c_0$. This implies that for all $t \geq 0$,

$$\psi(t)t \geq 4t - 4c_0. \tag{8.25}$$

In view of (8.1), choose $M_1 > 0$ such that

$$\psi(M_1/2) > L_0^{-1}M_0 + aL_1L_0^{-1} + 1, \; M_1 > 2(M_0 + aL_1 + 4c_0L_1). \tag{8.26}$$

Let $T \in [L_0, L_1]$ and an a. c. function $v : [0, T] \to R^n$ satisfy $I^f(0, T, v) \leq M_0$. Together with (8.3) this implies that

$$M_0 + aL_1 \geq \int_0^T \psi(|v(t)|)dt, \ \int_0^T \psi(|v'(t)|)|v'(t)|dt. \tag{8.27}$$

By (8.25) and (8.27),

$$M_0 + aL_1 \geq 4\int_0^T (|v'(t)| - c_0)dt \geq -4c_0L_1 + 4\int_0^T |v'(t)|dt,$$

$$\int_0^T (|v'(t)|)dt \leq M_0 + aL_1 + 4c_0L_1.$$

Together with (8.26) this implies that

$$\sup\{|v(t_2) - v(t_1)| : \ t_1, t_2 \in [0, T]\} \leq M_0 + aL_1 + 4c_0L_1 < M_1/2. \tag{8.28}$$

Assume that

$$\max\{|v(t)| : \ t \in [0, T]\} > M_1.$$

In view of (8.27) and (8.28),

$$|v(t)| \geq M_1/2 \text{ for all } t \in [0, T],$$

$$\psi(M_1/2)L_0 \leq \int_0^T \psi(|v(t)|)dt \leq M_0 + aL_1,$$

$$\psi(M_1/2) \leq L_0^{-1}M_0 + aL_1L_0^{-1}.$$

This contradicts (8.26). The contradiction we have reached proves Lemma 8.32.

8.5   Proof of Theorem 8.10

Theorem 8.10 easily follows from Lemma 8.28 and the following result.

Lemma 8.33. *Let* $M, \epsilon > 0$. *Then there exists* $L_0 > 0$ *such that for each number* $T \geq L_0$, *each a.c. function* $v : [0, T] \rightarrow R^n$ *satisfying*

$$I^f(0, T, v) \leq Tf(\bar{x}, 0) + M \tag{8.29}$$

and each number $s \in [0, T - L_0]$ *the inequality*

$$\min\{|v(t) - \bar{x}| : \ t \in [s, s + L_0]\} \leq \epsilon$$

holds.

Proof. In view of (8.1), there exists a number $M_0 > 0$ such that

$$\psi(M_0) > |f(\bar{x}, 0)| + a + 2. \tag{8.30}$$

By (A2), there exists $c_0 > 0$ such that

$$\sigma(f, T, x) \geq Tf(\bar{x}, 0) - c_0$$

for each $x \in R^n$ satisfying $|x| \leq M_0$ and each $T > 0$. $\hspace{2em}$ (8.31)

By Lemma 8.27, there exists $L_1 > 0$ such that the following property holds:

(Pi) for each number $T \geq L_1$, each a.c. function $u : [0, T] \to R^n$ satisfying

$$|u(0)| \leq M_0, \quad I^f(0, T, u) \leq Tf(\bar{x}, 0) + M$$

and each number $s \in [0, T - L_1]$,

$$\min\{|u(t) - \bar{x}| : t \in [s, s + L_1]\} \leq \epsilon.$$

Choose

$$L_0 > 4L_1 + M + c_0. \tag{8.32}$$

Assume that $T \geq L_0$ and an a. c. function $v : [0, T] \to R^n$ satisfies (8.29). Let us show that

$$\min\{|v(t)| : t \in [0, T]\} \leq M_0. \tag{8.33}$$

Assume the contrary. Then $|v(t)| > M_0$ for all $t \in [0, T]$. Together with (8.3), (8.29), and (8.30) this implies that

$$f(v(t), v'(t)) \geq f(\bar{x}, 0) + 2$$

for all $t \in [0, T]$ and

$$M \geq I^f(0, T, v) - Tf(\bar{x}, 0) \geq 2T \geq L_0.$$

This contradicts the choice of L_0 (see (8.32)). Thus (8.33) holds. Set

$$\tau_0 = \min\{t \in [0, T] : |v(t)| \leq M_0\}. \tag{8.34}$$

In view of (8.29) and (8.34),

$$M + Tf(\bar{x}, 0) \geq I^f(0, T, v) = I^f(0, \tau_0, v) + I^f(\tau_0, T, v). \tag{8.35}$$

By (8.3), (8.30), and (8.34),

$$I^f(0, \tau_0, v) \geq \tau_0(\psi(M_0) - a) \geq \tau_0(f(\bar{x}, 0) + 2). \tag{8.36}$$

It follows from (8.31) and (8.34) that

$$I^f(\tau_0, T, v) \geq (T - \tau_0)f(\bar{x}, 0) - c_0. \tag{8.37}$$

By (8.35)–(8.37),

$$M + Tf(\bar{x}, 0) \geq \tau_0(f(\bar{x}, 0) + 2) + (T - \tau_0)f(\bar{x}, 0) - c_0,$$

$$2\tau_0 \leq M + c_0. \tag{8.38}$$

In view of (8.29) and (8.36),

$$I^f(\tau_0, T, v) = I^f(0, T, v) - I^f(0, \tau_0, v)$$
$$\leq Tf(\bar{x}, 0) + M - \tau_0 f(\bar{x}, 0) \leq (T - \tau_0)f(\bar{x}, 0) + M. \tag{8.39}$$

By (8.32) and (8.38),

$$T - \tau_0 \geq L_0 - M - c_0 > 4L_1.$$

It follows from (8.34), (8.39), the inequality above, and property (Pi) that for each number S satisfying $[S, S + L_1] \subset [\tau_0, T]$,

$$\min\{|v(t) - \bar{x}| : t \in [S, S + L_1]\} \leq \epsilon. \tag{8.40}$$

Let a number S satisfy

$$[S, S + L_0] \subset [0, T].$$

By (8.32), (8.38), and the relation above, $E := [S, S + L_0] \cap [\tau_0, T]$ is a closed interval, $\mathrm{mes}(E) \geq L_0 - \tau_0 \geq 4L_1$. Together with (8.40) this implies that

$$\min\{|v(t) - \bar{x}| : t \in [S, S + L_0]\} \leq \min\{|v(t) - \bar{x}| : t \in E\} \leq \epsilon.$$

Lemma 8.33 is proved.

8.6 Proofs of Propositions 8.14, 8.15, 8.17, 8.19, 8.20 and 8.24

Proof of Proposition 8.14. By (8.14) and Theorem 8.8, there exists $\delta_0 > 0$ such that $\pi^f(x)$ is finite for any $x \in R^n$ satisfying $|x - \bar{x}| \le \delta_0$. Let us show that π^f is continuous at $\bar{x}$.

Let $\epsilon > 0$. By (A1) there exists $\delta \in (0, \delta_0)$ such that for each $(y, z) \in R^n \times R^n$ satisfying

$$|y - \bar{x}| + |z| \le 6\delta$$

we have

$$(y, z) \in \text{dom}(f), \ |f(y, z) - f(\bar{x}, 0)| \le \epsilon. \tag{8.41}$$

Let $x_1, x_2 \in R^n$ satisfy

$$|x_i - \bar{x}| \le \delta, \ i = 1, 2. \tag{8.42}$$

In order to complete the proof of proposition it is sufficient to show that $\pi^f(x_2) \le \pi^f(x_1) + \epsilon$. By the choice of δ_0 and (8.14), there exists an (f)-overtaking optimal function $v_0 : [0, \infty) \to R^n$ such that

$$v_0(0) = x_1, \tag{8.43}$$

$$\pi^f(x_1) = \liminf_{T \to \infty}[I^f(0, T, v_0) - Tf(\bar{x}, 0)]. \tag{8.44}$$

Define

$$v(t) = x_2 + t(x_1 - x_2), \ t \in [0, 1], \ v(t) = v_0(t - 1), \ t \in (1, \infty]. \tag{8.45}$$

In view of (8.43) and (8.45), $v : [0, \infty) \to R^n$ is an a. c. function. By (8.42) and (8.45), for all $t \in (0, 1)$,

$$|v'(t)| = |x_1 - x_2| \le 2\delta, \ |v(t) - \bar{x}| \le 3\delta. \tag{8.46}$$

By (8.46) and the choice of δ (see (8.41)), for all $t \in (0, 1)$,

$$|f(v(t), v'(t)) - f(\bar{x}, 0)| \le \epsilon. \tag{8.47}$$

It follows from (8.44), (8.45), (8.47), and Proposition 8.11 that

$$\pi^f(x_2) \le \liminf_{T \to \infty}[I^f(0, T, v) - Tf(\bar{x}, 0)]$$

$$= I^f(0, 1, v) - f(\bar{x}, 0) + \liminf_{T \to \infty}[I^f(0, T, v_0) - Tf(\bar{x}, 0)] \le \epsilon + \pi^f(x_1).$$

This completes the proof of Proposition 8.14.   □

Proof of Proposition 8.15. Let $M > 0$ and suppose that the set $\{x \in R^n : \pi^f(x) \leq M\}$ is nonempty. By Propositions 8.13 and 8.14, there exists $\delta > 0$ such that for each $x \in R^n$ satisfying $|x - \bar{x}| \leq \delta$, $\pi^f(x)$ is finite and

$$|\pi^f(x)| \leq 1. \tag{8.48}$$

By Lemma 8.33, there exists $L_0 > 0$ such that the following property holds:

(a) for each $T \geq L_0$, each a.c. function $v : [0, T] \to R^n$ satisfying

$$I^f(0, T, v) \leq Tf(\bar{x}, 0) + M + 1$$

and each $S \in [0, T - L_0]$,

$$\min\{|v(t) - \bar{x}| : t \in [S, S + L_0]\} \leq \delta.$$

By Lemma 8.32, there exists $M_1 > 0$ such that the following property holds:

(b) for each $S \in [1, L_0 + 1]$ and each a. c. function $v : [0, S] \to R^n$ satisfying

$$I^f(0, S, v) \leq (1 + L_0)|f(\bar{x}, 0)| + M + 1$$

the inequality $|v(t)| \leq M_1$ holds for all $t \in [0, S]$.
Assume that $x \in R^n$ satisfies

$$\pi^f(x) \leq M. \tag{8.49}$$

In view of (8.14) and (8.49),

$$\pi^f(x) = \liminf_{T \to \infty}[I^f(0, T, v) - Tf(\bar{x}, 0)] \leq M, \tag{8.50}$$

where

$$v \in \Lambda(f, x). \tag{8.51}$$

By (8.50) and property (a), there exists

$$t_0 \in [1, L_0 + 1] \tag{8.52}$$

such that

$$|v(t_0) - \bar{x}| \leq \delta. \tag{8.53}$$

It follows from (8.53) and the choice of δ (see (8.48)) that

$$|\pi^f(v(t_0))| \le 1. \tag{8.54}$$

By (8.49), (8.51), (8.54), and Proposition 8.12,

$$I^f(0, t_0, v) - t_0 f(\bar{x}, 0) = \pi^f(v(0)) - \pi^f(v(t_0)) \le M + 1. \tag{8.55}$$

In view of (8.52), (8.55), and property (b), $|v(t)| \le M_1$ for all $t \in [0, t_0]$ and in view of (8.51), $|x| = |v(0)| \le M_1$. Proposition 8.15 is proved. $\qquad\square$

Proof of Proposition 8.17. By Proposition 8.15, there exists $M_0 > 0$ such that

$$X_f \subset \{z \in R^n : |z| \le M_0\}. \tag{8.56}$$

By Lemma 8.29, there exists $M_1 > 0$ such that for each $T > 0$ and each a.c. function $v : [0, T] \to R^n$ which satisfies

$$|v(0)| \le M_0, \ I^f(0, T, v) \le Tf(\bar{x}, 0) + \inf(\pi^f) + 2$$

we have

$$|v(t)| \le M_1 \text{ for all } t \in [0, T]. \tag{8.57}$$

By (A2), there exists $c_1 > 0$ such that for each $T > 0$ and each $z \in R^n$ satisfying $|z| \le M_1$ we have

$$\sigma(f, T, z) \ge Tf(\bar{x}, 0) - c_1. \tag{8.58}$$

Let $x \in X_f$. Then

$$\pi^f(x) \le \inf(\pi^f) + 1. \tag{8.59}$$

By (8.59) and Proposition 8.16,

$$\pi^f(x) = \lim_{T \to \infty} [I^f(0, T, v) - Tf(\bar{x}, 0)], \tag{8.60}$$

where

$$v \in \Lambda(f, x). \tag{8.61}$$

By (8.59) and (8.60), there exists a number $T_0 > 0$ such that for all $T \ge T_0$,

$$I^f(0, T, v) - Tf(\bar{x}, 0) \le \inf(\pi^f) + 2. \tag{8.62}$$

By (8.56), (8.59), (8.61), (8.62), and the choice of M_1 (see (8.57)),

$$|v(t)| \leq M_1, t \in [0, \infty). \tag{8.63}$$

It follows from (8.62), (8.63), and the choice of c_1 (see (8.58)) that for any $T \in (0, T_0)$,

$$I^f(0, T, v) - Tf(\bar{x}, 0) = I^f(0, T_0, v) - T_0 f(\bar{x}, 0)$$
$$-(I^f(T, T_0, v) - (T_0 - T)f(\bar{x}, 0))$$
$$\leq \inf(\pi^f) + 2 + c_1. \tag{8.64}$$

In view of (8.61) and (8.63), $x = v(0) \in X_{M,f}$, where

$$M = |\inf(\pi^f)| + 2 + c_1 + M_1.$$

Proposition 8.17 is proved.   □

Proof of Proposition 8.19. Assume that $\{x_k\}_{k=1}^{\infty} \subset R^n$, $x \in R^n$ and that

$$\lim_{k \to \infty} x_k = x. \tag{8.65}$$

We show that $\pi^f(x) \leq \liminf_{k \to \infty} \pi^f(x_k)$. We may assume that there exists a finite

$$M_0 := \lim_{k \to \infty} \pi^f(x_k) \tag{8.66}$$

and that $\pi^f(x_k)$ is finite for all integers $k \geq 1$. By Proposition 8.16, for each integer $k \geq 1$,

$$\pi^f(x_k) = \lim_{T \to \infty} [I^f(0, T, v_k) - Tf(\bar{x}, 0)], \tag{8.67}$$

where

$$v_k \in \Lambda(f, x_k). \tag{8.68}$$

We may assume without loss of generality that for all integers $k \geq 1$,

$$\pi^f(x_k) \leq M_0 + 1. \tag{8.69}$$

By (8.65), (8.67)–(8.69), and Lemma 8.29, there exists $M_1 > 0$ such that for each integer $k \geq 1$,

$$|v_k(t)| \leq M_1, \ t \in [0, \infty). \tag{8.70}$$

By (A2) there exists $c_1 > 0$ such that for each $T > 0$ and each $z \in R^n$ satisfying $|z| \le M_1$,

$$\sigma(f, T, z) \ge Tf(\bar{x}, 0) - c_1. \tag{8.71}$$

Let $k \ge 1$ be an integer and $S > 0$. By (8.67) and (8.69), there exists $T > S$ such that

$$I^f(0, T, v_k) - Tf(\bar{x}, 0) \le M_0 + 2. \tag{8.72}$$

In view of the choice of c_1 (see (8.71)) and (8.70),

$$I^f(S, T, v_k) - (T - S)f(\bar{x}, 0) \ge -c_1. \tag{8.73}$$

Relations (8.72) and (8.73) imply that

$$I^f(0, S, v_k) - Sf(\bar{x}, 0) = I^f(0, T, v_k) - Tf(\bar{x}, 0)$$
$$-(I^f(S, T, v_k) - (T - S)f(\bar{x}, 0)) \le M_0 + c_1 + 2.$$

Thus for any integer $k \ge 1$ and any $S > 0$,

$$I^f(0, S, v_k) - Sf(\bar{x}, 0) \le M_2, \tag{8.74}$$

where

$$M_2 = M_0 + M_1 + c_1 + 2. \tag{8.75}$$

It follows from (8.68), (8.70), and (8.74) that for any integer $k \ge 1$,

$$x_k = v_k(0) \in X_{M_2, f}. \tag{8.76}$$

By (8.70), (8.74), and Proposition 8.30 using diagonalization process and re-indexing if necessary we may assume without loss of generality that there exists an a.c. function $v : [0, \infty) \to R^n$ such that for each natural number p,

$$v_k(t) \to v(t) \text{ as } k \to \infty \text{ uniformly on } [0, p], \tag{8.77}$$
$$I^f(0, p, v) \le \liminf_{k \to \infty} I^f(0, p, v_k). \tag{8.78}$$

Let $\epsilon > 0$. In view of Propositions 8.13 and 8.14, there exists a positive number

$$\delta < \min\{4^{-1}\epsilon, \ 4^{-1}\bar{r}\} \tag{8.79}$$

such that for each $y \in R^n$ satisfying $|y - \bar{x}| \leq \delta$,

$$|\pi^f(y)| = |\pi^f(y) - \pi^f(\bar{x})| \leq \epsilon/4. \tag{8.80}$$

By (8.69), (8.76), and Theorem 8.9, there exists $L > 0$ such that for each integer $k \geq 1$,

$$|v_k(t) - \bar{x}| \leq \delta, \ t \in [L, \infty). \tag{8.81}$$

In view of (8.80) and (8.81), for all integers $k \geq 1$,

$$|v_k(L) - \bar{x}| \leq \delta, \ |\pi^f(v_k(L))| \leq \epsilon/4. \tag{8.82}$$

It follows from (8.77), (8.80), and (8.82) that

$$|v(L) - \bar{x}| \leq \delta, \ |\pi^f(v(L))| \leq \epsilon/4. \tag{8.83}$$

Let $k \geq 1$ be an integer. By (8.68), (8.82), and Proposition 8.12,

$$I^f(0, L, v_k) = \pi^f(v_k(0)) - \pi^f(v_k(L)) + Lf(\bar{x}, 0) \leq \pi^f(v_k(0)) + Lf(\bar{x}, 0) + \epsilon/4.$$

Together with (8.66), (8.76), and (8.78), this implies that

$$I^f(0, L, v) \leq \liminf_{k \to \infty}[\pi^f(v_k(0)) + Lf(\bar{x}, 0) + \epsilon/4] = \lim_{k \to \infty} \pi^f(x_k) + Lf(\bar{x}, 0) + \epsilon/4. \tag{8.84}$$

By (8.65), (8.66), (8.76), (8.77), (8.83), (8.84), and Proposition 8.11,

$$\pi^f(x) = \pi^f(v(0)) \leq I^f(0, L, v) - Lf(\bar{x}, 0) + \pi^f(v(L)) \leq \lim_{k \to \infty} \pi^f(x_k) + \epsilon/2.$$

Since ϵ is any positive number this completes the proof of Proposition 8.19. □

Proof of Proposition 8.20. In view of (A4),

$$\lim_{t \to \infty} v(t) = \bar{x}. \tag{8.85}$$

Theorem 8.4 implies that there exists an (f)-overtaking optimal function $u :$ $[0, \infty) \to R^n$ such that

$$u(0) = v(0). \tag{8.86}$$

By (8.86) and Proposition 8.16,

$$\pi^f(v(0)) = \lim_{T \to \infty}[I^f(0, T, u) - Tf(\bar{x}, 0)]. \tag{8.87}$$

On the other hand, it follows from (8.19), (8.85), and Propositions 8.13 and 8.14 that for any $T > 0$,

$$I^f(0, T, v) - Tf(\bar{x}, 0) = \pi^f(v(0)) - \pi^f(v(T)) \to \pi^f(v(0)) \text{ as } T \to \infty. \qquad (8.88)$$

By (8.87) and (8.88),

$$\lim_{T \to \infty} [I^f(0, T, u) - I^f(0, T, v)] = 0.$$

This implies that v is an (f)-overtaking optimal function. Proposition 8.20 is proved.

$\square$

Proof of Proposition 8.24. First we prove assertion 1. Let $M > 0$. In view of (8.1) we may assume without loss of generality that

$$\psi(M) > |f(\bar{x}, 0)| + a + 2. \qquad (8.89)$$

By (A2), there exists $c_0 > 0$ such that for each $S > 0$ and each a. c. function $u : [0, S] \to R^n$ satisfying $|u(0)| \leq M$,

$$I^f(0, S, u) \geq Sf(\bar{x}, 0) - c_0. \qquad (8.90)$$

Assume that $T > 0$ and that $v : [0, T] \to R^n$ is an a. c. function satisfying

$$|v(0)| \leq M. \qquad (8.91)$$

In order to prove assertion 1 it is sufficient to show that

$$I^{\bar{f}}(0, T, v) \geq Tf(\bar{x}, 0) - c_0.$$

Let

$$u(t) = \bar{v}(t) = v(T - t), \ t \in [0, T]. \qquad (8.92)$$

In view of (8.23) and (8.92),

$$I^f(0, T, u) = I^{\bar{f}}(0, T, v). \qquad (8.93)$$

By (8.91) and (8.92),

$$|u(T)| \leq M. \qquad (8.94)$$

Let

$$\tau_0 = \min\{t \in [0, T] : |u(t)| \leq M\}. \qquad (8.95)$$

In view of (8.94), τ_0 is well defined. By (8.95) and the choice of c_0 (see (8.90)),

$$I^f(\tau_0, T, u) \geq (T - \tau_0)f(\bar{x}, 0) - c_0. \tag{8.96}$$

By (8.3), (8.89), and (8.95),

$$I^f(0, \tau_0, u) - \tau_0(\psi(M) - a) \geq \tau_0 f(\bar{x}, 0).$$

Together with (8.93) and (8.96) this implies that

$$I^{\bar{f}}(0, T, v) = I^f(0, \tau_0, u) + I^f(\tau_0, T, u) \geq T\bar{f}(\bar{x}, 0) - c_0.$$

Thus assertion 1 holds.

Let us prove assertion 2. Note that the integrand $\bar{f}$ satisfies (A1)–(A3). Let $v :$ $[0, \infty) \to R^n$ be an $(\bar{f})$-good function. Then there exists $M_1 > 0$ such that for each pair of integers $T_2 > T_1 \geq 0$,

$$\left| \int_{T_1}^{T_2} [\bar{f}(v(t), v'(t)) - f(\bar{x}, 0)]dt \right| \leq M_1. \tag{8.97}$$

In order to complete the proof of assertion 2 it is sufficient to show that

$$\lim_{t \to \infty} v(t) = \bar{x}.$$

Let $\epsilon > 0$. By Proposition 8.1 applied to the function $\bar{f}$, there exists $M_2 > 0$ such that

$$|v(t)| \leq M_2 \text{ for all } t \in [0, \infty). \tag{8.98}$$

By Lemma 8.28 and Proposition 8.23, there exists $\delta \in (0, \bar{r}/2)$ such that the following property holds:

(i) for each number $T \geq 2$ and each a.c. function $w : [0, T] \to R^n$ which satisfies

$$|w(0) - \bar{x}|, \ |w(T) - \bar{x}| \leq \delta,$$
$$I^{\bar{f}}(0, T, w) \leq \sigma(\bar{f}, T, w(0), w(T)) + \delta$$

the inequality $|w(t) - \bar{x}| \leq \epsilon$ is true for all numbers $t \in [0, T]$.

By Lemma 8.27 and (8.23), there exists $L_0 > 0$ such that the following property holds:

(ii) for each a.c. function $w : [0, L_0] \to R^n$ satisfying

$$|w(t)| \leq M_1 + M_2, \ t \in [0, L_0], \ I^{\bar{f}}(0, L_0, w) \leq L_0 f(\bar{x}, 0) + M_1 + M_2$$

we have

$$\min\{|w(t) - \bar{x}| : t \in [0, L_0]\} \le \delta.$$

Since the function v is $(\bar{f})$-good there exists $\tau_0 > 0$ such that for each pair of numbers $T_2 > T_1 \ge \tau_0$,

$$I^{\bar{f}}(T_1, T_2, v) \le \sigma(\bar{f}, T_2 - T_1, v(T_1), v(T_2)) + \delta. \tag{8.99}$$

It follows from (8.97), (8.98), and property (ii) that there exists a strictly increasing sequence of numbers $\{S_k\}_{k=1}^{\infty}$ such that

$$S_1 \ge \tau_0, \ S_{k+1} \ge S_k + 2 \text{ for all natural numbers } k,$$

$$|v(S_k) - \bar{x}| \le \delta \text{ for all integers } k \ge 1. \tag{8.100}$$

By (8.99), (8.100), and property (i), for each integer $k \ge 1$,

$$|v(t) - \bar{x}| \le \epsilon, \ t \in [S_k, S_{k+1}].$$

Therefore $|v(t) - \bar{x}| \le \epsilon$ for all $t \ge S_1$ and $\lim_{t\to\infty} v(t) = \bar{x}$. Proposition 8.24 is proved. □

8.7 The Basic Lemma for Theorem 8.25

Lemma 8.34. *Let $T_0 > 0$ and $\epsilon \in (0, 1)$. Then there exists $\delta \in (0, \epsilon)$ such that for each a. c. function $v : [0, T_0] \to R^n$ which satisfies*

$$\pi^f(v(0)) \le \inf(\pi^f) + \delta, \tag{8.101}$$

$$I^f(0, T_0, v) - T_0 f(\bar{x}, 0) - \pi^f(v(0)) + \pi^f(v(T_0)) \le \delta \tag{8.102}$$

there exists an (f)-overtaking optimal function $u : [0, \infty) \to R^n$ which satisfies

$$\pi^f(u(0)) = \inf(\pi^f), \tag{8.103}$$

$$|u(t) - v(t)| \le \epsilon \text{ for all } t \in [0, T_0].$$

Proof. Assume the contrary. Then there exist a sequence $\{\delta_k\}_{k=1}^{\infty} \subset (0, 1]$ and a sequence of a. c. functions $v_k : [0, T_0] \to R^n, k = 1, 2, \dots$ such that

$$\lim_{k\to\infty} \delta_k = 0 \tag{8.104}$$

and that for each integer $k \geq 1$ and each (f)-overtaking optimal function $u :$ $[0, \infty) \to R^n$ satisfying (8.103),

$$\pi^f(v_k(0)) \leq \inf(\pi^f) + \delta_k, \tag{8.105}$$

$$I^f(0, T_0, v_k) - T_0 f(\bar{x}, 0) - \pi^f(v_k(0)) + \pi^f(v_k(T_0)) \leq \delta_k, \tag{8.106}$$

$$\sup\{|u(t) - v_k(t)| : t \in [0, T_0]\} > \epsilon. \tag{8.107}$$

By (8.105) and Proposition 8.15, the sequence $\{v_k(0)\}_{k=1}^\infty$ is bounded. In view of (8.105) and (8.106), for each integer $k \geq 1$, $\pi^f(v_k(0))$ and $\pi^f(v_k(T_0))$ are finite and the sequence $\{I^f(0, T_0, v_k)\}_{k=1}^\infty$ is bounded. By Proposition 8.30 and the boundedness of the sequences $\{v_k(0)\}_{k=1}^\infty$ and $\{I^f(0, T_0, v_k)\}_{k=1}^\infty$, extracting a subsequence and re-indexing if necessary, we may assume without loss of generality that there exists an a.c. function $v : [0, T_0] \to R^n$ such that

$$v_k(t) \to v(t) \text{ as } k \to \infty \text{ uniformly on } [0, T_0], \tag{8.108}$$

$$I^f(0, T_0, v) \leq \liminf_{k \to \infty} I^f(0, T_0, v_k). \tag{8.109}$$

By (8.105), (8.108), and the lower semicontinuity of π^f (see Proposition 8.19),

$$\pi^f(v(0)) \leq \liminf_{k \to \infty} \pi^f(v_k(0)) = \inf(\pi^f)$$

and

$$\pi^f(v(0)) = \inf(\pi^f). \tag{8.110}$$

In view of (8.108) and Proposition 8.19,

$$\pi^f(v(T_0)) \leq \liminf_{k \to \infty} \pi^f(v_k(T_0)). \tag{8.111}$$

By (8.104)–(8.106), (8.109)–(8.111),

$$I^f(0, T_0, v) - T_0 f(\bar{x}, 0) - \pi^f(v(0)) + \pi^f(v(T_0))$$

$$\leq \liminf_{k \to \infty} I^f(0, T_0, v_k) - T_0 f(\bar{x}, 0) - \lim_{k \to \infty} \pi^f(v_k(0)) + \liminf_{k \to \infty} \pi^f(v_k(T_0))$$

$$\leq \liminf_{k \to \infty} [I^f(0, T_0, v_k) - T_0 f(\bar{x}, 0) - \pi^f(v_k(0)) + \pi^f(v_k(T_0))] \leq 0. \tag{8.112}$$

Relations (8.109), (8.110), and (8.112) imply that $\pi^f(v(T_0))$ is finite. Together with (8.110) and Proposition 8.11 this implies that

$$I^f(0, T_0, v) - T_0 f(\bar{x}, 0) - \pi^f(v(0)) + \pi^f(v(T_0)) \geq 0.$$

By the inequality above and (8.112),

$$I^f(0, T_0, v) - T_0 f(\bar{x}, 0) - \pi^f(v(0)) + \pi^f(v(T_0)) = 0. \tag{8.113}$$

Since $\pi^f(v(T_0))$ is finite, Theorem 8.4 implies that there is an (f)-overtaking optimal and (f)-good function $w : [0, \infty) \to R^n$ such that

$$w(0) = v(T_0). \tag{8.114}$$

For all $t > T_0$ set

$$v(t) = w(t - T_0). \tag{8.115}$$

Clearly, $v : [0, \infty) \to R^n$ is an a. c. function. Since the function w is (f)-good and (f)-overtaking it follows from Propositions 8.11 and 8.12 and (8.113) that for all $T > 0$,

$$I^f(0, T, v) - T f(\bar{x}, 0) - \pi^f(v(0)) + \pi^f(v(T)) = 0. \tag{8.116}$$

It is clear that the function v is (f)-good. In view of (8.116) and Proposition 8.20, v is an (f)-overtaking optimal function satisfying (8.110). By (8.108), for all sufficiently large natural numbers k, $|v_k(t) - v(t)| \leq \epsilon/2$ for all $t \in [0, T_0]$. This contradicts (8.107). The contradiction we have reached proves Lemma 8.34.

Since the function $\bar{f}$ satisfies the same assumptions as the function f Lemma 8.34 can be applied with the function $\bar{f}$.

8.8   Proof of Theorem 8.25

Recall (see (8.11) and (8.12)) that $\bar{r} \in (0, 1)$ and

$$\Omega_0 = \{(x, y) \in R^n \times R^n : |x - \bar{x}| \leq \bar{r} \text{ and } |y| \leq \bar{r}\} \subset \text{dom}(f). \tag{8.117}$$

By Lemma 8.34 applied to the function $\bar{f}$, there exists

$$\delta_1 \in (0, \min\{\epsilon, \bar{r}/2\})$$

such that the following property holds:

(Pii)   for each a. c. function $v : [0, \tau_0] \to R^n$ which satisfies

$$\pi^{\bar{f}}(v(0)) \leq \inf(\pi^{\bar{f}}) + \delta_1, \tag{8.118}$$

$$I^{\bar{f}}(0, \tau_0, v) - \tau_0 f(\bar{x}, 0) - \pi^{\bar{f}}(v(0)) + \pi^{\bar{f}}(v(\tau_0)) \leq \delta_1 \tag{8.119}$$

there exists an $(\bar{f})$-overtaking optimal function $u : [0, \infty) \to R^n$ such that

$$\pi^{\bar{f}}(u(0)) = \inf(\pi^{\bar{f}}), \qquad (8.120)$$

$$|u(t) - v(t)| \leq \epsilon \text{ for all } t \in [0, \tau_0]. \qquad (8.121)$$

By Propositions 8.13 and 8.14, (A1) and Lemma 8.31, there exists $\delta_2 \in (0, \delta_1)$ such that for each $z \in R^n$ satisfying $|z - \bar{x}| \leq 2\delta_2$,

$$|\pi^{\bar{f}}(z)| = |\pi^{\bar{f}}(z) - \pi^{\bar{f}}(\bar{x})| \leq \delta_1/8, \qquad (8.122)$$

for each $(x, y) \in R^n \times R^n$ satisfying $|x - \bar{x}| \leq 4\delta_2$, $|y| \leq 4\delta_2$,

$$|f(x, y) - f(\bar{x}, 0)| \leq \delta_1/8 \qquad (8.123)$$

and that the following property holds:

(Piii)   for each a.c. function $v : [0, 1] \to R^n$ satisfying $|v(0) - \bar{x}|, |v(1) - \bar{x}| \leq \delta_2$,

$$I^f(0, 1, v) \geq f(\bar{x}, 0) - \delta_1/8.$$

By Theorem 8.2 and Proposition 8.5, there exist $L \geq 1$ and $\delta_3 > 0$ such that the following property holds:

(Piv)   for each $T > 2L$ and each a.c. function $v : [0, T] \to R^n$ which satisfies

$$v(0) \in \bar{Y}_{L_0, f} \text{ and } I^f(0, T, v) \leq \sigma(f, T, v(0)) + \delta_3 \qquad (8.124)$$

we have

$$|v(t) - \bar{x}| \leq \delta_2 \text{ for all } t \in [L, T - L]. \qquad (8.125)$$

By Proposition 8.21 applied to the function $\bar{f}$ there exists $\tau_1 > 0$ such that for each $(\bar{f})$-overtaking optimal function $u : [0, \infty) \to R^n$ satisfying

$$\pi^{\bar{f}}(u(0)) = \inf(\pi^{\bar{f}}) \qquad (8.126)$$

we have

$$|u(t) - \bar{x}| \leq \delta_2 \text{ for all } t \geq \tau_1. \qquad (8.127)$$

Choose $\delta > 0$ and $T_0 > 0$ such that

$$\delta < (16(L + \tau_1 + \tau_0 + 6))^{-1} \min\{\delta_1, \delta_2, \delta_3\}, \qquad (8.128)$$

$$T_0 > 2L + 2\tau_0 + 2\tau_1 + 4. \qquad (8.129)$$

Assume that $T \geq T_0$ and that an a. c. function $v : [0, T] \to R^n$ satisfies

$$v(0) \in \bar{Y}_{L_0, f} \text{ and } I^f(0, T, v) \leq \sigma(f, T, v(0)) + \delta. \tag{8.130}$$

By (8.128)–(8.130) and property (Piv), relation (8.125) holds. In view of (8.129),

$$[T - L - \tau_0 - \tau_1 - 4, T - L - \tau_0 - \tau_1] \subset [L, T - L - \tau_0 - \tau_1]. \tag{8.131}$$

Relations (8.125) and (8.131) imply that

$$|v(t) - \bar{x}| \leq \delta_2 \text{ for all } t \in [T - L - \tau_0 - \tau_1 - 4, T - L - \tau_0 - \tau_1]. \tag{8.132}$$

By Theorem 8.4, there exists an $(\bar{f})$-overtaking optimal function $u : [0, \infty) \to R^n$ satisfying (8.120). Then (8.127) holds. Define

$$\tilde{v}(t) = v(t), \ t \in [0, T - L - \tau_0 - \tau_1 - 4],$$

$$\tilde{v}(t) = u(T - t), \ t \in [T - L - \tau_0 - \tau_1 - 3, T],$$

$$\tilde{v}(t + T - L - \tau_0 - \tau_1 - 4) = v(T - L - \tau_0 - \tau_1 - 4)$$

$$+ t[u(L + \tau_0 + \tau_1 + 3) - v(T - L - \tau_0 - \tau_1 - 4)], \ t \in (0, 1). \tag{8.133}$$

Clearly, $\tilde{v} : [0, T] \to R^n$ is an a. c. function. In view of (8.127) and (8.133),

$$|\tilde{v}(T - L - \tau_0 - \tau_1 - 3) - \bar{x}| = |u(L + \tau_0 + \tau_1 + 3) - \bar{x}| \leq \delta_2. \tag{8.134}$$

It follows from (8.127), (8.132), and (8.133) that for all $t \in (T - L - \tau_0 - \tau_1 - 4, T - L - \tau_0 - \tau_1 - 3)$,

$$|\tilde{v}'(t)| \leq |u(L + \tau_0 + \tau_1 + 3) - v(T - L - \tau_0 - \tau_1 - 4)| \leq 2\delta_2,$$

$$|\tilde{v}(t) - \bar{x}| \leq 3\delta_2. \tag{8.135}$$

By (8.123) and (8.135), for all $t \in (T - L - \tau_0 - \tau_1 - 4, T - L - \tau_0 - \tau_1 - 3)$,

$$|f(\tilde{v}(t), \tilde{v}'(t)) - f(\bar{x}, 0)| \leq \delta_1/8. \tag{8.136}$$

It follows from (8.130) and (8.133) that

$$\delta \geq I^f(0, T, v) - I^f(0, T, \tilde{v})$$

$$= I^f(T - L - \tau_0 - \tau_1 - 4, T, v) - I^f(T - L - \tau_0 - \tau_1 - 4, T, \tilde{v}).$$

Together with (8.136) this implies that

$$I^f(T - L - \tau_0 - \tau_1 - 4, T, v)$$
$$\leq \delta + I^f(T - L - \tau_0 - \tau_1 - 4, T - L - \tau_0 - \tau_1 - 3, \tilde{v})$$
$$+ I^f(T - L - \tau_0 - \tau_1 - 3, T, \tilde{v})$$
$$\leq \delta + \delta_1/8 + f(\bar{x}, 0) + I^{\bar{f}}(0, L + \tau_0 + \tau_1 + 3, u). \tag{8.137}$$

By (8.132) and property (Piii),

$$I^f(T - L - \tau_0 - \tau_1 - 4, T - L - \tau_0 - \tau_1 - 3, v) \geq \bar{f}(\bar{x}, 0) - \delta_1/8.$$

Together with (8.137) this implies that

$$I^f(T - L - \tau_0 - \tau_1 - 3, T, v) \leq \delta + \delta_1/4 + I^{\bar{f}}(0, L + \tau_0 + \tau_1 + 3, u). \tag{8.138}$$

Set

$$y(t) = v(T - t), \ t \in [0, L + \tau_0 + \tau_1 + 3]. \tag{8.139}$$

It follows from (8.23), (8.138), and (8.139) that

$$I^{\bar{f}}(0, L + \tau_0 + \tau_1 + 3, y)$$
$$= I^f(T - L - \tau_0 - \tau_1 - 3, T, v)$$
$$\leq \delta + \delta_1/4 + I^{\bar{f}}(0, L + \tau_0 + \tau_1 + 3, u). \tag{8.140}$$

By (8.120), (8.128)–(8.130), (8.132), (8.140), Propositions 8.11 and 8.12, and $(\bar{f})$-overtaking optimality of u,

$$\pi^{\bar{f}}(y(0)) - \inf(\pi^{\bar{f}}) + I^{\bar{f}}(0, \tau_0, y) - \tau_0 f(\bar{x}, 0) - \pi^{\bar{f}}(y(0)) + \pi^{\bar{f}}(y(\tau_0))$$
$$\leq \pi^{\bar{f}}(y(0)) - \pi^{\bar{f}}(u(0)) + I^{\bar{f}}(0, L + \tau_0 + \tau_1 + 3, y)$$
$$- (L + \tau_0 + \tau_1 + 3) f(\bar{x}, 0) - \pi^{\bar{f}}(y(0)) + \pi^{\bar{f}}(y(L + \tau_0 + \tau_1 + 3))$$
$$\leq \pi^{\bar{f}}(y(0)) - \pi^{\bar{f}}(u(0)) + 3\delta_1/8 + I^{\bar{f}}(0, L + \tau_0 + \tau_1 + 3, u)$$
$$- (L + \tau_0 + \tau_1 + 3) f(\bar{x}, 0) - \pi^{\bar{f}}(y(0)) + \pi^{\bar{f}}(y(L + \tau_0 + \tau_1 + 3))$$
$$= \pi^{\bar{f}}(y(0)) - \pi^{\bar{f}}(u(0)) + \pi^{\bar{f}}(u(0)) - \pi^{\bar{f}}(u(L + \tau_0 + \tau_1 + 3)) + 3\delta_1/8$$
$$- \pi^{\bar{f}}(y(0)) + \pi^{\bar{f}}(y(L + \tau_0 + \tau_1 + 3))$$
$$= \pi^{\bar{f}}(y(L + \tau_0 + \tau_1 + 3)) - \pi^{\bar{f}}(u(L + \tau_0 + \tau_1 + 3)) + 3\delta_1/8. \tag{8.141}$$

By (8.132) and (8.139), $|y(L + \tau_0 + \tau_1 + 3) - \bar{x}| \leq \delta_2$. Together with (8.134) and the choice of δ_2 (see (8.122)) this implies that

$$|\pi^{\bar{f}}(y(L + \tau_0 + \tau_1 + 3))|, \; |\pi^{\bar{f}}(u(L + \tau_0 + \tau_1 + 3))| \leq \delta_1/8.$$

By the inequalities above and (8.141),

$$\pi^{\bar{f}}(y(0)) - \inf(\pi^{\bar{f}}) + I^{\bar{f}}(0, \tau_0, y) - \tau_0 f(\bar{x}, 0) - \pi^{\bar{f}}(y(0)) + \pi^{\bar{f}}(y(\tau_0)) \leq \delta_1.$$

Combined with Proposition 8.11, (8.130), (8.132), and (8.139) the inequality above implies that

$$\pi^{\bar{f}}(y(0)) - \inf(\pi^{\bar{f}}) \leq \delta_1,$$
$$I^{\bar{f}}(0, \tau_0, y) - \tau_0 f(\bar{x}, 0) - \pi^{\bar{f}}(y(0)) + \pi^{\bar{f}}(y(\tau_0)) \leq \delta_1.$$

By the inequalities above and property (Pii) there exists an $(\bar{f})$-overtaking optimal function $w : [0, \infty) \to R^n$ such that $w(0) \in \mathcal{D}(\bar{f})$ and for all $t \in [0, \tau_0]$,

$$\epsilon \geq |y(t) - w(t)| = |v(T - t) - w(t)|.$$

Theorem 8.25 is proved. $\qquad\square$

8.9 Proof of Theorem 8.26

Theorems 8.10 and 8.25 and Lemma 8.32 imply the following result.

Theorem 8.35. *Let $\tau_0 > 0$ and $\epsilon > 0$. Then there exist $\delta > 0$ and $T_0 > \tau_0$ such that for each $T \geq T_0$ and each a. c. function $v : [0, T] \to R^n$ which satisfies*

$$I^f(0, T, v) \leq \sigma(f, T) + \delta$$

there exists an $(\bar{f})$-overtaking optimal function $v^* : [0, \infty) \to R^n$ such that $v^*(0) \in \mathcal{D}(\bar{f})$ and that for all $t \in [0, \tau_0]$,

$$|v(T - t) - v^*(t)| \leq \epsilon.$$

Theorem 8.26 follows from Theorem 8.35 applied to the functions f and $\bar{f}$.

8.10   Structure of Solutions of Problem (P_1)

We use the notation, definitions, and assumptions introduced in Sects. 8.1 and 8.2. The following theorem describes the structure of solutions of problems (P_1) in the regions closed to the endpoints of their domains.

Theorem 8.36. *Let $\tau_0 > 0$, $\epsilon > 0$, $x \in \cup\{\bar{Y}_{L,f} : L > 0\}$, and $y \in \cup\{Y_{L,f} : L > 0\}$. Then there exist $\delta > 0$ and $T_0 > \tau_0$ such that for each $T \geq T_0$ and each a. c. function $v : [0, T] \to R^n$ which satisfies*

$$v(0) = x, \ v(T) = y, \ I^f(0, T, v) \leq \sigma(f, T, x, y) + \delta$$

there exist an (f)-overtaking optimal function $\xi : [0, \infty) \to R^n$ and an $(\bar{f})$-overtaking optimal function $\eta : [0, \infty) \to R^n$ such that $\xi(0) = x$, $\eta(0) = y$ and

$$|\xi(t) - v(t)| \leq \epsilon, \ |v(T - t) - \eta(t)| \leq \epsilon$$

for all $t \in [0, \tau_0]$.

Theorem 8.36 follows from the next result applied to the functions f and $\bar{f}$.

Theorem 8.37. *Let $\tau_0 > 0$, $\epsilon > 0$, $L_0 > 0$, and $x \in \bar{Y}_{L_0,f}$. Then there exist $\delta > 0$ and $T_0 > \tau_0$ such that for each $T \geq T_0$ and each a. c. function $v : [0, T] \to R^n$ which satisfies*

$$v(0) = x, \ v(T) \in Y_{L_0,f}, \ I^f(0, T, v) \leq \sigma(f, T, x, v(T)) + \delta$$

there exists an (f)-overtaking optimal function $\xi : [0, \infty) \to R^n$ such that $\xi(0) = x$ and $|\xi(t) - v(t)| \leq \epsilon$ for all $t \in [0, \tau_0]$.

Theorem 8.37 is proved in Sect. 8.11. Theorems 8.2 and 8.37 and Proposition 8.5 imply the following result.

Theorem 8.38. *Let $\tau_0 > 0$, $\epsilon > 0$ and $x \in \cup\{\bar{Y}_{L,f} : L > 0\}$. Then there exist $\delta > 0$ and $T_0 > \tau_0$ such that for each $T \geq T_0$ and each a. c. function $v : [0, T] \to R^n$ which satisfies*

$$v(0) = x, \ I^f(0, T, v) \leq \sigma(f, T, x) + \delta$$

there exists an (f)-overtaking optimal function $\xi : [0, \infty) \to R^n$ such that $\xi(0) = x$ and $|\xi(t) - v(t)| \leq \epsilon$ for all $t \in [0, \tau_0]$.

8.11   Proof of Theorem 8.37

Assume that Theorem 8.37 does not hold. Then for each natural number k there exist

$$T_k \geq \tau_0 + k + 2L_0 \tag{8.142}$$

and an a. c. function $v_k : [0, T_k] \to R^n$ such that

$$v_k(0) = x, \ v_k(T_k) \in Y_{L_0,f}, \tag{8.143}$$

$$I^f(0, T_k, v_k) \leq \sigma(f, T_k, x, v_k(T_k)) + 1/k \tag{8.144}$$

and that for each a. c. function

$$\xi \in \Lambda(f, x) \tag{8.145}$$

we have

$$\sup\{|\xi(t) - v_k(t)| : \ t \in [0, \tau_0]\} > \epsilon. \tag{8.146}$$

By (8.142)–(8.144), Proposition 8.5, Theorem 8.3, and the inclusion

$$x \in \bar{Y}_{L_0,f} \tag{8.147}$$

the following property holds:

(Pv)   for each $\epsilon > 0$ there exist $L(\epsilon) > 0$ and a natural number $k(\epsilon)$ such that for each integer $k \geq k(\epsilon)$ and each $T_k > 2L(\epsilon)$,

$$|v_k(t) - \bar{x}| \leq \epsilon, \ t \in [L(\epsilon), T_k - L(\epsilon)].$$

Relations (8.142), (8.143), and (8.147) imply that for each integer $k \geq 1$,

$$\sigma(f, T_k, x, v_k(T_k)) \leq T_k f(\bar{x}, 0) + 2L_0 + 2L_0|f(\bar{x}, 0)|. \tag{8.148}$$

By (8.143), (8.144), (8.148), and Lemma 8.29, there exists $M_0 > 0$ such that for all integers $k \geq 1$,

$$|v_k(t)| \leq M_0 \text{ for all } t \in [0, T_k]. \tag{8.149}$$

(A2) implies that there exists $c_0 > 0$ such that for each $T > 0$ and each $z \in R^n$ satisfying $|z| \leq M_0$,

$$\sigma(f, T, z) \geq Tf(\bar{x}, 0) - c_0. \tag{8.150}$$

Let $p \geq 1$ be an integer. In view of (8.142), (8.144), (8.148)–(8.150), for each integer $k \geq p$,

$$I^f(0, p, v_k) = I^f(0, T_k, v_k) - I^f(p, T_k, v_k)$$
$$\leq T_k f(\bar{x}, 0) + 2L_0 + 2L_0|f(\bar{x}, 0)| + 1 - (T_k - p)f(\bar{x}, 0) + c_0$$
$$\leq pf(\bar{x}, 0) + 2L_0(1 + |f(\bar{x}, 0)|) + c_0 + 1. \tag{8.151}$$

By (8.143), (8.151), and Proposition 8.30, extracting subsequences and re-indexing if necessary, we may assume without loss of generality that there exists an a. c. function $v : [0, \infty) \to R^n$ such that for any integer $p \geq 1$,

$$v_k(t) \to v(t) \text{ as } k \to \infty \text{ uniformly on } [0, p], \tag{8.152}$$

$$I^f(0, p, v) \leq \liminf_{k \to \infty} I^f(0, p, v_k). \tag{8.153}$$

In view of (8.151) and (8.153), for any integer $p \geq 1$,

$$I^f(0, p, v) \leq pf(\bar{x}, 0) + 2L_0(1 + |f(\bar{x}, 0)|) + c_0 + 1. \tag{8.154}$$

We show that for each $T > 0$,

$$I^f(0, T, v) = \sigma(f, T, v(0), v(T)).$$

Assume the contrary. Then there exist $S > 0$ and $\Delta > 0$ such that

$$I^f(0, S, v) > \sigma(f, S, v(0), v(S)) + 7\Delta.$$

Clearly, there exists an a. c. function $\tilde{v} : [0, S] \to R^n$ such that

$$\tilde{v}(0) = x, \ \tilde{v}(S) = v(S), \ I^f(0, S, \tilde{v}) < I^f(0, S, v) - 6\Delta. \tag{8.155}$$

By (A1) and Lemma 8.31, there exists $\delta \in (0, \bar{r}/4)$ such that for each $(x, y) \in R^n \times R^n$ satisfying $|x - \bar{x}| \leq 2\delta, |y| \leq 2\delta$,

$$|f(x, y) - f(\bar{x}, 0)| \leq \Delta/2 \tag{8.156}$$

and that for each a.c. function $v : [0, 1] \to R^n$ satisfying $|v(0) - \bar{x}|, |v(1) - \bar{x}| \leq 2\delta$,

$$I^f(0, 1, v) \geq f(\bar{x}, 0) - \Delta/2. \tag{8.157}$$

By (Pv) and (8.142), there exist a number $L_1 > 0$ and an integer $k_1 > 2L_1$ such that for each integer $k \geq k_1$,

$$|v_k(t) - \bar{x}| \leq \delta, \ t \in [L_1, T_k - L_1]. \tag{8.158}$$

It follows from (8.142), (8.152), and (8.158) that

$$|v(t) - \bar{x}| \leq \delta \text{ for all } t \geq L_1. \tag{8.159}$$

Choose a natural number

$$k_2 > k_1 + S + 2L_1 + 9. \tag{8.160}$$

In view of (8.153),

$$I^f(0, k_1 + S + L_1 + 8, v) \leq \liminf_{k \to \infty} I^f(0, k_1 + S + L_1 + 8, v_k). \tag{8.161}$$

For each integer $k \geq k_2$ define

$$\hat{v}_k(t) = \tilde{v}(t), \ t \in [0, S], \ \hat{v}_k(t) = v(t), \ t \in (S, k_1 + S + L_1 + 8],$$

$$\hat{v}_k(k_1 + S + L_1 + 8 + t) = v(k_1 + S + L_1 + 8)$$
$$+ t[v_k(k_1 + S + L_1 + 9)$$
$$- v(k_1 + S + L_1 + 8)], \ t \in (0, 1]. \tag{8.162}$$

In view of (8.155) and (8.162), for each integer $k \geq k_2$, $\hat{v}_k : [0, k_1+S+L_1+9] \to R^n$ is an a. c. function which satisfies

$$\hat{v}_k(0) = x, \ \hat{v}_k(k_1 + S + L_1 + 9) = v_k(k_1 + S + L_1 + 9). \tag{8.163}$$

By (8.143), (8.144), and (8.163), for each integer $k \geq k_2$,

$$I^f(0, k_1 + S + L_1 + 9, v_k) \leq I^f(0, k_1 + S + L_1 + 9, \hat{v}_k) + 1/k. \tag{8.164}$$

In view of (8.162), there exists an integer $k \geq k_2$ such that

$$I^f(0, k_1 + S + L_1 + 8, v) \leq I^f(0, k_1 + S + L_1 + 8, v_k) + \Delta, \tag{8.165}$$

$$k^{-1}(k_1 + S + L_1 + 10) < \Delta. \tag{8.166}$$

It follows from (8.142) and (8.160) that

$$T_k - L_1 \geq k + 2L_0 - L_1 > k_1 + S + 2L_0 + L_1 + 9. \tag{8.167}$$

Relations (8.158), (8.159), (8.162), and (8.167) imply that for all $t \in (k_1+S+L_1+8, k_1+S+L_1+9)$,

$$|\hat{v}'_k(t)| \leq |v_k(k_1+S+L_1+9)-\bar{x}| + |\bar{x}-v(k_1+S+L_1+8)| \leq 2\delta,$$

$$|\hat{v}_k(t)-\bar{x}| \leq |v_k(k_1+S+L_1+9)-\bar{x}| + |\bar{x}-v(k_1+S+L_1+8)| \leq 2\delta. \qquad (8.168)$$

In view of (8.156) and (8.168),

$$|I^f(k_1+S+L_1+8, k_1+S+L_1+9, \hat{v}_k)-f(\bar{x},0)| \leq \Delta/2. \qquad (8.169)$$

By the choice of δ (see (8.157)), (8.158), and (8.167),

$$I^f(k_1+S+L_1+8, k_1+S+L_1+9, v_k) \geq f(\bar{x},0)-\Delta/2. \qquad (8.170)$$

By (8.155), (8.162), (8.164), (8.165), (8.169), and (8.170),

$$-1/k \leq I^f(0, k_1+S+L_1+9, \hat{v}_k) - I^f(0, k_1+S+L_1+9, v_k)$$
$$= I^f(0, k_1+S+L_1+8, \hat{v}_k) + I^f(k_1+S+L_1+8, k_1+S+L_1+9, \hat{v}_k)$$
$$\quad - I^f(0, k_1+S+L_1+8, v_k) - I^f(k_1+S+L_1+8, k_1+S+L_1+9, v_k)$$
$$\leq I^f(0, S, \tilde{v}) + I^f(S, k_1+S+L_1+8, v)$$
$$\quad - I^f(0, k_1+S+L_1+8, v_k) + f(\bar{x},0) + \Delta/2 - f(\bar{x},0) + \Delta/2$$
$$\leq I^f(0, S, v) - 6\Delta + I^f(S, k_1+S+L_1+8, v)$$
$$\quad - I^f(0, k_1+S+L_1+8, v_k) + \Delta \leq -4\Delta,$$
$$\Delta < 1/k.$$

This contradicts (8.166). The contradiction we have reached proves that

$$I^f(0, T, v) = \sigma(f, T, v(0), v(T))$$

for all $T > 0$. Together with (8.154) and Theorem 8.7 this implies that v is an (f)-overtaking optimal function satisfying $v(0) = x$. By (8.152) for all sufficiently large natural numbers k,

$$|v_k(t) - v(t)| \leq \epsilon/2, \ t \in [0, \tau_0].$$

This contradicts (8.146). The contradiction we have reached proves Theorem 8.37.

$\square$

Chapter 9
Dynamic Games with Extended-Valued Integrands

In this chapter we study a class of dynamic continuous-time two-player zero-sum unconstrained games with extended-valued integrands. We do not assume convexity-concavity assumptions and establish the existence and the turnpike property of approximate solutions.

9.1 Preliminaries and Main Results

In this chapter we continue to use the notation and definitions introduced in Chap. 8. We denote by $\text{mes}(E)$ the Lebesgue measure of a Lebesgue measurable set $E \subset R^1$, denote by $|\cdot|$ the Euclidean norm of the space R^q and by $\langle \cdot, \cdot \rangle$ the inner product of R^q where q is a natural number.

For each function $f : X \to \bar{R} := R^1 \cup \{-\infty, \infty\}$, where the set X is nonempty put

$$\text{dom}(f) = \{x \in X : -\infty < f(x) < \infty\}.$$

We suppose that $a > 0$, $\psi : [0, \infty) \to [0, \infty)$ is an increasing function such that

$$\lim_{t \to \infty} \psi(t) = \infty,$$

n, m be natural numbers and that $f : R^n \times R^n \times R^m \times R^m \to \bar{R}$ be a Borel measurable function.

We suppose that for all real numbers λ,

$$\lambda + \infty = \infty, \ \lambda - \infty = -\infty, \ -\infty < \lambda < \infty.$$

© Springer International Publishing Switzerland 2015
A.J. Zaslavski, *Turnpike Theory of Continuous-Time Linear Optimal Control Problems*, Springer Optimization and Its Applications 104,
DOI 10.1007/978-3-319-19141-6_9

For each $z \in R^1$ set

$$z^+ = \max\{z, 0\}, \ z^- = \max\{-z, 0\}.$$

Given $z_1, z_2 \in R^n$, $\xi_1, \xi_2 \in R^m$, and a positive number T we consider a continuous-time two-player zero-sum game over the interval $[0, T]$ denoted by $\Gamma(z_1, z_2, \xi_1, \xi_2, T)$. For this game the set of strategies for the first player is the set of all a. c. functions $x : [0, T] \to R^n$ satisfying $x(0) = z_1$ and $x(T) = z_2$, the set of strategies for the second player is the set of all a. c. functions $y : [0, T] \to R^m$ satisfying $y(0) = \xi_1$ and $y(T) = \xi_2$, and the objective function for the first player associated with the strategies $x : [0, T] \to R^n$, $y : [0, T] \to R^m$ is given by $\int_0^T f(x(t), x'(t), y(t), y'(t))dt$ if this integrand is well defined in the sense which is explained below.

Let

$$x_f \in R^n, \ y_f \in R^m.$$

Set

$$f^{(1)}(x, y) = f(x, y, y_f, 0), \ (x, y) \in R^n \times R^n,$$

$$f^{(2)}(x, y) = -f(x_f, 0, x, y), \ (x, y) \in R^m \times R^m.$$

We assume that

$$f^{(1)}(x, y) > -\infty \text{ for all } (x, y) \in R^n \times R^n,$$

$$f^{(2)}(x, y) > -\infty \text{ for all } (x, y) \in R^m \times R^m,$$

the functions $f^{(1)} : R^n \times R^n \to R^1 \cup \{\infty\}, f^{(2)} : R^m \times R^m \to R^1 \cup \{\infty\}$ are lower semicontinuous functions, the sets

$$\mathrm{dom}(f^{(1)}) = \{(x, y) \in R^n \times R^n : f(x, y, y_f, 0) < \infty\},$$

$$\mathrm{dom}(f^{(2)}) = \{(x, y) \in R^m \times R^m : f(x_f, 0, x, y) > -\infty\}$$

are nonempty, convex, and closed and

$$f(x, y, y_f, 0) = f^{(1)}(x, y) \geq \max\{\psi(|x|), \psi(|y|)|y|\} - a$$

$$\text{for each } x, y \in R^n, \tag{9.1}$$

$$f(x_f, 0, x, y) = -f^{(2)}(x, y) \leq -\max\{\psi(|x|), \psi(|y|)|y|\} + a$$

$$\text{for each } x, y \in R^m. \tag{9.2}$$

For each $T^1 \in R^1$, each $T_2 > T_1$, and each pair of functions $v : [T_1, T_2] \to R^n$, $u : [T_1, T_2] \to R^m$ set

$$I^{f^{(1)}}(T_1, T_2, v) = \int_{T_1}^{T_2} f^{(1)}(v(t), v'(t))dt,$$

$$I^{f^{(2)}}(T_1, T_2, v) = -\int_{T_1}^{T_2} f^{(2)}(u(t), u'(t))dt.$$

We suppose that

$$\text{dom}(f) = \text{dom}(f^{(1)}) \times \text{dom}(f^{(2)}), \tag{9.3}$$

the function f is bounded on all bounded subsets of $\text{dom}(f)$,

$$f(x_f, 0, y_f, 0) \leq f(x, 0, y_f, 0) \text{ for each } x \in R^n, \tag{9.4}$$

$$f(x_f, 0, y_f, 0) \geq f(x_f, 0, y, 0) \text{ for each } y \in R^m \tag{9.5}$$

and that the following assumptions hold:

(C1) $(x_f, 0)$ is an interior point of $\text{dom}(f^{(1)})$ and $f^{(1)}$ is continuous at the point $(x_f, 0)$; $(y_f, 0)$ is an interior point of $\text{dom}(f^{(2)})$ and $f^{(2)}$ is continuous at the point $(y_f, 0)$;

(C2) for each $M > 0$ there is a number $c_M > 0$ such that $f(x, u, y, v) \geq -c_M$ for each $(x, u) \in R^n \times R^n$ and each $(y, v) \in \text{dom}(f^{(2)})$ satisfying $|y|, |v| \leq M$, and $f(x, u, y, v) \leq c_M$ for each $(y, v) \in R^m \times R^m$ and each $(x, u) \in \text{dom}(f^{(1)})$ satisfying $|x|, |u| \leq M$;

(C3) for each $M > 0$ there exists $c_M > 0$ such that for each $T > 0$ and each a.c. function $x : [0, T] \to R^n$ satisfying $|x(0)| \leq M$,

$$\int_0^T f(x(t), x'(t), y_f, 0)dt \geq Tf(x_f, 0, y_f, 0) - c_M$$

and for each $T > 0$ and each a.c. function $y : [0, T] \to R^m$ satisfying $|y(0)| \leq M$,

$$\int_0^T f(x_f, 0, y(t), y'(t))dt \leq Tf(x_f, 0, y_f, 0) + c_M;$$

(C4) for each $x \in R^n$ the function $f(x, \cdot, y_f, 0) : R^n \to R^1 \cup \{\infty\}$ is convex and for each $y \in R^m$ the function $f(x_f, 0, y, \cdot) : R^m \to R^1 \cup \{-\infty\}$ is concave.

By (9.4) and (9.5), the pair (x_f, y_f) is a saddle point for the function $\bar{f}(x, y) := f(x, 0, y, 0)$, $x \in R^n$, $y \in R^m$.

Assumption (C3) imply that for each a. c. function $v : [0, \infty) \to R^n$ the function

$$T \to \int_0^T f(v(t), v'(t), y_f, 0)dt - Tf(x_f, 0, y_f, 0), \quad T \in (0, \infty)$$

is bounded from below and that for each a. c. function $u : [0, \infty) \to R^m$ the function

$$T \to \int_0^T f(x_f, 0, u(t), u'(t))dt - Tf(x_f, 0, y_f, 0), \quad T \in (0, \infty)$$

is bounded from above.

For all $x_1, x_2 \in R^n$, $y_1, y_2 \in R^m$ define

$$f^+(x_1, x_2, y_1, y_2) = \max\{f(x_1, x_2, y_1, y_2), 0\},$$
$$f^-(x_1, x_2, y_1, y_2) = \max\{-f(x_1, x_2, y_1, y_2), 0\}.$$

Let $-\infty < T_1 < T_2 < \infty$, $x : [T_1, T_2] \to R^n$, $y : [T_1, T_2] \to R^m$ be a. c. functions. The pair (x, y) is called admissible if at least one of the integrals

$$\int_{T_1}^{T_2} f^+(x(t), x'(t), y(t), y'(t))dt, \quad \int_{T_1}^{T_2} f^-(x(t), x'(t), y(t), y'(t))dt$$

is finite. If (x, y) is admissible, then we set

$$I^f(T_1, T_2, x, y) = \int_{T_1}^{T_2} f(x(t), x'(t), y(t), y'(t))dt$$

$$= \int_{T_1}^{T_2} f^+(x(t), x'(t), y(t), y'(t))dt$$

$$- \int_{T_1}^{T_2} f^-(x(t), x'(t), y(t), y'(t))dt.$$

It should be mentioned that analogs of assumption (C3) are used in the infinite horizon optimal control and they are usually posed, when one obtains a turnpike result where the turnpike is a singleton. See, for example, [53].

Let us now define approximate solutions (saddle points) of games

$$\Gamma(z_1, z_2, \xi_1, \xi_2, T)$$

with $z_1, z_2 \in R^n$, $\xi_1, \xi_2 \in R^m$ and a positive constant T.

Let $M \geq 0$, $-\infty < T_1 < T_2 < \infty$, $x : [T_1, T_2] \to R^n$, $y : [T_1, T_2] \to R^m$ be a.c. functions such that the pair (x, y) is admissible. The pair (x, y) is called (M)-good if the integral $\int_{T_1}^{T_2} f(x(t), x'(t), y(t), y'(t))dt$ is finite and the following properties hold:

for each a. c. function $z : [T_1, T_2] \to R^n$ such that the pair (z, y) is admissible and $z(T_i) = x(T_i), i = 1, 2,$

$$\int_{T_1}^{T_2} f(z(t), z'(t), y(t), y'(t))dt \geq \int_{T_1}^{T_2} f(x(t), x'(t), y(t), y'(t))dt - M;$$

for each a. c. function $\xi : [T_1, T_2] \to R^m$ such that the pair (x, ξ) is admissible and $\xi(T_i) = y(T_i), i = 1, 2,$

$$\int_{T_1}^{T_2} f(x(t), x'(t), \xi(t), \xi'(t))dt \leq \int_{T_1}^{T_2} f(x(t), x'(t), y(t), y'(t))dt + M.$$

If (x, y) is (0)-good, then (x, y) is called a saddle point of the game $\Gamma(x(T_1), x(T_2), y(T_1), y(T_2), T_2 - T_1)$.

Let $L > 0$. Denote by $X_{L,1}$ the set of all points $x \in R^n$ for which there exist a number $T \in (0, L]$ and an a. c. function $v : [0, T] \to R^n$ such that

$$v(0) = x, \ v(T) = x_f,$$

$$|v(t)|, |v'(t)| \leq L, \ t \in [0, L]$$

and $I^{f^{(1)}}(0, T, v) < \infty$ and by $X_{L,2}$ the set of all points $y \in R^m$ for which there exist a number $T \in (0, L]$ and an a. c. function $v : [0, T] \to R^m$ such that

$$v(0) = y, \ v(T) = y_f,$$

$$|v(t)|, |v'(t)| \leq L, \ t \in [0, L]$$

and $I^{f^{(2)}}(0, T, v) < \infty.$

Denote by $\bar{X}_{L,1}$ the set of all points $x \in R^n$ for which there exist a number $T \in (0, L]$ and an a. c. function $v : [0, T] \to R^n$ such that

$$v(0) = x_f, \ v(T) = x,$$

$$|v(t)|, |v'(t)| \leq L, \ t \in [0, L]$$

and $I^{f^{(1)}}(0, T, v) < \infty$ and by $\bar{X}_{L,2}$ the set of all points $y \in R^m$ for which there exist a number $T \in (0, L]$ and an a. c. function $v : [0, T] \to R^m$ such that

$$v(0) = y_f, \ v(T) = y,$$

$$|v(t)|, |v'(t)| \leq L, \ t \in [0, L]$$

and $I^{f^{(2)}}(0, T, v) < \infty.$

Note that the existence of a saddle point of the game $\Gamma(z_1, z_2, \xi_1, \xi_2, T)$ with $z_1, z_2 \in X$, $\xi_1, \xi_2 \in Y$ and $T > 0$ is not guaranteed. Nevertheless, the next result is proved in Sect. 9.3.

Theorem 9.1. *Let $M > 0$. Then there exists $M_* > 0$ such that for each $T_1 \in R^1$, each $T_2 > T_1 + 2M$, each $z_1 \in X_{M,1}$, each $z_2 \in \bar{X}_{M,1}$, each $\xi_1 \in X_{M,2}$ and each $\xi_2 \in \bar{X}_{M,2}$ there exists an (M_*)-good pair of a.c. functions $x : [T_1, T_2] \to R^n$, $y : [T_1, T_2] \to R^m$ such that*

$$x(T_i) = z_i, \; y(T_i) = \xi_i, \; i = 1, 2,$$
$$x(t) = x_f, \; y(t) = y_f, \; t \in [T_1 + M, T_2 - M],$$
$$|x(t)|, |x'(t)|, |y(t)|, |y'(t)| \le M, \; t \in [T_1, T_1 + M] \cup [T_2 - M, T_2]$$

and

$$I^{f^{(1)}}(T_1, T_1 + M, x), \; I^{f^{(1)}}(T_2 - M, T_2, x),$$
$$I^{f^{(2)}}(T_1, T_1 + M, y), \; I^{f^{(2)}}(T_2 - M, T_2, y) < \infty.$$

It should be mentioned that in Theorem 9.1 the constant M_* does not depend on the length of the interval $T_2 - T_1$.

In this chapter we establish a turnpike property of good pairs of a. c. functions which means that they spend most of the time in a small neighborhood of the pair (x^*, y^*). It is known in the optimal control theory that turnpike properties of approximately optimal solutions are deduced from an asymptotic turnpike property of solutions of corresponding infinite horizon optimal control problems [53].

We say that f possesses the asymptotic turnpike property (or briefly ATP), if the following properties hold:
for each a. c. function $x : [0, \infty) \to R^n$ such that

$$\sup \left\{ \int_0^T f(x(t), x'(t), y_f, 0)dt - Tf(x_f, 0, y_f, 0) : \; T \in (0, \infty) \right\} < \infty$$

we have $\lim_{t \to \infty} |x(t) - x_f| = 0$;
for each a. c. function $y : [0, \infty) \to R^m$ such that

$$\inf \left\{ \int_0^T f(x_f, 0, y(t), y'(t))dt - Tf(x_f, 0, y_f, 0) : \; T \in (0, \infty) \right\} > -\infty$$

we have $\lim_{t \to \infty} |y(t) - y_f| = 0$.
The following theorem is our main result of this chapter.

Theorem 9.2. *Let f possess (ATP) and $M, M_1, \epsilon > 0$. Then there exist $l > 0$ and an integer $Q \ge 1$ such that for each $T > Ql$ and each (M_1)-good admissible pair of a.c. functions $x : [0, T] \to R^n$, $y : [0, T] \to R^m$ such that*

$$x(0) \in X_{M,1}, \; x(T) \in \bar{X}_{M,1}, \; y(0) \in X_{M,2}, \; y(T) \in \bar{X}_{M,2}$$

there exist a natural number $q \le Q$ and a sequence of closed intervals $[a_i, b_i] \subset$ $[0, T]$, $i = 1, \ldots, q$ such that

$$0 \le b_i - a_i \le l, \ i = 1, \ldots, q,$$
$$|x(t) - x_f| \le \epsilon, \ |y(t) - y_f| \le \epsilon$$

for all $t \in [0, T] \setminus \cup_{i=1}^{q} [a_i, b_i]$.

Chapter 9 is organized as follows. Section 9.2 contains an auxiliary result. Theorem 9.1 is proved in Sect. 9.3. Auxiliary results for Theorem 9.2 are given in Sect. 9.4. Theorem 9.2 is proved in Sect. 9.5.

9.2 An Auxiliary Result

Lemma 9.3. *Let $\{g, s, z_g\}$ be either $\{f^{(1)}, n, x_f\}$ or $\{f^{(2)}, m, y_f\}$. Then there exists $c_* > 0$ such that for each $T > 0$ and each a.c. function $x : [0, T] \to R^s$,*

$$\int_0^T g(x(t), x'(t)) dt \ge Tg(z_g, 0) - c_*. \tag{9.6}$$

Proof. Clearly, there is $M > 0$ such that

$$\psi(M) > a + 4 + |f(x_f, 0, y_f, 0)|. \tag{9.7}$$

By (C3), there exists $c_* > 0$ such that for each $T > 0$ and each a.c. function $x : [0, T] \to R^s$ satisfying $|x(0)| \le M$,

$$\int_0^T g(x(t), x'(t)) dt \ge Tg(z_g, 0) - c_*. \tag{9.8}$$

Let $T > 0$ and $x : [0, T] \to R^s$ be an a. c. function. We show that (9.6) holds. If $|x(t)| > M$ for all $t \in [0, T]$, then relations (9.1), (9.2), and (9.7) imply (9.6).

Assume that

$$\{t \in [0, T] : |x(t)| \le M\} \ne \emptyset.$$

Set

$$\tau_0 = \min\{t \in [0, T] : |x(t)| \le M\}.$$

It is clear that

$$|x(\tau_0)| \leq M, \tag{9.9}$$

$$|x(t)| > M \text{ for all } t \text{ satisfying } 0 \leq t < \tau_0. \tag{9.10}$$

By (9.1), (9.2), (9.7)–(9.10), and the choice of c_*,

$$\int_0^T g(x(t), x'(t))dt - Tg(z_g, 0)$$

$$\geq (\psi(M) - a)\tau_0 + \int_{\tau_0}^T g(x(t), x'(t))dt - Tg(z_g, 0)$$

$$\geq \psi(M) - a - |f(x_f, 0, y_f, 0)|\tau_0 - c_* \geq -c_*.$$

Lemma 9.3 is proved.

9.3 Proof of Theorem 9.1

We may assume that

$$M > 4 + |x_f| + |y_f|.$$

Since the function f is bounded on bounded subsets of $\mathrm{dom}(f^{(1)}) \times \mathrm{dom}(f^{(2)})$ there is a constant $M_0 > 0$ such that

$$|f(z_1, z_2, \xi_1, \xi_2)| \leq M_0 \text{ for all } (z_1, z_2) \in \mathrm{dom}(f^{(1)})$$

and all $(\xi_1, \xi_2) \in \mathrm{dom}(f^{(2)})$ satisfying $|z_i|, |\xi_i| \leq M, \ i = 1, 2.$ \qquad (9.11)

By Lemma 9.3, there exists $c_* > 0$ such that:
for each $T > 0$ and each a.c. function $v : [0, T] \to R^n$,

$$\int_0^T f(x(t), x'(t), y_f, 0)dt \geq Tf(x_f, 0, y_f, 0) - c_*; \tag{9.12}$$

for each $T > 0$ and each a.c. function $u : [0, T] \to R^m$,

$$\int_0^T f(x_f, 0, u(t), u'(t))dt \leq Tf(x_f, 0, y_f, 0) + c_*. \tag{9.13}$$

By (C2), there is a number $M_1 > 0$ such that

$$f(x, u, y, v) \geq -M_1 \text{ for each } (x, u) \in R^n \times R^n$$

$$\text{and each } (y, v) \in \text{dom}(f^{(2)}) \text{ satisfying } |y|, |v| \leq M; \qquad (9.14)$$

$$f(x, u, y, v) \leq M_1 \text{ for each } (y, v) \in R^m \times R^m$$

$$\text{and each } (x, u) \in \text{dom}(f^{(1)}) \text{ satisfying } |x|, |u| \leq M. \qquad (9.15)$$

Fix

$$M_* \geq 2M_1 M + c_* + 1 + 2MM_0 + 4M|f(x_f, 0, y_f, 0)|. \qquad (9.16)$$

Let $T_1 \in R^1$, $T_2 > T_1 + 2M$,

$$z_1 \in X_{M,1}, \ z_2 \in \bar{X}_{M,1}, \ \xi_1 \in X_{M,2}, \ \xi_2 \in \bar{X}_{M,2}. \qquad (9.17)$$

In view of (8.17), there exist a. c. functions $x : [T_1, T_2] \to R^n$, $y : [T_1, T_2] \to R^m$ such that

$$x(T_1) = z_1, \ x(t) = x_f, \ t \in [T_1 + M, T_2 - M], \ x(T_2) = z_2, \qquad (9.18)$$

$$|x(t)|, |x'(t)| \leq M, \ t \in [T_1, T_1 + M] \cup [T_2 - M, T_2], \qquad (9.19)$$

$$y(T_1) = \xi_1, \ y(t) = y_f, \ t \in [T_1 + M, T_2 - M], \ y(T_2) = \xi_2, \qquad (9.20)$$

$$|y(t)|, |y'(t)| \leq M, \ t \in [T_1, T_1 + M] \cup [T_2 - M, T_2], \qquad (9.21)$$

$$I^{f^{(1)}}(T_1, T_1 + M, x), \ I^{f^{(1)}}(T_2 - M, T_2, x),$$

$$I^{f^{(2)}}(T_1, T_1 + M, y), \ I^{f^{(2)}}(T_2 - M, T_2, y) < \infty. \qquad (9.22)$$

It follows from (9.11), (9.19), (9.21), and (9.22) that for a. e. $t \in [T_1, T_1 + M] \cup [T_2 - M, T_2]$,

$$(x(t), x'(t)) \in \text{dom}(f^{(1)}), \ (y(t), y'(t)) \in \text{dom}(f^{(2)}), \qquad (9.23)$$

$$|f(x(t), x'(t), y(t), y'(t))| \leq M_0. \qquad (9.24)$$

By (9.18), (9.20), and (9.24),

$$\left| \int_{T_1}^{T_2} f(x(t), x'(t), y(t), y'(t)) dt - (T_2 - T_1) f(x_f, 0, y_f, 0) \right|$$

$$\leq 2MM_0 + 2M|f(x_f, 0, y_f, 0)|. \qquad (9.25)$$

We show that (x, y) is an (M_*)-good pair of a.c. functions.

Assume that $\xi : [T_1, T_2] \to R^n$ is an a.c. function. By (9.12), (9.16), (9.20), (9.21), (9.23)–(9.25), and (C2)

$$\int_{T_1}^{T_2} f(\xi(t), \xi'(t), y(t), y'(t)) dt$$

$$= \int_{T_1}^{T_1+M} f(\xi(t), \xi'(t), y(t), y'(t)) dt + \int_{T_1+M}^{T_2-M} f(\xi(t), \xi'(t), y_f, 0) dt$$

$$+ \int_{T_2-M}^{T_2} f(\xi(t), \xi'(t), y(t), y'(t)) dt$$

$$\geq -2M_1 M - c_* + (T_2 - T_1 - 2M) f(x_f, 0, y_f, 0)$$

$$\geq \int_{T_1}^{T_2} f(x(t), x'(t), y(t), y'(t)) dt - 2MM_0 - 4M|f(x_f, 0, y_f, 0)| - 2M_1 M - c_*$$

$$\geq \int_{T_1}^{T_2} f(x(t), x'(t), y(t), y'(t)) dt - M_*.$$

Assume that $\xi : [T_1, T_2] \to R^m$ is an a.c. function. By (9.10), (9.13), (9.15), (9.18), (9.19), (9.23), (9.25), and (C2),

$$\int_{T_1}^{T_2} f(x(t), x'(t), \xi(t), \xi'(t)) dt$$

$$= \int_{T_1}^{T_1+M} f(x(t), x'(t), \xi(t), \xi'(t)) dt + \int_{T_1+M}^{T_2-M} f(x_f, 0, \xi(t), \xi'(t)) dt$$

$$+ \int_{T_2-M}^{T_2} f(\xi(t), \xi'(t), x(t), x'(t)) dt$$

$$\leq 2M_1 M + c_* + (T_2 - T_1 - 2M) f(x_f, 0, y_f, 0)$$

$$\leq \int_{T_1}^{T_2} f(x(t), x'(t), y(t), y'(t)) dt + 2MM_0 + 4M|f(x_f, 0, y_f, 0)| + 2M_1 M + c_*$$

$$\leq \int_{T_1}^{T_2} f(x(t), x'(t), y(t), y'(t)) dt + M_*.$$

Theorem 9.1 is proved. □

9.4 Auxiliary Results for Theorem 9.2

Let

$$(g, s, \bar{x}) \in \{(f^{(1)}, n, x_f), (f^{(2)}, m, y_f)\}.$$

It is not difficult to see that the triplet $(g, s, \bar{x})$ satisfies all the assumptions of Sect. 3.1 of Chap. 3 of [53] and the results stated there are true for this triplet.

Clearly, there is $r \in (0, 1)$ such that

$$D := \{(x, y) \in R^s \times R^s : |x - \bar{x}| \le \bar{r} \text{ and } |y| \le \bar{r}\} \subset \text{dom}(g),$$

$$|g(x, y)| - g(\bar{x}, 0)| \le 1 \text{ for all } (x, y) \in D.$$

For each pair of points $x, y \in R^n$ and each positive number T define

$$\sigma(g, T, x, y) = \inf \left\{ \int_0^T g(v(t), v'(t))dt : v : [0, T] \to R^s \right.$$

$$\left. \text{is an a. c. function satisfying } v(0) = x, \ v(T) = y \right\}.$$

For each pair of number $T_1 \in R^1$, $T_2 > T_1$ and each a.c. function $v : [T_1, T_2] \to R^n$ put

$$I^g(T_1, T_2, v) = \int_{T_1}^{T_2} g(v(t), v'(t))dt.$$

Theorem 3.4 of [53] implies the following result.

Proposition 9.4. *Let $\epsilon, M > 0$. Then there exist an integer $Q \ge 1$ and a positive number l such that for each number $T > lQ$ and each a.c. function $v : [0, T] \to R^n$ which satisfies*

$$|v(0) - \bar{x}|, \ |v(T) - \bar{x}| \le r,$$
$$I^g(0, T, v) \le \sigma(g, T, v(0), v(T)) + M$$

there exists a finite sequence of closed intervals $[a_i, b_i] \subset [0, T]$, $i = 1, \ldots, q$ such that

$$q \le Q, \ b_i - a_i \le l, \ i = 1, \ldots, q,$$
$$|v(t) - \bar{x}| \le \epsilon, \ t \in [0, T] \setminus \cup_{i=1}^q [a_i, b_i].$$

In this chapter we use the following Lemma 3.10 of [53] (see also Lemma 5.1 of Chap. 8).

Lemma 9.5. *Let M, ϵ be positive numbers. Then there exists a positive number L_0 such that for each number $T \ge L_0$, each a.c. function $v : [0, T] \to R^s$ satisfying*

$$|v(0)| \le M, \ I^g(0, T, v) \le Tg(\bar{x}, 0) + M$$

and each number $s \in [0, T - L_0]$ the inequality

$$\min\{|v(t) - \bar{x}| : t \in [s, s + L_0]\} \le \epsilon$$

holds.

Lemma 9.6. *Let M be a positive number. Then there exists a positive number M_0 such that for each number $T \ge 1$, each a.c. function $v : [0, T] \to R^s$ satisfying*

$$I^g(0, T, v) \le Tg(\bar{x}, 0) + M$$

the inequality

$$\min\{|v(t)| : t \in [0, 1]\} \le M_0$$

holds.

Proof. By Lemma 9.3, there exists $c_* > 0$ such that for each $T > 0$ and each a.c. function $x : [0, T] \to R^s$,

$$I^g(0, T, x) \ge Tg(\bar{x}, 0) - c_*. \tag{9.26}$$

Choose a number $M_0 > 0$ such that

$$\psi(M_0) > |a| + |g(\bar{x}, 0)| + 2M + 2c_* + 1. \tag{9.27}$$

Assume that $T \ge 1$ and an a.c. function $v : [0, T] \to R^s$ satisfies

$$I^g(0, T, v) \le Tg(\bar{x}, 0) + M. \tag{9.28}$$

If $|v(t)| > M_0$ for all $t \in [0, T]$, then by (9.1) and (9.2), for a. e. $t \in [0, T]$,

$$g(v(t), v'(t)) \ge \psi(M_0) - a.$$

Together with (9.27) and the inequality $T \ge 1$ this implies that

$$I^g(0, T, v) - Tg(\bar{x}, 0) \ge T(\psi(M_0) - a - g(\bar{x}, 0))$$
$$\ge \psi(M_0) - a - |g(\bar{x}, 0)| > M + 1.$$

This contradicts (9.28). The contradiction we have reached proves

$$\{t \in [0, T] : |v(t)| \le M_0\} \ne \emptyset.$$

Set

$$\tau_0 = \min\{t \in [0, T] : |v(t)| \le M_0\}. \tag{9.29}$$

By (9.1), (9.2), (9.26), (9.28), and (9.29),

$$M \geq I^g(0, T, v) - Tg(\bar{x}, 0)$$
$$\geq \tau_0 \psi(M_0) + I^g(\tau_0, T, v) - Tg(\bar{x}, 0) - \tau_0 a$$
$$\geq \tau_0(\psi(M_0) - g(\bar{x}, 0) - a) + I^g(\tau_0, T, v) - (T - \tau_0)g(\bar{x}, 0)$$
$$\geq \tau_0(\psi(M_0) - g(\bar{x}, 0) - a) - c_*$$

and

$$\tau_0(\psi(M_0) - |g(\bar{x}, 0)| - a) \leq c_* + M.$$

Together with (9.27) this implies that $\tau_0 \leq 1$. Lemma 9.6 is proved.

Lemma 9.7. *Let M, ϵ be positive numbers. Then there exists a positive number L_0 such that for each number $T \geq L_0$, each a.c. function $v : [0, T] \rightarrow R^n$ satisfying*

$$I^g(0, T, v) \leq Tg(\bar{x}, 0) + M$$

and each number $s \in [0, T - L_0]$ the inequality

$$\min\{|v(t) - \bar{x}| : t \in [s, s + L_0]\} \leq \epsilon$$

holds.

Proof. By Lemma 9.6, there exists a positive number M_0 such that the following property holds:

(i) for each number $T \geq 1$, each a.c. function $v : [0, T] \rightarrow R^s$ satisfying

$$I^g(0, T, v) \leq Tg(\bar{x}, 0) + M$$

we have

$$\min\{|v(t)| : t \in [0, 1]\} \leq M_0.$$

By Lemma 9.5, there exists a number $L_0 > 1$ such that the following property holds:

(ii) for each number $T \geq L_0 - 1$, each a.c. function $v : [0, T] \rightarrow R^n$ satisfying

$$|v(0)| \leq M_0,$$
$$I^g(0, T, v) \leq Tg(\bar{x}, 0) + M + a + |g(\bar{x}, 0)|$$

and each $S \in [0, T - L_0 + 1]$, we have

$$\min\{|v(t) - \bar{x}| : t \in [S, S + L_0 - 1]\} \le \epsilon.$$

Assume that $T \ge L_0$ and an a. c. function $v : [0, T] \to R^n$ satisfies

$$I^g(0, T, v) \le Tg(\bar{x}, 0) + M. \tag{9.30}$$

In view of (9.30) and property (i), there is

$$\tau_0 \in [0, 1] \tag{9.31}$$

such that

$$|v(\tau_0)| \le M_0. \tag{9.32}$$

It follows from (9.1), (9.2), (9.30), and (9.31) that

$$I^g(\tau_0, T, v) = I^g(0, T, v) - I^g(0, \tau_0, v) \le Tg(\bar{x}, 0) + M + a$$
$$\le (T - \tau_0)g(\bar{x}, 0) + M + a + |g(\bar{x}, 0)|. \tag{9.33}$$

By (9.31)–(9.33), property (ii), and the relation $T \ge L_0 > 1$, for each number S satisfying

$$[S, S + L_0 - 1] \subset [\tau_0, T]$$

we have

$$\min\{|v(t) - \bar{x}| : t \in [S, S + L_0 - 1]\} \le \epsilon.$$

This implies that for each number S satisfying

$$[S, S + L_0] \subset [0, T]$$

we have

$$[S + 1, (S + 1) + L_0 - 1] \subset [0, T],$$
$$\min\{|v(t) - \bar{x}| : t \in [S, S + L_0]\} \le \epsilon.$$

Lemma 9.7 is proved.

Lemma 9.8. *Let $M > 0$. Then there exist $L_0, M_0 > 0$ such that for each number $T \ge 2L_0$ and each a.c. function $v : [0, T] \to R^s$ which satisfies*

$$I^g(0, T, v) \leq Tg(\bar{x}, 0) + M$$

there are

$$\tau_1 \in [0, L_0], \ \tau_2 \in [T - L_0, T]$$

such that

$$|v(\tau_i) - \bar{x}| \leq r, \ i = 1, 2 \tag{9.34}$$

and

$$I^g(\tau_1, \tau_2, v) \leq \sigma(g, v(\tau_1), v(\tau_2), \tau_2 - \tau_1) + M_0.$$

Proof. Let $L_0 > 0$ be as guaranteed by Lemma 9.7 with $\epsilon = r$. By Lemma 9.3, there exists $c_* > 0$ such that for each $T > 0$ and each a.c. function $v : [0, T] \to R^s$,

$$I^g(0, T, v) \geq Tg(\bar{x}, 0) - c_*. \tag{9.35}$$

Choose a number

$$M_0 \geq c_* + 2 + 2L_0 + M + 2L_0|g(\bar{x}, 0)| + 2L_0 a. \tag{9.36}$$

Assume that $T \geq 2L_0$ and an a.c. function $v : [0, T] \to R^s$ satisfies

$$I^g(0, T, v) \leq Tg(\bar{x}, 0) + M. \tag{9.37}$$

It follows from (9.37), the choice of L_0, and Lemma 9.7 that there are

$$\tau_1 \in [0, L_0], \ \tau_2 \in [T - L_0, T]$$

such that (9.34) holds. By (9.35) and (9.37),

$$\sigma(g, v(\tau_1), v(\tau_2), \tau_2 - \tau_1) \geq (\tau_2 - \tau_1)g(\bar{x}, 0) - c_*. \tag{9.38}$$

By (9.1), (9.2), (9.36), (9.37), and (9.38),

$$\begin{aligned}
I^g(\tau_1, \tau_2, v) &= I^g(0, T, v) - I^g(0, \tau_1, v) - I^g(\tau_2, T, v) \\
&\leq Tg(\bar{x}, 0) + M + 2L_0 a \\
&\leq (\tau_2 - \tau_1)g(\bar{x}, 0) + M + 2L_0 a + 2L_0|g(\bar{x}, 0)| \\
&\leq \sigma(g, v(\tau_1), v(\tau_2), \tau_2 - \tau_1) + c_* + M + 2L_0 a + 2L_0|g(\bar{x}, 0)| \\
&\leq \sigma(g, v(\tau_1), v(\tau_2), \tau_2 - \tau_1) + M_0.
\end{aligned}$$

Lemma 9.8 is proved.

Proposition 9.4 and Lemma 9.8 imply the following result.

Proposition 9.9. *Let $\epsilon, M > 0$. Then there exist an integer $Q \geq 1$ and a positive number l such that for each number $T > lQ$ and each a.c. function $v : [0, T] \to R^s$ which satisfies*

$$I^g(0, T, v) \leq Tg(\bar{x}, 0) + M$$

there exists a finite sequence of closed intervals $[a_i, b_i] \subset [0, T]$, $i = 1, \ldots, q$ such that

$$q \leq Q, \ 0 \leq b_i - a_i \leq l, \ i = 1, \ldots, q,$$
$$|v(t) - \bar{x}| \leq \epsilon, \ t \in [0, T] \setminus \cup_{i=1}^{q} [a_i, b_i].$$

9.5 Proof of Theorem 9.2

By Theorem 9.1, there exists $M_2 > 0$ such that the following property holds:

(iii) for each $T_1 \in R^1$, each $T_2 > T_1 + 2M$, each $z_1 \in X_{M,1}$, each $z_2 \in \bar{X}_{M,1}$, each $\xi_1 \in X_{M,2}$ and each $\xi_2 \in \bar{X}_{M,2}$ there exists an (M_2)-good pair of a.c. functions $x : [T_1, T_2] \to R^n$, $y : [T_1, T_2] \to R^m$ such that

$$x(T_i) = z_i, \ y(T_i) = \xi_i, \ i = 1, 2,$$
$$x(t) = x_f, \ y(t) = y_f, \ t \in [T_1 + M, T_2 - M],$$
$$|x(t)|, |x'(t)|, |y(t)|, |y(t)| \leq M, \ t \in [T_1, T_1 + M] \cup [T_2 - M, T_2]$$

and

$$I^{f^{(1)}}(T_1, T_1 + M, x), \ I^{f^{(1)}}(T_2 - M, T_2, x),$$
$$I^{f^{(2)}}(T_1, T_1 + M, y), \ I^{f^{(2)}}(T_2 - M, T_2, y) < \infty.$$

By (C2), there exists $M_3 > 0$ such that:

$$f(z_1, z_2, \xi_1, \xi_2) \geq -M_3 \text{ for each } (z_1, z_2) \in R^n \times R^n$$

and each $(\xi_1, \xi_2) \in \text{dom}(f^{(2)})$ satisfying $|\xi_1|, |\xi_2| \leq M;$ (9.39)

$$f(z_1, z_2, \xi_1, \xi_2) \leq M_3 \text{ for each } (\xi_1, \xi_2) \in R^m \times R^m$$

and each $(z_1, z_2) \in \text{dom}(f^{(1)})$ satisfying $|z_1|, |z_2| \leq M.$ (9.40)

By Lemma 9.3, there exists $c_* > 0$ such that for each $T > 0$ and each pair of a. c. functions $v : [0, T] \to R^n$ and $u : [0, T] \to R^m$,

$$\int_0^T f(v(t), v'(t), y_f, 0)dt \geq Tf(x_f, 0, y_f, 0) - c_*, \tag{9.41}$$

$$\int_0^T f(x_f, 0, u(t), u'(t))dt \leq Tf(x_f, 0, y_f, 0) + c_*. \tag{9.42}$$

By Proposition 9.9 applied to the functions $f^{(i)}$, $i = 1, 2$, there exist integer $Q_1, Q_2 \geq 1$ and positive numbers l_1, l_2 such that the following properties hold:

(iv) for each number $T > l_1 Q_1$ and each a. c. function $u : [0, T] \to R^n$ which satisfies

$$I^{f^{(1)}}(0, T, u) \leq Tf^{(1)}(x_f, 0) + 2M_1 + c_* + 4MM_3 + 8M$$

there exists a finite sequence of closed intervals $[a_i, b_i] \subset [0, T]$, $i = 1, \ldots, q$ such that

$$q \leq Q_1, \ b_i - a_i \leq l_1, \ i = 1, \ldots, q,$$
$$|u(t) - x_f| \leq \epsilon, \ t \in [0, T] \setminus \cup_{i=1}^q [a_i, b_i];$$

(v) for each number $T > l_2 Q_2$ and each a. c. function $u : [0, T] \to R^m$ which satisfies

$$I^{f^{(2)}}(0, T, u) \leq Tf^{(2)}(y_f, 0) + 2M_1 + c_* + 4MM_3 + 8M$$

there exists a finite sequence of closed intervals $[a_i, b_i] \subset [0, T]$, $i = 1, \ldots, q$ such that

$$q \leq Q_2, \ b_i - a_i \leq l_2, \ i = 1, \ldots, q,$$
$$|u(t) - y_f| \leq \epsilon, \ t \in [0, T] \setminus \cup_{i=1}^q [a_i, b_i].$$

Set

$$Q = Q_1 + Q_2 + 4,$$

$$l = \max\{l_1, l_2, M\}.$$

Assume that $T > Ql$ and $x : [0, T] \to R^n$ and $y : [0, T] \to R^m$ is an (M_1)-good admissible pair such that

$$x(0) \in X_{M,1}, \ x(T) \in \bar{X}_{M,1}, \ y(0) \in X_{M,2}, \ y(T) \in \bar{X}_{M,2}. \tag{9.43}$$

By (9.43) and property (ii), there exists an (M_2)-good pair of a.c. functions $\tilde{x} :$ $[0, T] \to R^n, \tilde{y} : [0, T] \to R^m$ such that

$$\tilde{x}(0) = x(0), \ \tilde{x}(T) = x(T), \ \tilde{y}(0) = y(0), \ \tilde{y}(T) = y(T), \tag{9.44}$$

$$\tilde{x}(t) = x_f, \ \tilde{y}(t) = y_f, \ t \in [M, T - M], \tag{9.45}$$

$$|\tilde{x}(t)|, |\tilde{x}'(t)|, |\tilde{y}(t)|, |\tilde{y}'(t)| \leq M, \ t \in [0, M] \cup [T - M, T] \tag{9.46}$$

and

$$I^{f^{(1)}}(0, M, \tilde{x}), \ I^{f^{(1)}}(T - M, T, \tilde{x}),$$

$$I^{f^{(2)}}(0, M, \tilde{y}), \ I^{f^{(2)}}(T - M, T_2, \tilde{y}) < \infty. \tag{9.47}$$

In view of (9.45)–(9.47), the pairs $(x, \tilde{y})$ and $(\tilde{x}, y)$ are admissible. Since the pair (x, y) is (M_1)-good it follows from (9.44) and (9.45) that

$$-M_1 + \int_0^M f(x(t), x'(t), \tilde{y}(t), \tilde{y}'(t)) dt$$

$$+ \int_M^{T-M} f(x(t), x'(t), y_f, 0) dt + \int_{T-M}^T f(x(t), x'(t), \tilde{y}(t), \tilde{y}'(t)) dt$$

$$= -M_1 + \int_0^T f(x(t), x'(t), \tilde{y}(t), \tilde{y}'(t)) dt$$

$$\leq \int_0^T f(x(t), x'(t), y(t), y'(t)) dt$$

$$\leq M_1 + \int_0^T f(\tilde{x}(t), \tilde{x}'(t), y(t), y'(t)) dt$$

$$= M_1 + \int_0^M f(\tilde{x}(t), \tilde{x}'(t), y(t), y'(t)) dt + \int_M^{T-M} f(x_f, 0, y(t), y'(t)) dt$$

$$+ \int_{T-M}^T f(\tilde{x}(t), \tilde{x}'(t), y(t), y'(t)) dt. \tag{9.48}$$

By (9.39)–(9.42), (9.46), and (9.48),

$$-M_1 - 2MM_3 - c_* + (T - 2M) f(x_f, 0, y_f, 0)$$

$$\leq -M_1 - 2MM_3 + \int_M^{T-M} f(x(t), x'(t), y_f, 0) dt$$

$$\leq \int_0^T f(x(t), x'(t), y(t), y'(t)) dt$$

$$\leq M_1 + 2MM_3 + \int_M^{T-M} f(x_f, 0, y(t), y'(t))dt$$

$$\leq M_1 + 2MM_3 + c_* + (T - 2M)f(x_f, 0, y_f, 0).$$

The relation above implies that

$$\int_M^{T-M} f(x_f, 0, y(t), y'(t))dt \geq -2M_1 - 4MM_3 - c_* + (T - 2M)f(x_f, 0, y_f, 0),$$

$$(9.49)$$

$$\int_M^{T-M} f(x(t), x'(t), y_f, 0)dt \leq 2M_1 + 4MM_3 + c_* + (T - 2M)f(x_f, 0, y_f, 0).$$

$$(9.50)$$

By (9.49), (9.50), and properties (iv) and (v), there exist finite sequence of closed intervals

$$[a_{i,1}, b_{i,1}] \subset [M, T - M], \ i = 1, \dots, q_1$$

and

$$[a_{i,2}, b_{i,2}] \subset [M, T - M], \ i = 1, \dots, q_2$$

such that

$$q_1 \leq Q_1, \ q_2 \leq Q_2,$$
$$0 \leq b_{i,1} - a_{i,1} \leq l_1, \ i = 1, \dots, q_1,$$
$$0 \leq b_{i,2} - a_{i,2} \leq l_2, \ i = 1, \dots, q_2,$$
$$|x(t) - x_f| \leq \epsilon, \ t \in [M, T - M] \setminus \cup_{i=1}^{q_1}[a_{i,1}, b_{i,1}],$$
$$|y(t) - y_f| \leq \epsilon, \ t \in [M, T - M] \setminus \cup_{i=1}^{q_2}[a_{i,2}, b_{i,2}].$$

This completes the proof of Theorem 9.2.   □

9.6 Examples

Example 9.10. Assume that $f : R^n \times R^n \times R^m \times R^m \to R^1$ is a Borel measurable function which is bounded on all bounded subsets of $R^n \times R^n \times R^m \times R^m$ and satisfies (C2). Let

$$x_f \in R^n, \ y_f \in R^m$$

and set

$$f^{(1)}(x, y) = f(x, y, y_f, 0), \quad (x, y) \in R^n \times R^n,$$

$$f^{(2)}(x, y) = -f(x_f, 0, x, y), \quad (x, y) \in R^m \times R^m.$$

Let a_0 be a positive number, $\psi_0 : [0, \infty) \to [0, \infty)$ be an increasing function satisfying

$$\lim_{t \to \infty} \psi_0(t) = \infty,$$

$l_1 \in R^n$, $l_2 \in R^m$ and let $L_1 : R^n \times R^n \to [0, \infty)$ and $L_2 : R^m \times R^m \to [0, \infty)$ be lower semicontinuous functions such that

$$L_1(x, y) \geq \max\{\psi_0(|x|), \ \psi_0(|y|)|y|\} - a_0 + |l_1||y| \text{ for each } x, y \in R^n,$$

$$L_2(x, y) \geq \max\{\psi_0(|x|), \ \psi_0(|y|)|y|\} - a_0 + |l_2||y| \text{ for each } x, y \in R^m,$$

for each $x, y \in R^n$,

$$L_1(x, y) = 0 \text{ if and only if } (x, y) = (x_f, 0),$$

for each $x, y \in R^m$,

$$L_2(x, y) = 0 \text{ if and only if } (x, y) = (y_f, 0),$$

the function L_1 is continuous at $(x_f, 0)$, the function L_2 is continuous at $(y_f, 0)$, for each point $x \in R^n$ the function $L_1(x, \cdot) : R^n \to R^1$ is convex, for each point $x \in R^m$ the function $L_2(x, \cdot) : R^m \to R^1$ is convex, and that for each $x, y \in R^n$,

$$f^{(1)}(x, y) = f(x, y, y_f, 0) = L_1(x, y) + f(x_f, 0, y_f, 0) + \langle l_1, y \rangle,$$

for each $x, y \in R^m$,

$$-f^{(2)}(x, y) = f(x_f, 0, x, y) = -L_2(x, y) + f(x_f, 0, y_f, 0) - \langle l_2, y \rangle.$$

We show that f satisfies all the assumptions made in Sect. 9.1. It is not difficult to see that (9.1) and (9.2) hold under the appropriate choice of $a > 0$, ψ. Clearly, (9.3)–(9.5), (C1), and (C4) hold. Now we need only to show that (C3) and (ATP) hold.

Evidently, all the assumptions made in Example 1.14 hold for the functions $f^{(1)}$, $f^{(2)}$. In view of Example 1.14, all the assumptions made in Sect. 1.4 (including (A2) and (A4)) hold for the functions $f^{(1)}, f^{(2)}$. This implies that (C3) and (ATP) hold for the function f. Therefore Theorems 9.1 and 9.2 are true for f.

Example 9.11. Assume that $f : R^n \times R^n \times R^m \times R^m \to R^1$ is a Borel measurable function which is bounded on all bounded subsets of $R^n \times R^n \times R^m \times R^m$ and satisfies (C2). Let

$$x_f \in R^n, \; y_f \in R^m$$

and set

$$f^{(1)}(x,y) = f(x,y,y_f,0), \; (x,y) \in R^n \times R^n,$$
$$f^{(2)}(x,y) = -f(x_f,0,x,y), \; (x,y) \in R^m \times R^m.$$

Assume that

$$f(x_f,0,y,0) \le f(x_f,0,y_f,0) \le f(x,0,y_f,0)$$

for each $x \in R^n$ and each $y \in R^m$.

Let a be a positive number, $\psi : [0,\infty) \to [0,\infty)$ be an increasing function satisfying

$$\lim_{t\to\infty} \psi(t) = \infty.$$

Assume that (9.1) and (9.2) hold and the functions $f^{(1)}$, $f^{(2)}$ are continuous and strictly convex. We claim that f satisfies all the assumptions made in Sect. 9.1. Clearly, (9.3)–(9.5), (C1), and (C4) hold. Now we need only to show that (C3) and (ATP) hold. Evidently, all the assumptions made in Example 1.15 hold for the functions $f^{(1)}$, $f^{(2)}$. In view of Example 1.15, all the assumptions made in Sect. 1.4 (including (A2) and (A4)) hold for the functions $f^{(1)}$, $f^{(2)}$. This implies that (C3) and (ATP) hold for the function f. Therefore Theorems 9.1 and 9.2 are true for f.

References

1. Artstein Z, Leizarowitz A (1985) Tracking periodic signals with the overtaking criterion. IEEE Trans Autom Control AC-30:1123–1126
2. Aseev SM, Kryazhimskiy AV (2004) The Pontryagin maximum principle and transversality conditions for a class of optimal control problems with infinite time horizons. SIAM J Control Optim 43:1094–1119
3. Aseev SM, Veliov VM (2012) Maximum principle for infinite-horizon optimal control problems with dominating discount. Dyn Continuous Discrete Impuls Syst Ser B 19:43–63
4. Aseev SM, Veliov VM (2012) Necessary optimality conditions for improper infinite-horizon control problems. In: Operations research proceedings, pp 21–26
5. Aubry S, Le Daeron PY (1983) The discrete Frenkel-Kontorova model and its extensions I. Physica D 8:381–422
6. Baumeister J, Leitao A, Silva GN (2007) On the value function for nonautonomous optimal control problem with infinite horizon. Syst Control Lett 56:188–196
7. Blot J (2009) Infinite-horizon Pontryagin principles without invertibility. J Nonlinear Convex Anal 10:177–189
8. Blot J, Cartigny P (2000) Optimality in infinite-horizon variational problems under sign conditions. J Optim Theory Appl 106:411–419
9. Blot J, Hayek N (2000) Sufficient conditions for infinite-horizon calculus of variations problems. ESAIM Control Optim Calc Var 5:279–292
10. Blot J, Hayek N (2014) Infinite-horizon optimal control in the discrete-time framework. SpringerBriefs in optimization. Springer, New York
11. Bright I (2012) A reduction of topological infinite-horizon optimization to periodic optimization in a class of compact 2-manifolds. J Math Anal Appl 394:84–101
12. Carlson DA (1990) The existence of catching-up optimal solutions for a class of infinite horizon optimal control problems with time delay. SIAM J Control Optim 28:402–422
13. Carlson DA, Haurie A, Leizarowitz A (1991) Infinite horizon optimal control. Springer, Berlin
14. Cartigny P, Michel P (2003) On a sufficient transversality condition for infinite horizon optimal control problems. Automatica J IFAC 39:1007–1010
15. Coleman BD, Marcus M, Mizel VJ (1992) On the thermodynamics of periodic phases. Arch Ration Mech Anal 117:321–347
16. Gaitsgory V, Rossomakhine S, Thatcher N (2012) Approximate solution of the HJB inequality related to the infinite horizon optimal control problem with discounting. Dyn Continuous Discrete Impuls Syst Ser B 19:65–92
17. Gale D (1967) On optimal development in a multi-sector economy. Rev Econ Stud 34:1–18

© Springer International Publishing Switzerland 2015

A.J. Zaslavski, *Turnpike Theory of Continuous-Time Linear Optimal Control Problems*, Springer Optimization and Its Applications 104, DOI 10.1007/978-3-319-19141-6

18. Guo X, Hernandez-Lerma O (2005) Zero-sum continuous-time Markov games with unbounded transition and discounted payoff rates. Bernoulli 11:1009–1029
19. Hayek N (2011) Infinite horizon multiobjective optimal control problems in the discrete time case. Optimization 60:509–529
20. Jasso-Fuentes H, Hernandez-Lerma O (2008) Characterizations of overtaking optimality for controlled diffusion processes. Appl Math Optim 57:349–369
21. Kolokoltsov V, Yang W (2012) The turnpike theorems for Markov games. Dyn Games Appl 2:294–312
22. Leizarowitz A (1985) Infinite horizon autonomous systems with unbounded cost. Appl Math Optim 13:19–43
23. Leizarowitz A (1986) Tracking nonperiodic trajectories with the overtaking criterion. Appl Math Optim 14:155–171
24. Leizarowitz A, Mizel VJ (1989) One dimensional infinite horizon variational problems arising in continuum mechanics. Arch Ration Mech Anal 106:161–194
25. Lykina V, Pickenhain S, Wagner M (2008) Different interpretations of the improper integral objective in an infinite horizon control problem. J Math Anal Appl 340: 498–510
26. Makarov VL, Rubinov AM (1977) Mathematical theory of economic dynamics and equilibria. Springer, New York
27. Malinowska AB, Martins N, Torres DFM (2011) Transversality conditions for infinite horizon variational problems on time scales. Optim Lett 5:41–53
28. Marcus M, Zaslavski AJ (1999) On a class of second order variational problems with constraints. Israel J Math 111:1–28
29. Marcus M, Zaslavski AJ (1999) The structure of extremals of a class of second order variational problems. Ann Inst H Poincaré Anal Non linéaire 16:593–629
30. Marcus M, Zaslavski AJ (2002) The structure and limiting behavior of locally optimal minimizers. Ann Inst H Poincaré, Anal Non linéaire 19:343–370
31. McKenzie LW (1976) Turnpike theory. Econometrica 44:841–866
32. Mordukhovich BS (1990) Minimax design for a class of distributed parameter systems. Autom Remote Control 50:1333–1340
33. Mordukhovich BS (2011) Optimal control and feedback design of state-constrained parabolic systems in uncertainly conditions. Appl Anal 90:1075–1109
34. Mordukhovich BS, Shvartsman I (2004) Optimization and feedback control of constrained parabolic systems under uncertain perturbations. In: Optimal control, stabilization and nonsmooth analysis. Lecture notes in control and information science. Springer, Berlin, pp 121–132
35. Ocana Anaya E, Cartigny P, Loisel P (2009) Singular infinite horizon calculus of variations. Applications to fisheries management. J Nonlinear Convex Anal 10:157–176
36. Pickenhain S, Lykina V, Wagner M (2008) On the lower semicontinuity of functionals involving Lebesgue or improper Riemann integrals in infinite horizon optimal control problems. Control Cybern 37:451–468
37. Porretta A, Zuazua E (2013) Long time versus steady state optimal control. SIAM J Control Optim 51:4242–4273
38. Rockafellar RT (1970) Convex analysis. Princeton University Press, Princeton, NJ
39. Rubinov AM (1984) Economic dynamics. J Sov Math 26:1975–2012
40. Samuelson PA (1965) A catenary turnpike theorem involving consumption and the golden rule. Am Econ Rev 55:486–496
41. von Weizsacker CC (1965) Existence of optimal programs of accumulation for an infinite horizon. Rev Econ Stud 32:85–104
42. Zaslavski AJ (1987) Ground states in Frenkel-Kontorova model. Math USSR Izvestiya 29:323–354
43. Zaslavski AJ (1999) Turnpike property for dynamic discrete time zero-sum games. Abstr Appl Anal 4:21–48
44. Zaslavski AJ (2006) Turnpike properties in the calculus of variations and optimal control. Springer, New York

45. Zaslavski AJ (2007) Turnpike results for a discrete-time optimal control systems arising in economic dynamics. Nonlinear Anal 67:2024–2049

46. Zaslavski AJ (2008) A turnpike result for a class of problems of the calculus of variations with extended-valued integrands. J Convex Anal 15:869–890

47. Zaslavski AJ (2009) Structure of approximate solutions of variational problems with extended-valued convex integrands. ESAIM Control Optim Calc Var 15:872–894

48. Zaslavski AJ (2010) Optimal solutions for a class of infinite horizon variational problems with extended-valued integrands. Optimization 59:181–197

49. Zaslavski AJ (2011) Two turnpike results for a continuous-time optimal control systems. In: Proceedings of an international conference, complex analysis and dynamical systems IV: function theory and optimization, vol 553, pp 305–317

50. Zaslavski AJ (2011) The existence and structure of approximate solutions of dynamic discrete time zero-sum games. J Nonlinear Convex Anal 12:49–68

51. Zaslavski AJ (2013) Structure of solutions of variational problems. SpringerBriefs in optimization. Springer, New York

52. Zaslavski AJ (2013) Structure of approximate solutions of optimal control problems. SpringerBriefs in optimization. Springer, New York

53. Zaslavski AJ (2014) Turnpike phenomenon and infinite horizon optimal control. Springer optimization and its applications. Springer, New York

54. Zaslavski AJ (2014) Structure of approximate solutions of dynamic continuous-time zero-sum games. J Dyn Games 1:153–179

55. Zaslavski AJ (2014) Structure of solutions of variational problems with extended-valued integrands in the regions close to the endpoints. Set-Valued Var Anal 22:809–842

56. Zaslavski AJ, Leizarowitz A (1997) Optimal solutions of linear control systems with nonperiodic integrands. Math Oper Res 22: 726–746

57. Zaslavski AJ, Leizarowitz A (1998) Optimal solutions of linear periodic control systems with convex integrands. Appl Math Optim 37:127–150

Index

© Springer International Publishing Switzerland 2015
A.J. Zaslavski, *Turnpike Theory of Continuous-Time Linear Optimal Control Problems*, Springer Optimization and Its Applications 104, DOI 10.1007/978-3-319-19141-6

Printed in the United States
By Bookmasters